Media and Communications – Technologies, Policies and Challenges

Reality Television – Merging the Global and the Local

MEDIA AND COMMUNICATIONS – TECHNOLOGIES, POLICIES AND CHALLENGES

Additional books in this series can be found on Nova's website under the Series tab.

Additional E-books in this seriers can be found on Nova's website under the E-books tab.

Copyright © 2010 by Nova Science Publishers, Inc.

All rights reserved. No part of this book may be reproduced, stored in a retrieval system or transmitted in any form or by any means: electronic, electrostatic, magnetic, tape, mechanical photocopying, recording or otherwise without the written permission of the Publisher.

For permission to use material from this book please contact us:
Telephone 631-231-7269; Fax 631-231-8175
Web Site: http://www.novapublishers.com

NOTICE TO THE READER

The Publisher has taken reasonable care in the preparation of this book, but makes no expressed or implied warranty of any kind and assumes no responsibility for any errors or omissions. No liability is assumed for incidental or consequential damages in connection with or arising out of information contained in this book. The Publisher shall not be liable for any special, consequential, or exemplary damages resulting, in whole or in part, from the readers' use of, or reliance upon, this material.

Independent verification should be sought for any data, advice or recommendations contained in this book. In addition, no responsibility is assumed by the publisher for any injury and/or damage to persons or property arising from any methods, products, instructions, ideas or otherwise contained in this publication.

This publication is designed to provide accurate and authoritative information with regard to the subject matter covered herein. It is sold with the clear understanding that the Publisher is not engaged in rendering legal or any other professional services. If legal or any other expert assistance is required, the services of a competent person should be sought. FROM A DECLARATION OF PARTICIPANTS JOINTLY ADOPTED BY A COMMITTEE OF THE AMERICAN BAR ASSOCIATION AND A COMMITTEE OF PUBLISHERS.

LIBRARY OF CONGRESS CATALOGING-IN-PUBLICATION DATA
Reality television : merging the global and the local / editor: Amir
Hetsroni.
p. cm.
Includes bibliographical references and index.
ISBN 978-1-62100-068-6 (softcover)
1. Reality television programs--Social aspects. 2. Reality television
programs--Political aspects. I. Hetsroni, Amir.
PN1992.8.R43R43 2010
791.45'6--dc22
2010016721

Published by Nova Science Publishers, Inc. ✦ *New York*

MEDIA AND COMMUNICATIONS – TECHNOLOGIES, POLICIES AND CHALLI

REALITY TELEVISION – MERGING THI GLOBAL AND THE LOCAL

AMIR HETSRONI
EDITOR

Nova
Nova Science Publishers, Inc.
New York

CONTENTS

Preface		vii
Introduction		1
Section I: North America		5
Chapter 1	Do You Know who your Friends are? An Analysis of Voting Patterns and Alliances on the Reality Television Show *Survivor* *Erich M. Hayes and Norah E. Dunbar*	7
Chapter 2	Reality Television and Computer-Mediated Identity: Offline Exposure and Online Behavior *Michael A. Stefanone, Derek Lackaff and Devan Rosen*	25
Section II: Europe		45
Chapter 3	You'll See, You'll Watch: The Success of *Big Brother* in Post-Communist Bulgaria *Maria Raicheva-Stover*	47
Chapter 4	Gok Wan and the Magical Wardrobe *Gareth Palmer*	65
Chapter 5	Slovene Reality Television: The Commercial Re-Inscription of the National *Zala Volcic and Mark Andrejevic*	79
Chapter 6	Talking about Big Brother: Interpersonal Communication about a Controversial Television Format *Helena Bilandzic and Matthias R. Hastall*	95
Section III: The Middle East		113
Chapter 7	Reality vs. Reality TV: News Coverage in Israeli Media at the Time of Reality TV *Dror Abend-David*	115

vi   Contents

Chapter 8   Real Love Has No Boundaries? Dating Reality TV Shows
between Global Format And Local-Cultural Conflicts   **123**
Motti Neiger

Chapter 9   The Seal of Culture in Format Adaptations:
Singing for a Dream on Turkish Television   **137**
Sevilay Celenk

Chapter 10   The Praise and the Critique of a Nasty Format:
An Analysis of the Public Debate Over Reality TV in Israel   **151**
Amir Hetsroni

Section IV: Cross-Cultural Studies   **163**

Chapter 11   Performing the Nation: A Cross-Cultural Comparison
of *Idol* Shows in Four Countries   **165**
Oren Livio

Chapter 12   Mobile Makeovers: Global and Local Lifestyles
and Identities in Reality Formats   **189**
Tania Lewis

Chapter 13   From Reality TV To Coaching TV: Elements of Theory
and Empirical Findings Towards Understanding the Genre   **211**
Jürgen Grimm

Chapter 14   Reality Nations: An International Comparison
of the Historical Reality Genre   **259**
Emily West

About the Contributors   **279**

Index   **283**

PREFACE

Reality television has become a worldwide phenomenon which has the capability to crossover cultural boundaries and appeal to distinctly different markets. Drawing theories from media studies, economics, cultural studies and social science, this new book reviews how reality TV has conquered the world and has the potential to remove successful dramatic genres from the prime-time lineup.

Chapter 1 - There has been a boom in the popularity of reality television programming among the large U.S. networks because it is a cost-efficient way to produce popular programming without the need to employ writers to develop scripts or pay actors to portray fictional characters (Poniewozik, 2009). In reality shows like *Survivor,* drama is created by putting interesting people into unique situations so the audience can then imagine themselves in those situations. *Survivor*, created by Mark Burnett and currently in its twentieth season, has become the model for many other reality television shows where contestants are isolated and are eliminated competitively each week until there is a single winner remaining. Shows such as *The Amazing Race, Big Brother,* and even *Project Runway* or *HGTV's Design Star* have been modeled after the *Survivor* formula. By studying the behavior of the contestants in the show, our goal is to examine the reasons for the show's success and discuss the impact that the decisions made by the contestants have on the audience.

Chapter 2 - Life on the screen makes it very easy to present oneself as other than one is in real life. And although some people think that representing oneself as other than one is always a deception, many people turn to online life with the intention of playing it in precisely this way. (Turkle, 1995, p. 228.)

In her now-classic work *Life on the Screen*, sociologist Sherry Turkle (1995) effectively captured the radical zeitgeist of the early public internet: absent physical cues in the text-based medium, individuals were free to construct and deconstruct identity as they saw fit. Gender, race, and ability only became a component of social exchange to the degree that individuals chose to introduce it. "We are creating a world that all may enter without privilege or prejudice accorded by race, economic power, military force, or station of birth," optimistically declared another early commentator (Barlow, 1996). Significant amounts of subsequent research energy have been devoted to exploring how computer mediation affects personal identity construction and social interaction (e.g. Donath, 1999; Ellison, Heino, and Gibbs, 2006; Walther, 2007).

Chapter 3 - Until recently, Bulgarians thought of Big Brother as the embodiment of a totalitarian government capable of subjecting everybody to an uninterrupted surveillance apparatus. Having survived 45 years of communist control, Bulgarians were all too aware of George Orwell's gripping description of the totalitarian state in his *Nineteen-Eighty-Four* novel. Not a lot of people would have heard of the reality show by the same name. The introduction of *Big Brother* as the first Bulgarian reality show, however, marked a sharp transition to a new media reality in this post-communist country. Almost overnight, *Big Brother* became a big media event, attracting not only unparalleled attention from audiences, but dramatically changing the television landscape of the country. According to official data, 2.1 million viewers (62.5 percent of all viewers) were glued to the screen for the first season finale (Antonova and Kandov, 2005).

Chapter 4 - Lifestyle Television can be profitably analysed by considering the way in which magical symbolism combine with commercial interests to sell the ideology of consumerism. In this paper I begin by discussing Lifestyle Television's relationship to Reality Television and trends in consumerism in work very much informed by my reading of Adorno. I then follow with a detailed analysis of Gok's Fashion Fix. In this I draw on the mythical studies of Joseph Campbell to show how the Individual-as-hero and their mythic Quest to find the self presents a magical sheen to what is essentially a drive to align the individual with products more suited to their personality-type. I follow this by analysing the role of the Supernatural aid or Magical Helper. The gifts of this individual to befriend and help the hero-questor achieve self realization has to be considered in the light of the advertising that goes in and around the programme which in turn connects to the distinctly unmagical business of commerce. I then conclude by considering 'Entering the Belly of the Whale' - the Transformation – the onstage/catwalk fulfillment which is presented as a self-overcoming and a re-birth but is also a loud and aggressive celebration of what consumerism can do for the individual. In short I want to consider how the rational and the irrational are blended together to create a productive consumer whose self-fulfilment has a fortunate connection with the world of goods and services.

Chapter 5 - In the fall of 2007 Slovenia found its most popular reality format to date, a show called simply The Farm, which tapped directly into the country's self-identification with its rural, agricultural history. The Farm easily outperformed all competitors, including locally produced familiar global formats like The Bachelor, PopStars, Big Brother, and Who Wants to be a Millionaire, earning high ratings among the coveted 18-49 demographic, and breaking ratings records with its finale. On average, the show's 2007 season regularly drew almost half of the viewers watching TV during its primetime slot (45%) in a county of 2 million people (POP TV, 2007). This success was even more striking, given the fact that Slovenia is a nation with relatively high penetration of cable television (57%) and a wide selection of regional and international channels. In short, The Farm was a national phenomenon and, as we will argue, a nationalist one, insofar as it tapped into a deep vein of rural nostalgia for Slovene folk culture, complete with traditional costumes, accordian-centred folk music, and a celebration of the country's agricultural way of life.

Chapter 6 - At the beginning of the 21st century, episodes of the reality television show *Big Brother* were watched by millions of viewers worldwide and became the subject of countless media and interpersonal debates (Bignell, 2005). *Big Brother* was "in many ways a watershed for our understanding of media audiences" (Ross and Nightingale, 2003, p. 3), as it provoked unprecedented levels of audience ratings and audience involvement. This chapter explores the relationship between a media spectacle like *Big Brother* and interpersonal communications about such events by viewers and non-viewers. We follow Hartley's (1999) understanding of interpersonal communication as a face-to-face communication from one individual to another, in which personal characteristics, social roles and social relationships of the communicating individuals are reflected by form and content of the communication. Our perspective is not restricted to family communication (e.g., Larson, 1993), but encompasses all situations and locations in which interpersonal communication about television programs occurs.

Chapter 7 - On Monday, April 20, 2009, the Israeli press highlighted an important announcement by the Governor of Israel's National Bank, Professor Stanley Fischer: One or two major companies might find themselves in bankruptcy by the end of the week (Avriel, 2009). The press was immediately buzzing with the names of possible Israeli companies and individuals who might be candidates for bankruptcy. However, by the end of the week, the press was already preoccupied with the hostilities between Fischer and Bank Hapoalim owner Shari Arison, and the identity of the Israeli mogul to be banished from the circles of economic powers was never revealed.

Chapter 8 - Globalization has become one of the most debated topics in social science research since the 1960s (Appadurai, 1990; Ritzer, 1995). During the late 1980s, mainly due to the collapse of the Soviet Union, the concept began to gain momentum in the scientific community (Srebreny-Mohammadi et al., 1997), and for over a decade a significant growth can be observed both in the research literature that refers to the phenomenon and in the number of researchers whose work deals with it (Guillén, 2001). One of the key questions regarding Globalization – the process that integrates different societies/cultures into a single cross-national unit – is whether local/national cultures are able to challenge the process of Globalization/Americanization ("The McDonaldization of Society", Ritzer, 1993) in a combined dynamics of "Glocalization" and Hybridity (Robertson, 1994; Ritzer and Ryan, 2003; Kraidy, 2002) or they must surrender their local identity in order to become integrated and compete in the capitalist market.

Chapter 9 - My father never wore a suit. He watches you on television and says "Ibrahim Tatlises is wearing such a nice suit." May I ask you, Mr. Ibrahim, to give the suit that you are wearing to my father? (Peker and Peker, 2007, episode 8).

These words are from a viewer letter sent to the popular Turkish television show *Hayalin Icin Soyle (Singing for a Dream)*. No one, neither the participants, nor the viewers watching the program on the television, were expecting these words when the presenter announced that she would be reading a letter from a viewer. Nevertheless, this unusual request by a 10-year-old viewer was not so astonishing as it was only one of the many unexpected demands by the viewers, who wanted their own share of the glamorous world of Turkish television. In other words, this request was being expressed through such a locality and culturality, that it rendered ordinary its own strangeness. Ibrahim Tatlises, who is one of the most popular singers of Turkey, received many requests during this reality show, where he served as a jury member. In the letter mentioned above, he was being asked to take off what he is wearing and to give it to someone else. Still, this was not the only interesting expression of culture and identity, which rendered *Singing for a Dream* reality television talent competition overly local. The show, which had been adapted from a foreign format, had even established locality as an "excessiveness." At one point, Muazzez Abaci, who has been one of the most famous singers of Turkey for the last 50 years -though she was not as popular at all times- cheered a disabled contestant by saying "kurban olurum sana"—a traditional Turkish expression, which means "I would sacrifice myself for you." Neither was this one of the most interesting expressions of cultural difference. The tension between the content of *Singing for a Dream* and its foreign format forced it to become overly local and national—so as to eliminate any concerns of "foreignness." It even forced the program to drift into adopting a populist nationalist discourse.

Chapter 10 - Like in many other countries the success of reality TV formats such as "Big Brother" and "Survivor," which are based on closed-circuit television (CCTV) and feature constant surveillance of the contestants' daily life, performance of humiliating and sometimes explicit tasks, and unscripted battles between participants brought with it a heated public debate concerning the moral legality of the genre and a high pitched discourse revolving around the impact of the broadcasts. This paper reviews the controversy, which occupied a significant portion of the mass-media related public agenda in Israel over the years 2005-2008. Through a thematic content analysis of the arguments articulated in praise and critique of reality TV, I map the public discourse and contextualize the claims in more general longstanding debates. Thus, even though the data on which the analysis is based was collected in a single country, the mapping portrays the public discourse about reality TV in a way that can be relevant to various cultures.

Chapter 11 - Reality television provides wonderfully fertile grounds for investigating contemporary processes of globalization and localization, as well as their implications for the modern-day nation-state, due to the genre's simultaneous reliance on both universal, cross-national formats and the particular, localized customization of these formats for specific national contexts. As noted by Darling-Wolf (in press), "the genre's attraction is predicated on its successful adaptation of global formats to local environments," rendering it "the perfect exemplar of twenty-first century capitalism at its best – in all its glocalized, deterritorialized, indigenized and disjunctive messiness." This study examines the tensions and ambivalences of globalization and localization in their practical manifestations within the context of one of the most popular televisual formats in recent years, the *Idol* singing competition – focusing on the complex ways in which four different versions of the show (from the United Kingdom, the United States, Canada, and Israel) adapt the universal formula to accommodate different local cultural nuances and project divergent imaginings of a shared national identity. Rather than attempting to resolve the longstanding debate regarding the relative dominance of the universalizing versus particularizing impulses associated with modernity, I more modestly aim to identify some of the ways in which the tensions and struggles that accompany processes of globalization and localization are materialized in concrete cultural practice, and, more importantly, the ways in which these materializations are closely related to specific power relations and to the market-based motivations of show producers, who strategically appeal to perceived local, national identities in order to achieve popular and commercial success.

Chapter 12 - 'These twelve people have one thing in common—they're about to change their lives forever!': so goes the dramatic voiceover for *The Biggest Loser* (TBL), one of the recent big success stories in reality TV formats. A competitive weight-loss show, the format has captured the imagination of audiences worldwide with local versions of the US program broadcast in the UK, Australia and the Middle East (where the show has been rebadged *The Biggest Winner*). While the format has varied somewhat over different seasons and between countries, the basic premise involves a group of overweight people being brought together in a *Big Brother*-style house to compete for a large sum of money by losing the highest percentage of their starting body weight, initially as part of competing teams and later as individuals.

Chapter 13 - This chapter has a twofold goal: theoretical clarification of the reality TV genre and testing the theoretical essentials by empirical findings referring to the *Supernanny* formats in five countries. Part 1 tries to define a basic framework of reality TV and aims at answering the questions which social developments add to the popularity of this truly global television genre and which criteria form its smallest common denominator. Based on Alfred Schutz' sociological phenomenology of everyday life (called *"lifeworld"* theory), reality TV is analyzed with regards to its specific contributions to the recipient's everyday living environment. Especially the latest trends towards coaching TV (lifestyle, upbringing) show clearer than before what makes up reality TV's extra-medial reference and what lines of development its sub-genres follow. The second part presents results of a comparative international study of *Supernanny* programs in England, Germany, Austria, Spain and Brazil gained by a research project at the University of Vienna. Together with the survival and celebrity shows, the *Supernanny* format marks the most dynamic area of development in the post-*Big Brother*-era. The interpretation of results on British edutainment television is governed by the question whether and if so, to what degree global marketing and stable formats work well with the adaptation to the respective countries' parenting traditions and which different upbringing problems on the one hand are visible and what different parenting recommendations are being given on the other hand in the various countries. The data of the content analysis are supplemented by results from a survey of 1611 *Supernanny* viewers in Germany and Austria that allow to double-check some of the theoretical essentials on reality TV concerning the audience's motives. Finally, I rely on the findings to evaluate and forecast the future perspectives of reality TV.

Chapter 14 - When 1900 House (Hoppe, 2000) premiered in the UK in 2000, a hybrid television form was born that would spawn spin-offs and imitators over the next several years in several other countries. These series place people in historical settings, asking them to leave their 21st century lives behind, and live within the material and social constraints of the past for a period of three or four months. Part historical documentary, part re-enactment, part gamedoc - like Survivor, and part observational reality show or docusoap - like The Real World, the new historical reality genre drew upon a number of formulae. From the historical documentary tradition it inherited the pedagogical mission of addressing historical ignorance and shoring up national collective memory; from reality genres it drew emphases on entertainment and putting "real people" in visually and emotionally interesting situations.

In: Reality Television-Merging the Global and the Local
Editor: Amir Hetsroni, pp. 1-4

ISBN 978-1-62100-068-6
© 2010 Nova Science Publishers, Inc.

INTRODUCTION:
TOWARDS A CULTURALLY SENSITIVE AND CROSS-CULTURAL RESEARCH OF REALITY TV

Amir Hetsroni
Ariel University Center, Israel

Is this book just another collection of essays about reality TV? Technically, the answer is yes; in practice, hopefully, the answer is no. Since the genre first appeared on the TV screen and conquered a significant share of the networks' lineup as well as a lion's share of specialty channels (e.g. "Reality Central") it moved further to capture other media conduits like the internet and tabloids and has become a staple of poplar media and the subject of a number of edited scholarly anthologies (e.g. Holmes & Jermyn, 2004; Murray & Ouellette, 2009). Most of these essays collections were, in my opinion, quite good (or at least featured some noteworthy contributions), but they were also too general in their approach. They featured articles that dealt with nearly any possible aspect of reality TV. Even a relatively focused collection like the volume edited by Escoffery (2006) used a very comprehensive term such "the representation in reality TV" as anchor.

The current collection aims to be different. The articles here do not attempt to encompass all the various facets of reality TV. They admittedly avoid one million dollar questions such as "what is reality TV all about" for a number of reasons. First, these questions have been discussed intensively in previous books and articles. Second, right from the start of the discourse regarding this genre it seemed to go without saying that reality TV is "something of a catch all phrase" which contains features that originate from different formats (Kilborn, 1994, p.423). While certain attributes - such as minimal writing, the use of nonactors, the constant surveillance of daily life, the inclusion of eye witness testimonials, the reconstruction of storylines as a reflection of real life, the employment of experts as referees who – together with the audience at home – determine the outcome of shows are widely accepted in the literature as characteristics of reality TV (see, for example, Dovey, 2001), it seems to me that above all reality TV is a primitive concept, which means that we know it when we see it, but it is difficult to write down a definition upon which the scientific community and the public agree.

So this collection does not attempt to redefine the field. What it does try to provide is a culturally sensitive examination and a cross-cultural view – things which I find dearly missing in the literature. There are plentiful of reasons why a culturally sensitive investigation and a cross-cultural view are important. First, reality programs are global media products that need at least some cultural adaptation to succeed outside their habitués. In the vast majority of cases, franchised or non-franchised formats are locally produced rather than shows are imported and broadcast in their original version - as often happens with other TV genres such as drama. Second, reality TV includes some of the very few globally successful formats that do not originate in the USA or the UK. For instance, one of the most successful developers and exporters of reality formats, Endemol, which is solely responsible for the worldwide success of "Fear Factor", "Big Brother" and "Extreme Makeover", is Dutch. Finally, most of the previous studies about reality TV have concentrated on America and Great Britain and ignored the potentially different impact and significantly dissimilar content that shows may have in other countries. While this book does not ignore the USA or the British Isles, it does place reality TV a more international context.

Carrying the flag raised by Cooper-Chen (1994) in her classic study of game shows throughout the world the entries in this book, which shed a light on the content of reality shows, describe their audience, and offer explanations as to their appeal in different countries, are sectioned geographically according to the prime foci of their analysis.

The first section presents studies from North America. The opening article, chapter one by Hayes and Dunbar, examines voting patterns of contestants in the American version of Survivor. The authors draw on theories of interpersonal communication to predict voting patterns that represent the contestants' popularity and anchor their findings in some of the well known attributes of American culture and US mass media. The second article in this section, chapter two by Michael Stefanone, Derek Lackaff and Devan Rosen, is an empirical testing of the various relationships that exist between reality TV viewing and activity in Web2.0 sites such as Facebook and Myspace. This study questions the extent to which reality television consumption explains activity in the context of social network sites and detects a significant correlation between the two.

The second section presents studies from Europe. The first article in this section, chapter three by Maria Raicheva-Stover, is a review of the content and reception of reality programming in Bulgaria. By comparing the acceptance of the genre in that Balkan country with the situation in Russia and Poland, the author constructs a comprehensive picture of the way reality TV has been adapted and adopted in the former Soviet block. Chapter four, written by Gareth Palmer, is an insightful essay about the messages embedded in lifestyle programming in Britain. Using the show Fashion Fix as a case study, the author shows how magical symbolism is combined with commercial interests to promote consumerism. The makeover plotline, which is essential to the show, is read as reality mini-program. This entry places lifestyle programming, a sub-genre of reality TV, in the context of present day British mass culture. Volvic and Andrejevic's article (chapter five) starts with a comprehensive review of the history of reality TV in Slovenia and continues with a close analysis of "The Farm" – an adaptation of a Dutch format that became the most successful reality show in Slovenia. The article shows how the show, which mixes elements of traditional Slovenian folk culture with post-modern individualistic western competitiveness, epitomizes the current intermittent status of Slovenian society as a culture in transition. The third article in the second section, chapter six by Helena Bilandzic and Matthias Hastall, explores the relationships between the initial appearance of Big Brother as a controversial media event and interpersonal communication on the topic. Using Germany of the early 2000s as a study field and survey research as methodology, the authors prove that while involved communication about Big Brother (e.g., talking about the inhabitants) was strongly related to watching the show, reflective communication about it (e.g., talking about moral concerns) was not. The two step flow model of communication is used to explicate the findings.

The third part of the book presents studies from the Middle East. This section starts with chapter seven, an article by Dror Abend-David who points at some of the intriguing ways in which reality programming has been influencing news coverage in Israel, and consequently, some of the different ways in which news events have been treated, and perhaps even shaped by models of reporting developed in reality programs.

In the second article in this section (chapter eight), Motti Neiger offers a refreshing reading of an Israeli dating program. The author explains why some romantic conflicts that relate to religion and national belonging are insolvable in the context of a reality show, whereas other conflicts (e.g. age gap) have a happy ending. The third article in this section (chapter nine) is a textual analysis of the Turkish version of the show "Pop Idol. The author, Sevilay Celenk, argues that the salient characteristics of national identity and of local culture can fully conquer the adaptations of foreign formats and that the tension that emerges due to a confrontation between local and global leads to excessiveness which is a prime feature of talent shows. The final article in this section, chapter ten for which I am responsible, explores the public debate over the positive and negative values embedded in reality TV. Relying on a combination of quantitative content analysis of newspaper articles on the topic and a closer reading of a few representative examples, my study maps the arguments articulated in praise and critique of reality TV and contextualizes the different claims in more general cultural debates.

The last part of this book is devoted to cross-cultural studies. The section's opening chapter, chapter eleven, is Oren Livio's study of "Pop Idol" in four different countries - USA, UK, Canada and Israel. This study employs a close discursive analysis of a large number of programs to demonstrate the commonalities and disparities in the shows' respective constructions of perceptions of national unity, solidarity, identity, and space, as well as abstract conceptions of individuality, social mobility, civic participation, and democracy. The paper exemplifies how producers in different countries adapt the program's universal formula to accommodate different local cultural nuances and project divergent imaginings of their countries' national identities. Tania Lewis' article (chapter twelve) focuses on reality-based makeover programs in Australia and Singapore to demonstrate the ways in which international formats are shaped by both globalizing forces and domestic concerns. Lewis argues against a one-size-fits-all approach to reality TV and brings together a variety of critical frameworks for understanding the global spread of the format. Chapter thirteen is a study of "Supernanny" in Austria, Germany, Spain, Brazil and Britain. Jürgen Grimm's work is a fascinating multi-level study which includes both a content analysis of "Supernanny" programs and a viewers' survey. The author questions people's motives to watch the program and points at cross-cultural differences. Uses and gratifications models are used to connect content findings with public expectations from the show, whereas the philosophical ideas of Alfred Schütz are used to explain viewers' motivations to watch "Supernanny" as these motivations relate to searching for meaningful life experiences. The final article, chapter fourteen by Emily West, is a comparative textual analysis of historical reality series broadcast in the USA, UK, Canada and Australia. This chapter argues that adaptations of the reality historical format in different countries are remarkably similar. Whether they show turn-of-the-century middle-class Brits, pioneers on the American frontier, or Newfoundlanders eking out an existence in a remote fishing village, these programs ultimately purport to revisit the past through the daily life experiences in order to shed light on the origins, and the "true meaning," of the nation today. The subtext, and sometimes the explicit text, is that such an exercise will recapture the essence of the nation. The programs seek to "know" the national past, but the game they feature pairs this nationalist project with another, more personal, search of our private identity. West argues that this pairing is the key to the success of the historical reality format. Perhaps, this is - at least to some extent - also the key to the success of reality TV in general in so many different cultures. While this book does not aim to proclaim that we have found a perfect recipe to a successful reality format that can cross over cultures, I hope that the articles give at the very least some helpful hints.

Let me end with a few acknowledgements: I would like to thank all the scholars who expressed interest in contributing to this book. Due to space limitations and the need to keep a cohesive content framework and a high level of scholarship, I was able to publish only 14 of 29 submissions that were sent to me; however, I do thank all the 29 submitters for the vote of confidence and patience during endless revision iterations. I also thank the people at Nova Science for their willingness to publish this book. A different kind of gratefulness is sent from here to Professor Hanna Adoni, my all-time and old-time mentor from the Hebrew University of Jerusalem, whose guidance helped me to become a better writer. I am also indebted to my research assistant at Ariel University Center, Naz Kameli, who helped in formatting the book chapters. Finally, I would like to thank my parents – Sima and David – for allowing me to get the education that is needed to complete this book and my significant other, Katarazyna (Kasia) Markiewicz, for providing emotional support, when the work became overly exhausting. See you all on the next book!

REFERENCES

Cooper-Chen, A. (1994) *Games in the global village: A 50 nation study of entertainment television*. Bowling-Green, OH: Bowling-Green State University Popular Culture Press.

Dovey, J. (2001). Reality TV. In C. Creeber (ed.), *The television genre book* (pp. 134-137). London: British Film Institute.

Escoffery, D. S. (2006). *How real is reality TV: Essays on representation and truth*. Jefferson, NC: McFarland & Co.

Holmes, S., & Jermyn, D. (2004). *Understanding reality television*. London: Routledge.

Kilborn, R. (1994). How real can you get? Recent developments in reality television. *European Journal of Communication, 9*(4), 421-439.

Murray, S., & Ouellette, L. (2009). *Reality TV: Remaking television culture*. NY: New York University Press.

SECTION I: NORTH AMERICA

In: Reality Television-Merging the Global and the Local
Editor: Amir Hetsroni, pp. 7-24

ISBN 978-1-62100-068-6
© 2010 Nova Science Publishers, Inc.

Chapter 1

DO YOU KNOW WHO YOUR FRIENDS ARE? AN ANALYSIS OF VOTING PATTERNS AND ALLIANCES ON THE REALITY TELEVISION SHOW *SURVIVOR*

Erich M. Hayes and Norah E. Dunbar
University of Oklahoma, USA

There has been a boom in the popularity of reality television programming among the large U.S. networks because it is a cost-efficient way to produce popular programming without the need to employ writers to develop scripts or pay actors to portray fictional characters (Poniewozik, 2009). In reality shows like *Survivor,* drama is created by putting interesting people into unique situations so the audience can then imagine themselves in those situations. *Survivor,* created by Mark Burnett and currently in its twentieth season, has become the model for many other reality television shows where contestants are isolated and are eliminated competitively each week until there is a single winner remaining. Shows such as *The Amazing Race, Big Brother,* and even *Project Runway* or *HGTV's Design Star* have been modeled after the *Survivor* formula. By studying the behavior of the contestants in the show, our goal is to examine the reasons for the show's success and discuss the impact that the decisions made by the contestants have on the audience.

Survivor places several strangers in a remote area of the world and asks them to compete against the environment and each other. The goal is to "outwit, outplay, and outlast" your opponents and win the ultimate prize of one million dollars and the title of sole survivor. The contestants spend 40 days in a remote location, are divided into tribes, and vote one player out of the game at "tribal council" every three days. The producers create challenges in which one can win immunity from the votes at tribal council or rewards such as food, comforts, and adventures away from camp. Initially, the tribes compete against one another, but at some point during the game, normally with 10 to 12 players left, the tribes are merged and the players play the game as individuals, vying for individual immunity and individual rewards. Ostensibly, the game is one of survival because the players must forage for their own food, build their own shelters, and make fires. However, even more important than one's survival skills against the environment are one's skills in creating interpersonal connections with the other players. A successful player must vie for popularity among their fellow participants in the game and create alliances to protect themselves from opponents.

Survivor and reality television fans tune in primarily to watch the relationships among the people featured on the shows develop. There are many unofficial fan websites that discuss *Survivor* and the topics addressed in those forums are often about the personalities of the players themselves. For example, on the Survivorskills.com website, one fan posted the question, "Is Debbie smarter than we think?" about the *Survivor: Tocantins* player Debra Beebe. That post had 44 responses and was viewed 1460 times. Another post called "the immunity challenge" had just 12 responses and was viewed 565 times, but many of the posts were actually about the contestants' responses to losing the immunity challenge rather than the challenge itself. *Survivor* fans remember the relationships that are formed and the unscripted volcanic arguments more than they do the immunity and reward challenges created by the producers. The social relationships, although somewhat contrived by the producers, are worth examining to see how people interact when confronted with adverse situations. Determining how the interpersonal relationships, player voting patterns, and their depiction on the television show will help fans appreciate the show and help scholars understand how the social interactions impact popularity among fellow players and ultimately success in the game.

Part of the appeal of *Survivor* is watching the players form a new society within the context of the game. The people are randomly placed together and have to adapt to each other and their environment. Players take on specific roles within the context of their tribe – leaders, followers, outsiders, jocks, etc. These interactions and dialogue are a part of what make the television show popular for many viewers. The inherent conflict associated with the creation of this new society makes for interesting television drama. The way that the new society plays out is different each season. Show host Jeff Probst and the producers Charlie Parsons and Mark Burnett play a role in creating this society by establishing rules for the new society and, in greater part during some seasons, by the way tribes are assigned. For example, the producers limit access to food, provide immunity from votes to certain tribes or members of a tribe, isolate some contestants from their tribes, and provide rewards to certain tribes or members of the tribes. In some seasons, tribes have been assigned based upon race, gender, or other characteristics that dictated how the tribes have recreated their society. Seeing how the new society evolves over time and how successful the contestants become while facing adverse conditions is likely one reason for the popularity of *Survivor*.

This project will focus on analyzing the voting patterns exhibited in *Survivor* in fifteen seasons of the show for which complete voting records are publicly available on the official CBS *Survivor* website and unofficial fan websites. We will also examine detailed summaries of the shows available from Survivor fan sites (listed below) to explain the reasons why a player is voted out—is it due to a particular event or does it have to do with the strength of their alliance? The summaries provide accounts of what happens in each season as well as individual episodes. Specific examination will center on the how voting with the majority of players and receiving votes predicts the success of a player on *Survivor*. Based upon the assessment of these voting patterns, conclusions may be drawn of how gaming strategies affect a player's success on *Survivor*. Further, this project will show via uses and gratifications media theory (Katz, Blumler, and Gurevitch, 1974; Lundy, Ruth, and Park, 2008) why the drama and dialogue associated voting makes people watch the television show

Survivor. Uses and gratifications theory argues that people consume media that satisfies their own needs or explains a motivation behind their behavior. Audiences attend to programming that accounts for a personal stimulus. It will be shown that the voting and alliances analyzed in this chapter and the player manipulations associated with these aspects of the game make *Survivor* popular for its viewers and fans.

REALITY TELEVISION

Defining reality television is a somewhat difficult task. It is considered "reality" because the performances are unscripted and are not portrayed by actors (Smith and Wood, 2003). Some shows blur the line between fact and fiction by scripting some parts of the action. The appeal of reality television for many viewers is that they get to watch people in what seems like a more natural setting than a show with actors and written scripts. The whole point of reality TV seems to be to allow the audience to think that the people on the show are similar to them and they can imagine themselves in the situations faced by the participants. However, the contestants chosen to take part in *Survivor* and other similar shows are often not representative of the mainstream audience, but are chosen for their quirkiness or peculiar personality. Many of the participants are trying to make a career in the entertainment industry and the producers emphasize their unordinary occupations and genuine hobbies (Patkin, 2003). The players are chosen for their unique characteristics·that will provide the conflict and entertainment that many viewers enjoy (Nabi, Stitt, Halford, and Finnerty, 2003). Players often represent cultural stereotypes that are conducive to conflict and television viewership.

Reality television continues to have a loyal following among fans and the finales continue to attract record ratings. In its eighteenth season, *Survivor* rated in the top 20 television shows in the U.S., still dominates its time slot on Thursday nights, and drew over 13 million viewers for the most recent premiere of *Survivor: Tocantins* (Gorman, 2009). In a representative survey of adults in Tucson Arizona, 71% of respondents said they regularly watched at least one reality program and the respondents had watched four of sixteen reality programs included in the survey (Nabi et al., 2006). The poor reputation of the quality of reality programming is pervasive, but it doesn't appear to affect its appeal to general television viewing audiences. A recent *Time* magazine article suggests that many networks are responding to the splintering of television audiences by abandoning the costly scripted dramas that have traditionally dominated prime-time television in favor of more cost-effective reality programming (Poniewozik, 2009).

THEORETICAL FRAMEWORK

Power and Social Capital

Social capital is the idea that people allied together can form power in certain situations. In terms of *Survivor*, the power of allied people comes from the voting outcomes that determine which player leaves the game. Knowledge acquisition is a benefit of social capital (Inkpen and Tsang, 2005). Knowing who is voting for whom is one of the more important aspects to being successful in *Survivor* so developing trust and open communication with fellow contestants is crucial. Trust is a primary ingredient in maintaining and supporting the development of social capital. Trust is gained by the belief in reciprocity (Laser and Leibowitz, 2009). Trust is the linking element in creating high or low levels of social capital. As an extension, players in *Survivor* with a higher level of trust will have more social capital than players with lower levels of trust. Not surprisingly, those players with the social skills necessary to establish trust and detect deception accurately are more successful than players without those skills (Boone, 2003).

One way to examine whether or not a player is knowledgeable about the strategies of the other players in the game is to examine whether or not they vote with the majority at tribal council. Oftentimes, players conspire to "blindside" an opponent by undercutting their alliance and voting them out without the player realizing it is happening. Our first research question examines how often a player votes with the majority of the other players in relation to their success in the game:

RQ1: Does Voting with the Majority Predict where Players will Place in the Television Show *Survivor*?

The obligation and trust created through reciprocity is an important element in creating cohesiveness for a social group and dividends for individuals later (Laser and Leibowitz, 2009). In the case of *Survivor*, players establish trust within their alliances because they trade voting favor in exchange for voting with the alliance. Choosing which alliance to be a part of is a big decision for any *Survivor* player as they advance in the game because only truly cohesive alliances survive. As groups become larger, solidarity becomes more difficult to maintain (Laser and Leibowitz, 2009). Larger alliances in *Survivor* eventually must splinter as those who remain in the more cohesive part of the alliance are voted out of the game and those in the alliance try to improve their position in the pecking order within the alliance. Even those alliances that survive and manage to eliminate all their opponents must eventually turn inward and start voting off their own members. If an alliance member is not committed to the group and realizes they are the lowest status alliance member, it is to their benefit to leave the alliance and try establishing a connection with others outside the alliance. This is where sub-alliances within the larger alliance begin to become important. These smaller alliances are held together by what Laser and Leibowitz (2009) refer to as bridges. Bridging smaller groups into larger groups occurs in *Survivor* when the larger alliances break into smaller alliances as the game progresses.

Trust plays an important part in the *Survivor* alliances. Truth-telling (or at least the perception of truthfulness) is important in the success of the game's alliances. A player's ability to maintain their integrity is difficult in the game because deception and manipulation are necessary to "outwit" one's opponents. However, players that are caught lying usually don't last long in the game because the other players cannot trust them to keep promises or remain loyal to their alliance. For example, in season 18, *Survivor: Tocantins*, Debra attempted to flip on her long-time ally Coach. Rather than seeing her as a new ally and welcoming into their alliance, JT and Stephen saw her as untrustworthy because of her lack of loyalty to Coach and voted her out. In general, friends have higher levels of trust than nonacquaintances, something called the "relationship truth bias" in the deception literature. In general, we trust people with whom we have formed relationships more than strangers and the more vulnerable we make ourselves to them, the more we believe they are truthful to us (Vrij, 2006). Friends involved in a negotiation have a higher level of trust compared with acquaintances and non-friends in which discretion is important for trust expectations (Olk and Marta, 2001). The concept that friends trust each other more than non-friends is intuitive, but is important to understanding the need for players to form friendships and alliances quickly in the game of *Survivor*.

Alliances and Friendships

A key component of the game of *Survivor* is the formation of friendships that eventually become alliances. Much of the television depiction of the show focuses upon this element. Although, it is difficult to know how much of the actual game play is focused upon alliance building, it seems intuitive that making friends in *Survivor* would benefit players as they navigate the game. There are several reasons that players appear to be voted off *Survivor*. Some are voted off because they are considered weak players and their tribe thinks that they will lose challenges because of them. Some players do not bond well with the other participants early in the game and do not establish friendships or alliances with others. After the tribes are joined, it is often the stronger or smarter members that are voted off because the other players fear that they will win the challenges. Others are voted off because they are especially nice or popular and would be difficult to beat in the finals. Our second research question calls for a systematic examination of the reasons why players are voted off the game:

RQ2: What are the Reasons Players are Voted out of the Television Show *Survivor*?

While initial impressions are important for establishing a quick connection with other players and avoiding early votes, they are also important in forming long-term relationships that last throughout the game as well. Marek, Knapp, and Wanzer (2004) found that roommates that formed a positive first impression were more likely to have a long-term positive roommate relationship. Although the players on the game of *Survivor* are not quite the same as semester-long roommates because they have limited privacy and a more competitive relationship, their ability to form relationships is just as crucial as in longer-term relationships. For example, Marek et al. (2004) found that those roommates who formed positive initial impressions were more likely to use healthier conflict management strategies. The televised depictions of *Survivor* emphasize this conflict because it makes for better ratings, but much of the time spent on the island is spent talking and doing tasks together. Initial impressions are especially important because the relationships must develop quickly and alliances must be formed before the first vote takes place. Lewandowski and Le (2007) suggest that the basis for these first impressions on *Survivor* is largely attractiveness or appearance, similarity, and one's performance in the early physical challenges because these can be used to create schemas with relatively little other information. Our third research question addresses how important the initial impressions formed are to success in the game:

RQ3: How Important are First Impressions in Creating Friendships or Alliances in the Television Show *Survivor*?

Dominance and Power

Although social capital is created when people band together to form alliances, the dominance expressed by people in the show is one strategy used to attempt to outlast one's opponents. Dunbar and Burgoon (2005) differentiate between the ability to influence the decisions of another, which they call "power" and the behavioral tactics used to maintain or create the impression of power including dominance, manipulation, deception, or even non-negotiation. Perceptions of having power are just as important as the actual power that one holds because it is often the perception of power that creates the impetus to act (Dunbar, 2004; Neff and Harter, 2002). For example, Teven (2004) analyzed the final four contestants of one season of *Survivor* and found they were willing to deceive and manipulate their fellow competitors and build alliances with others to gain an advantage.

Certain traits may also play a role in one's ability to play the game. Dominance can be discussed both as a trait (Schmidt Mast and Hall, 2003) and the actual behaviors performed in interactions (Dunbar and Burgoon, 2005). Dominant individuals achieve influence because they are perceived to be more competent by fellow group members. Although this doesn't necessarily mean that those group members were more competent at those tasks, rather that the perception of competence made them a dominant individual (Anderson and Kilduff, 2009). The primary function of dominance in the context of *Survivor* is to gain control of the social setting. Dominance in the social setting could be beneficial if it means controlling votes and it could also be detrimental if it means becoming a target for others in the tribe. Often, competent players are voted out by their tribe-mates if they are seen as overbearing or bossy. In *Survivor*, one way to prove your dominance or competence is to win the immunity or reward challenges in the game. Although this strategy can be risky because other players may see you as a threat later on, other players may also see you as a more attractive ally because of your skill in winning challenges and may build their alliance around you. A player can also improve one's chances in the game by providing physical comfort and food to allies that are won in rewards as well as immunity from votes at tribal council. Patkin (2003) notes that Colby's streak of reward challenges in *Survivor: The Australian Outback* gave him superior nourishment, health, and strength, which gave him an advantage over the other players. In *Survivor: Panama*, Terry survived to take third place in the game despite the fact that he did not have a single ally because he won an unprecedented five immunity challenges in a row. Our next research question addresses the importance of winning challenges in the game:

RQ4: How Important is Winning Immunity and Reward Challenges to Players' Success in the Television Show *Survivor*?

Dominance traits are not only a function of a player's personality, but are also a function of their demographic traits. It is well established that sex, age, race, and class have created inequalities within U.S. culture and the producers of *Survivor* have exploited these inequalities in creating the show. In *Survivor: Panama*, the tribes were divided by age and sex and in *Survivor: Cook Islands,* the tribes were divided by race. The majority of contestants are single and those who have spouses or children are sometimes seen as a threat because they can use their family to garner sympathy with the final jury (made up of contestants previously voted out). The producers have tried to maintain diversity in the cast by including disabled participants, gay and lesbian participants, participants from every geographic region in the U.S., and participants with a wide range of occupations, but only four of the eighteen winners listed in Table 1 have been racial minorities (Vecepia from *Survivor: Marquesas* and Earl from *Survivor: Fiji* are African-American, Sandra from *Survivor: Pearl Islands* is Latina, and Yul from *Survivor: Cook Islands* is Asian-American). The elderly are often the first to be voted out because of their difficulty keeping up in the physical challenges early on. Only one winner has been over 40: Bob, the winner of *Survivor: Gabon,* was 57, but very physically fit.

Sex is an interesting case because it appears that women are succeeding well in the game, certainly better than anticipated, when the show first aired, in light of the emphasis put on physical challenges (Wolgast and Lanza, 2007). Seven of the eighteen (38%) winners have been female and some of the early winners like Tina, Jenna, and Sandra appeared to be unlikely winners because they lacked outdoor survival skills. Wolgast and Lanza (2007) make the case that the unassuming female is a more dangerous threat in *Survivor* than the burly alpha-male because they can fly "under the radar" while quietly making friends and joining alliances without appearing threatening. They argue that women are socialized to hone their skills at making connections and social networks at a young age and they use those skills to improve their chances at winning the game. At the time of their analysis, six of the twelve winners they examined were female but only one woman has won in the six seasons that have followed since (Parvarti, who won in her second appearance on *Survivor: Fans vs. Favorites*). Perhaps the contestants have stopped underestimating women and now see their social skills as a threat as well. The influence of sex on an individual's ability to win remains to be seen.

The geographic home or occupation of the contestants may also be a factor because they can be used as a way to establish connections with others or may be used to portray a non-threatening image to fellow participants. For example, Tom, the 46-year-old goat farmer from Virginia with overalls and a thick southern accent in *Survivor: Africa* was seen as friendly and likable and survived well into the game despite his age and lack of physical ability. He placed fourth. Rodger, age 53, in *Survivor: The Australian Outback* was described as a Kentucky farmer and high school shop teacher, but his position as a business owner and CEO of a bank was downplayed by producers to make him more likeable (Patkin, 2007). He established a paternal relationship with some of the other contestants and placed fifth in his season. Conversely, ex-NFL quarterback Gary Hogeboom concealed his true profession from his tribe so as not to appear intimidating or so that other players would not vote him out early assuming he was not in need of the prize money (Lewandowski and Le, 2007). Our interest in the demographics of the contestants leads us to the following research question:

RQ5: What Role do Players' Sex, Age, Race, Marital Status, Occupation, or Regional Home Play in Determining their Success in the Television Show *Survivor*?

Method

Participants

We analyzed fifteen seasons of the CBS Television show *Survivor*, aired between 2000 and 2009, which are summarized in Table 1. Our data for the show comes from four sources: (1) The *Survivor* Episode Guide on TV.com, (2) the Wikipedia.org website for *Survivor* (Survivor: U.S. TV series), (3) A fan-based *Survivor* wiki found at survivorwiki.wetpaint.com and (4) The Official CBS website for *Survivor*. The four sources of information were triangulated and cross-checked to ensure accuracy. Of the eighteen seasons that have aired at the time of this writing, three seasons were excluded from the analyses because they departed from the basic show format of including strangers who competed in the game until they were eliminated. In Season 7, participants who had been voted off were allowed to return to the game, which gave them a second chance at being eliminated. One contestant who had been eliminated ended up being the runner-up in that season. In Seasons 8 and 16, participants who had previously appeared on the show were included in *Survivor: All-Stars* and *Survivor: Fans vs. Favorites*. These two seasons were excluded because previous contestants may have an advantage in building alliances with other contestants they already knew and were skilled at the challenges. This resulted in 257 *Survivor* contestants, whom were included in the final analyses. Most seasons have 16-18 participants although a few had 20 (see Table 1) and one season, season 14 filmed in Fiji, had 19 players because one participant quit the game before it started filming.

Table 1. Summary of 18 seasons of Survivor

Season	*N*	Location	Premier Air Date	Winner
1	16	Borneo	May 2000	Richard Hatch
2	16	Australia	Jan. 2001	Tina Wesson
3	16	Africa	Oct. 2001	Ethan Zohn
4	16	Marquesas	Feb. 2002	Vecepia Towery
5	16	Thailand	Sept. 2002	Brian Heidik
6	16	Amazon	Feb. 2003	Jenna Morasca
7*	16	Pearl Islands	Sept. 2003	Sandra Diaz-Twine
8*	18	Pearl Islands (All-Stars)	Feb. 2004	Amber Brkich
9	18	Vanuatu	Sept. 2004	Chris Daugherty
10	20	Palau	Feb. 2005	Tom Westman
11	18	Guatemala	Sept. 2005	Danni Boatwright
12	16	Panama	Feb. 2006	Aras Baskauskas
13	20	Cook Islands	Sept. 2006	Yul Kwon
14	19	Fiji	Feb. 2007	Earl Cole
15	16	China	Sept. 2007	Todd Herzog
16*	20	Micronesia (Fans Vs. Favorites)	Feb. 2008	Parvati Shallow
17	18	Gabon	Sept. 2008	Robert Crowley
18	16	Tocantins	Feb. 2009	James Thomas

*Excluded from analyses.

In order to collect the demographics of our sample, we examined the bios of the contestants on the Official CBS *Survivor* website. The race of the participants was determined by examining the photos of the participants along with their bios, but the participants could only be grossly classified as white (including Hispanics)[1], African-American, or Asian-American using this method. The participants ranged in age from 19 to 72 years old (M = 33.51, SD = 10.61) and were evenly split between males and females. The majority of participants on the show are white or Hispanic (84%) with 10.5% African-American and 5.4% Asian-American. This compares with the general US population of 75.1% white or Hispanic, 12.3% African-American, and 3.6% Asian-American (Grieco and Cassidy, 2001). The participants were mostly single ($n = 175$) with a few reported to be married ($n = 68$), engaged ($n = 8$), or divorced ($n = 6$). In order to determine the participant's region, the contestants' home city listed on the CBS website was compared with the four U.S. regions according to the U.S. Census (2001). They were largely from the Western (33%) or Southern (31.1%) United States, with the remainder coming from the Northeast (21.7%) or Midwest (14%). This compares with the general US population of 22.4% Western, 35.6% South, 19% Northeast and 22.8% Midwest. Our research of the internet sources listed above found that the occupations of the participants were diverse including working in corporate positions ($n = 37$), education ($n = 31$), retail or sales ($n = 30$), law or law enforcement ($n = 25$), food service ($n = 21$), medical/dental ($n = 14$), real estate ($n = 8$) or "other" ($n = 70$). Some participants were full-time students ($n = 15$) or retired ($n = 6$). Most of the participants were selected to appear on the show though a nationwide application process, but according to the show's fan website on Wikipedia.org (n.d.) the contestants who appeared on season 14 were reportedly recruited by the production staff to increase the show's diversity.

Measures

Dependent Variables: Success in the game. The internet sources listed above contained information we used to create variables for our analysis. We counted the number of episodes in a season and then assigned numbers to those episodes. The *episode voted out* refers to the number of the episode in which a participant was voted out (a higher number means the participant stayed in the game longer and was more successful). Because the final three or four contestants are usually voted out in the same final episode, we also created the variable called *place* which refers to the player's final place in the game (first place means the participant won the game and the grand prize of $1,000,000).

Independent Variables: Alliances, Friendships, and Popularity. We counted the total number of *votes* a player received during the course of the game (which means other players voted to have him or her removed from the game either successfully or unsuccessfully). We also recorded the episode in which the player received their *first vote* from fellow participants. The number of *jury votes* refers to how many jury members (usually the final 9 or 10 players voted out before the final episode) voted to award the finalists the grand prize. We also counted how many times a participant voted for the person who was actually voted

---

1 The U.S. Census does not include "Hispanic" as a racial category. Whether or not a person is of Hispanic origin is addressed in a separate question after the racial question. The only races in the US Census are "White", "Black or African American", "American Indian", "Asian", "Native Hawaiian or other Pacific Islander," or "Some other race" (Grieco & Cassidy, 2001). We coded for White, Black, and Asian, and combined all the participants of Hispanic origin with the White category.

out. In other words, the number of times a player votes with the *majority* of the tribe, either by accident or because they are allied with the majority.

Challenges. Immunity refers to the number of times a player won an individual immunity challenge to protect themselves from being voted out. *Reward* refers to the number of times a player won a reward challenge.

RESULTS

In order to test our research questions, we conducted two linear regression analyses with the indicators of success in the game (*episode voted out* or *place in the game*) as the criterion variables and game-related variables (voting with the majority, immunity and reward challenges won) and the demographic variables (age, sex, race, marital status, occupation and regional home) as the predictors. The game predictors were entered as a block in step one and the demographic predictors were entered in step 2. The results are summarized in Table 2, but will be discussed according to each research question that follows. Our first research question asked: Does voting with the majority predict where players will place in the television show *Survivor*? The regression analysis produced a significant regression model for both *episode voted out* $F(9, 37) = 4.47$, $p < .01$ and *place in the game* $F(9, 37) = 5.93$, $p < .01$. For both outcomes, voting with the majority was a significant predictor suggesting that players that vote with the other players will be voted out later and place higher in the game.

Table 2.Summary of Hierarchical Regression Analysis for Variables Episode Voted Out and Place

Variable	Episode Voted Out			Place		
	B	$SE\ B$	β	B	$SE\ B$	β
Step 1						
Voting with the majority	0.31	0.07	.58**	-0.56	0.10	-.65**
Immunity	-0.02	0.14	-.02	-0.23	0.22	-.13
Reward	0.44	0.21	.28*	-0.10	0.31	-.04
Step 2						
Age	0.04	0.04	.26	0.01	0.05	.03
Sex	-0.57	0.37	-.19	-0.07	0.54	-.02
Race	-0.72	0.33	-.27*	0.44	0.49	.10
Marital status	-0.11	0.36	-.07	-0.48	0.53	-.19
Occupation	-0.01	0.04	-.03	-0.06	0.06	-.10
Regional home	-0.15	0.15	-.12	0.15	0.23	.08

Note. For episode voted out $R^2 = .37$ for Step 1; $\Delta R^2 = .16$ for Step 2 (p <.05).
For place in the game $R^2 = .55$ for Step 1; $\Delta R^2 = .04$ for Step 2 (p <.05). *p < .05. **p < .01.

Our second research question asked: What are the reasons players are voted out of *Survivor*? We summarized the reasons given from the episode guides available on our source websites in Table 3 resulting in 220 different reasons. The two researchers identified themes among the reasons and discussed each reason and grouped the reasons according to theme until 100% agreement was reached. Of these, the most common reasons for being voted out in the game of *Survivor* are a lack of a strong alliance, being perceived as too weak or too strong, and personality conflicts with other members of the tribe. Of these, the lack of a strong alliance was clearly the most common, accounting for 30.5% of the reasons given for a contestant to be voted out. If a player does not have an alliance to back them when others consider voting them out behind their back, the result is that the player leaves the game. Often, the other players in the game attempt to "blindside" a strong player by conspiring to vote for them when they are not expecting it. It is this situation where having the backing of a strong alliance is crucial because they can defend and protect a player.

Table 3. Reasons why people are voted out of Survivor

Reason for voting out a player	Number of times reason given
No strong alliance	67 (30.5%)
Seen as a threat (mentally/physically strong or too popular)	31 (14.1%)
Having personality clashes with other members of the tribe (seen by others as annoying, overly talkative, stubborn, bossy, weird, etc.)	24 (10.9%)
Perceived as a liability in challenges or a weak tribe member	23 (10.5%)
Being uncommitted to an alliance or untrustworthy	15 (6.8%)
Being lazy	11 (5%)
Causing conflicts with others	11 (5%)
Losing tie-breaker vote or challenge at tribal council	9 (4.1%)
Removed from game for injury, left voluntarily, or asked to be voted out	9 (4.1%)
Being physically sick or injured at camp	6 (2.7%)
Lost final immunity challenge or was next in line when immunity idol was played at tribal council	4 (1.8%)
Not bonding with other members of the tribe	4 (1.8%)
Weakest person in alliance once opponents are gone	4 (1.8%)
Stealing or hiding food	2 (1.0%)
Total	220

Our third research question asked: How important are first impressions in creating friendships or alliances in the television show *Survivor*? We analyzed this by examining the episode in which participants received their first vote as a measure of their first impressions because receiving votes earlier means that you have made a negative first impression to your tribe-mates. We conducted a linear regression analysis in which *first vote* was the criterion variable and the *episode voted out*, the *place in the game*, and *voting with the majority* were the predictor variables. The predictors were all entered simultaneously. The result was a significant regression equation in which the episode voted out and voting with the majority were significant predictors (see Table 4). This suggests that receiving a first vote later means a player is more likely to succeed and make alliances with others.

Table 4. Summary of Multiple Regression Analysis for First Vote Received

	First Vote		
Variable	B	$SE\ B$	β
Episode voted out	1.04	0.15	1.22**
Place in the game	.22	0.13	.29
Voting with the majority	-.55	0.12	-.46**
R^2	.61		
$F_{(3,253)}$	50.56*		

Note. $*p < .05.$ $**p < .01.$

Our fourth research question asked: How important is winning immunity and reward challenges to players' success in the television show *Survivor*? We examined the linear regression analysis reported earlier (see Table 2) with the indicators of success in the game (*episode voted out* or *place*) as the criterion variables and the number of immunity and reward challenges won as predictor variables (as part of step 1). Only the reward challenges proved to be a significant predictor of episode voted out suggesting that winning more reward challenges translated into being voted out later in the game.

When considering success in the game, however, it is important to consider not only the final result (such as one's place in the game) but also how a player succeeds along the way to the finale. We also tested whether winning reward or immunity challenges played a role in determining the number of votes they received overall, the number of jury votes that finalists received, and whether the player voted with the majority. *First vote* was not included because individual immunity challenges do not take place until later in the game. We ran a series of multiple linear regression analyses with the measures of popularity (*votes, jury votes, majority*) as the criterion variables and the number of immunity and reward challenges won as the predictor variables. All predictor variables were entered simultaneously. The resulting regression models for votes received $F(8, 38) = .87$, $p = .55$, $R^2 = .39$, jury votes received, $F(8, 33) = 1.64$, $p = .15$, $R^2 = .53$, and voting with the majority $F(8, 38) = 2.08$, $p = .06$, $R^2 = .55$ were all non-significant, suggesting that winning challenges does not affect your ability to receive votes or make alliances.

Our fifth research question asked: What roles do players' sex, age, race, marital status, occupation, or regional home play in determining their success in the television show *Survivor*? Once again, we examine the linear regression reported in table 2 where *episode voted out* and *place* are the criterion and sex, age, race, marital status, occupation, and regional home are the predictor variables (in step 2 of the regression reported earlier). Only race was a significant predictor of episode voted out. We also tested whether players' sex, age, race, marital status, occupation, or regional home played a role in determining the number of votes they received overall, the number of jury votes that finalists received, when the player received their first vote, and whether the player voted with the majority. We ran a series of linear regression analyses with the measures of popularity (*votes, jury votes, first vote, and majority*) as the criterion variables and sex, age, race, marital status, occupation, and regional home as the predictor variables. All predictor variables were entered simultaneously. The analysis for *votes* produced a significant regression model reported in Table 5 with both race and regional home emerging as significant predictors. The analysis also produced a non-significant regression model for *jury votes*, $F(6, 223) = .62$, $p = .62$, $R^2 = .13$ and for *majority* $F(6, 250) = .73$, $p = .73$, $R^2 = .13$ and none of the predictors were significant.

Table 5. Summary of Multiple Regression Analysis for Number of Votes Received

Variable	Votes		
	B	$SE\ B$	β
Age	-0.02	0.02	-.06
Sex	-0.01	0.32	.00
Race	0.61	0.31	.12*
Marital status	0.15	0.24	.06
Occupation	-0.01	0.04	-.01
Regional home	-0.38	0.14	-.17**
R^2	.23		
$F_{(6, 250)}$	2.30*		

Note. $*p < .05.\ **p < .01.$

A follow-up comparison of the means using an ANOVA for race $F(2, 254) = 3.21$, $p < .05$, partial $\eta^2 = .03$ revealed that white participants averaged fewer elimination votes throughout the game than did African-Americans or Asian-Americans (see Table 6). A follow-up comparison of the means using an ANOVA for regional home $F(3, 253) = 3.34$, $p < .05$, partial $\eta^2 = .04$ revealed that those from the West and Midwest received more elimination votes than those from the South or Northeast (see Table 7).

Table 6. Means and Standard Deviation for Votes Received According to Race

	N	Mean	SD	SE
White	216	5.25	2.61	.18
Black	27	6.26	3.01	.58
Asian	14	6.57	1.34	.36
Total	257	5.43	2.62	.16

Table 7. Means and Standard Deviation for Votes Received According to Regional Home

	N	Mean	SD	SE
West	85	5.96	2.87	.31
Midwest	36	5.94	2.84	.47
South	80	5.10	1.98	.22
Northeast	56	4.77	2.73	.36
Total	257	5.43	2.62	.16

DISCUSSION

Alliances and Voting Patterns

The results of our research project examining the voting patterns of players in *Survivor* over the many seasons have shown a few trends. Clearly, players must establish their social capital to create power over their opponents and establish alliances to help them ward off threats in the game. Research question one asked whether voting with the majority predicted success on *Survivor*. Our analysis of RQ1 demonstrates unequivocally that one should vote with the majority in order to place well in the game and avoid being eliminated early. We tested the effect that receiving a vote early in the game has on the chances that a player will make it far into the game of *Survivor* and eventually win. This means being part of the group, establishing bonds with other players, and creating trust so that others will tell you how they are voting. While some players may vote with the majority by accident, even when they are not part of an alliance, this is unlikely to proceed throughout the game. At the very minimum, these players are plugged into the information-sharing network enough to know who the rest of the tribe is voting for.

Additionally, based upon the reasons that players were voted out of *Survivor* analyzed using the descriptive statistics from summaries of the show for RQ2, it becomes important to have other players watching your back because being blindsided by the tribe or not having enough people to support you when your name comes up as someone to vote out, means you will be eliminated earlier. The most frequent reason for a player to be voted out at any point of the game was because they had no strong alliance (mentioned 30.4% of the time). This reason was twice as common as the next most common reason, "being seen as a threat". From the name *Survivor*, it might be expected that weaker players would be voted out first as in the concept of survival of the fittest, but players were voted out for being the weakest link of the tribe only 23 times (10.5%). It is clear from the reasons that the players are voted out that having a strong alliance is the key to being successful, staying in the game, and eventually winning. It is somewhat surprising that being too strong is more of a voting liability than being too weak as a tribe member, but this fits with Wolgast and Lanza's (2007) analysis that the weaker (and largely female) members can "fly under the radar" and make more social connections than the dominant males. Personality clashes do not seem to play as large a role as one's social connections implying that tribe-mates are willing to put up with quirky or even annoying players as long as they have a strong alliance.

Our third research question asked how important first impressions are towards success in *Survivor*. We examined how quickly someone received their first vote. Players that are able to quickly acquire social capital in the form of friends and allies appear to have more success in the game of *Survivor*. Rarely did a player receiving votes early in the game make it far into the game of *Survivor* and six of the winners never received a single elimination vote in the game. There were a few notable exceptions to this including Rudy from season one. As a retired Navy Seal well into his 70's, many didn't expect much from him. He turned out to be one of the most skilled, fit, and popular Survivors. However, the impression that he would be a weak player was somewhat confirmed when he suffered health problems in *Survivor: All-Stars*, which made him an early-exit in that season. Forming good first impressions appears to be a good indicator of likely success in the game of *Survivor*.

Our fourth research question looked to discover how winning immunity and reward challenges would translate into success in the game of *Survivor*. On the show, the host, Jeff Probst, makes it seem as though winning challenges are crucial to success. Reward challenges provide a way for players to acquire social capital in the form of inviting their friends and allies with them to the rewards. They also earn rest and nourishment that the other players do not receive. Winning reward challenges was a predictor of being voted out later in the game, which supports the concept that reward challenges are important. On the other hand, winning rewards could be a detriment as players that didn't win the reward may become jealous of the other player's success in the game or it establishes the player as a future threat in winning rewards. Players should be cautious about gloating openly about rewards won.

Compared to rewards, immunity transcends social capital in that someone with no social capital can still be successful by virtue of winning immunity. The most unpopular player can continue on simply by winning an immunity challenge, forcing an established alliance to vote out one of their own and change the dynamics of the tribe. Winning immunity obviously means that a player can't be voted out and therefore will make it at least one more week in the game, but it did not predict overall success in the game. It is possible though, that winning immunity challenges translates into likely voting with the majority, which was an important predictor of success in the game. Immune players who cannot receive votes are likely to be sought as allies for majority votes and it would not make much sense for a player that cannot receive votes to ally themselves with a doomed alliance.

Our final research question looked to analyze the demographic characteristics of players that determine their success in the television show *Survivor*. As discussed in the review of literature, people with specific characteristics are more capable of acquiring social capital than others. Determining what characteristics are important towards success in *Survivor* may provide an understanding of how those types of players acquire social capital. We examined how age, sex, race, marital status, occupation and regional home influenced players' success in the game. Of these characteristics, only two of them were significant: race and regional home of origin. The race of the participants played a role in determining where contestants placed in the game and how many votes one receives in the elimination rounds of the game. The means presented in Table 6 suggest that white participants received fewer votes on average than African-American or Asian-American players. While the producers have made strides to increase the diversity of the game by creating four equal-sized tribes of different ethnicities in *Survivor: Cook Islands* (although the tribes were mixed after only two episodes) and by recruiting more minorities for *Survivor: Fiji* (in which the top four final contestants were all minorities), most seasons had only one or two minority members. In a game where people need to make connections quickly and look for like-minded individuals with which to form an alliance, the race or ethnicity of their fellow competitors can play a role. Some minority players such as Cirie who finished fourth in *Survivor: Panama* and Vecepia who won *Survivor: Marquesas* were able to transcend race and make connections with other players on the basis of their personality or other features than race.

The geographical region from which *Survivor* participants hailed also related to total elimination votes received throughout the game. Players from the Midwest and West received more votes than players from the South or Northeast. It is difficult to account for why a player's region would impact voting other than by suggesting that players use the geographical region of one another to look for similarities between them and form alliances. It may also be that cultural stereotypes about a certain region influence expectations for other players. It might also make sense that rural players might have more survival skills than city-dwellers, but this would only be partially accounted for in region.

Resonating with Audiences

All of the characteristics that impact voting patterns are also what make the television show of *Survivor* so appealing to viewers. Uses and gratifications media theory (Katz, Blumler, and Gurevitch, 1974) postulates that people will consume media that satisfies a need or accounts for an internal motivation. It would be difficult to argue that viewers consciously consider the number of votes each player has received while deciding to watch programming. The vote itself is only a small part of each episode's programming. However, everything that occurs in the game is focused on the outcome of the vote. Players are at all times immersed in the game, jockeying for position and favor within their tribe and building friendships and alliances with their tribe-mates. These behaviors ultimately impact voting and these behaviors are what appeals to audiences.

An analysis of the voting patterns on *Survivor* doesn't necessarily explain why people continue to watch *Survivor*. Rarely does the vote itself come as a complete surprise for the viewers. Everything leading up to the vote is what people watch. Viewers like watching the characters form a new society from scratch. And, the friendships within that society make the television show worth viewing. The voting is, for the most part, the major source of drama and compelling viewing in the television show *Survivor*. This drama is caused by the need to recreate a society for survival in the wilderness and alliances for survival in the game. Watching people's lives unfold in front of viewers may fulfill a voyeuristic need to see how others react to troubling situations or it may just be an interest in what would happen if people were placed in a survival scenario. The vote is a small part of each episode and it's likely not the most interesting part of the episode, but it is the driving force for everything that goes on during the show and makes viewers watch every week and every season.

Overall, this project has shown that the key to winning the game of *Survivor* is to acquire friends and make alliances. The ways that players do this vary widely and the process of acquiring social capital is different every season. The game has created some interesting social connections such as friendships between a retired Navy Seal and a gay nudist, a high school physics teacher and a swimsuit model, and a cattle rancher and a corporate consultant. It is these strange and evolving friendships that keep audiences tuning in to more than the challenges or physical aspects of *Survivor*.

CONCLUSION

The producers of *Survivor* like to portray the game as something that anyone can win. This is not entirely the case. While the winners have come from a wide range of occupations, geographical regions, and ethnic backgrounds, only players who are good at making social connections and establishing trust early on with their competitors truly have the ability to win. While being in good physical shape certainly is an asset, the game also takes mental sharpness and social skills. No player can expect to win all the immunity challenges and survive without the help of friends. The key to winning the game of *Survivor* is to make friends with others in the game and make them early. Making these alliances translates into staying in the game longer and increasing a player's chances of becoming the ultimate *Survivor*. Our research shows that these alliances are more important than saying the right things, winning rewards, performing at immunity challenges, and any other form of game success. As Jeff Probst would say, the key to winning the game of *Survivor* is to "outwit, outplay and outlast".

REFERENCES

Anderson, C., and Kilduff, G. (2009). Why do dominant personalities attain influence in face-to-face groups? The competence-signaling effects of trait dominance. *Journal of Personality and Social Psychology, 96,* 491-503.

Bhandari, H., and Yasunobu, K. (2009). What is social capital? A comprehensive review of the concept. *Asian Journal of Social Science, 37,* 480-510.

Boone, R. T. (2003). The nonverbal communication of trustworthiness: A necessary *Survivor* skill. In M. J. Smith and A. F. Wood (Eds.), *Survivor lessons: Essays on Communication and Reality Television* (pp. 97-110). Jefferson, NC: McFarland and Company.

Chambliss, W. (1965). The selection of friends. *Social Forces, 43,* 370-380.

Dunbar, N.E., and Burgoon, J. (2005). Perceptions of power and interactional dominance in interpersonal relationships. *Journal of Social and Personal Relationships, 22,* 207-233.

Ellis, W. and, Zarbatany, L. (2007). Explaining friendship formation and friendship stability. Merrill-Palmer Quarterly, *53,* 79-104.

Gorman, B. (13 February, 2009). Thursday ratings: Survivor returns Well, Grey's boosts practice to demo beatdown. *TV by the Numbers.* Retrieved July 12, 2009 from http://tvbythenumbers.com/2009/02/13/thursday-ratings-survivor-returns-well-greys-and-practice-deliver-demo-beatdown/12787.

Grieco, E.M. and Cassidy, R. C. (2001, March). Overview of race and Hispanic origin. Census 2000 Brief. U.S. Census Bureau. Retrieved August 30, 2009 from http://www.census.gov/prod/2001pubs/c2kbr01-1.pdf.

Guinote, A. (2008). Power and affordances: When the situation has more power over powerful than powerless individuals. *Journal of Personality and Social Psychology, 95,* 237-252.

Inkpen, A. and Tsang, E. (2005). Social capital, networks, and knowledge transfer. *Academy of Management Review, 30,* 146-165.

Katz, E., Blumler, J. and Gurevitch, M. (1974). Uses and gratifications research. *Public Opinion Quarterly.* 37, 509-524.

Laser, J., and Leibowitz, G. (2009). Promoting positive outcomes for healthy youth development: Utilizing social capital theory. *Journal of Sociology and Social Welfare, 36,* 87-102.

Lewandowski, G., and Le, B. (2007). Outwitting to outlast. In R. J. Gerrig (Ed.). *The psychology of Survivor: Leading psychologists take an unauthorized look at the most elaborate psychological experiment ever conducted...Survivor!* (pp.57-70). Dallas, TX: Benbella Books.

Lundy, L., Ruth, A., and Park, T. (2008). Simply irresistible: Reality TV consumption patterns. *Communication Quarterly, 56*, 208-225.

Marek, C., Wanzer, M., and Knapp, J. (2004). An exploratory investigation of the relationship between roommates' first impressions and subsequent communication patterns. *Communication Research Reports, 21*, 210-220.

Nabi, R.L., Biely, E.N., Morgan, S.J., and Stitt, C. (2003). Reality-based television programming and the psychology of its appeal. *Media Psychology, 5*, 303-330.

Nabi, R.L., Stitt, C., Haliford, J., and Finnerty, K., (2006). Emotional and cognitive predictors of the enjoyment of reality-based and fictional television programming: An elaboration of the uses and gratifications perspective. *Media Psychology, 8*, 421-447.

Neff, K. and Harter, S. (2002). The role of power and authenticity in relationship styles emphasizing autonomy, connectedness, or mutuality among adult couples. *Journal of Social and Personal Relationships, 19*, 835-857.

Olk, P. and Elvira, M., (2001). Friends and strategic agents: The role of friendship and discretion in negotiating strategic alliances. *Group Organization Management, 26*, 124-164.

Papacharissi, Z., and Mendelson, A., (2007). An exploratory study of reality appeal: Uses and gratifications of reality TV shows. *Journal of Broadcasting and Electronic Media, 51*, 355-370.

Patkin, T. T. (2003). Individual and cultural identity in the world of reality television. In M. J. Smith and A. F. Wood (Eds.), *Survivor lessons: Essays on Communication and Reality Television* (pp. 13-26). Jefferson, NC: McFarland and Company.

Poniewozik, J. (2009, September 14). Jay Leno is shrinking your TV. *Time, 174*(10), pp. 40-47.

Reiss, S., and Wiltz, J. (2004). Why people watch reality TV. *Media Psychology, 6*, 363-378.

Raven, B. Centers, R., and Rodrigues, A. (1975). The bases of conjugal power. In R. E. Cromwell and D. H. Olson (Eds.), *Power in Families* (pp. 217-232). New York: Sage.

Schlenker, B., Lifka, A., and Wowra, S. (2004). Helping new acquaintances make the right impression: Balancing image concerns of others and self. *Self and Identity, 3*, 191-206.

Schmid Mast, M. and Hall, J. A. (2003). Anybody can be a boss but only certain people make good subordinates: Behavioral impacts of striving for dominance and dominance aversion. *Journal of Personality, 71*, 871-891.

Squires, C. (2008). Race and reality TV: Tryin' to make it real – but real compared to what? *Critical Studies in Media Communication, 25*, 434-440.

Teven, J. (2004). Survivor the Amazon: An examination of the persuasive strategies used to outwit, outplay, and outlast the competition. *Texas Speech Communication Journal, 29*, 52-64.

U.S. Cenus. (2001, July). Census 2000, geographic definitions. Retrieved August 30, 2009 from http://www.census.gov/geo/www/geo_defn.html#CensusDivision.

Vrij, A. (2006). Nonverbal Communication and Deception. In V. Manusov and M. L. Patterson (Eds.), *The Sage handbook of nonverbal communication.* (pp. 341-359). Thousand Oaks, CA US: Sage Publications, Inc.

Wolgast, B., and Lanza, M. J. (2007). "What? How did SHE win?" In R. J. Gerrig (ed.), *The psychology of Survivor: Leading psychologists take an unauthorized look at the most elaborate psychological experiment ever conducted...Survivor!,* (pp.177-196). Dallas, TX: Benbella Books.

In: Reality Television-Merging the Global and the Local
Editor: Amir Hetsroni, pp. 25-44

ISBN 978-1-62100-068-6
© 2010 Nova Science Publishers, Inc.

Chapter 2

REALITY TELEVISION AND COMPUTER-MEDIATED IDENTITY: OFFLINE EXPOSURE AND ONLINE BEHAVIOR

Michael A. Stefanone(1), Derek Lackaff (2) and Devan Rosen(3)

1. State University of New York at Buffalo
2. University of Texas at Austin, USA
3. University of Hawaii at Manoa, USA

> Life on the screen makes it very easy to present oneself as other than one is in real life. And although some people think that representing oneself as other than one is always a deception, many people turn to online life with the intention of playing it in precisely this way. (Turkle, 1995, p. 228.)

In her now-classic work *Life on the Screen*, sociologist Sherry Turkle (1995) effectively captured the radical zeitgeist of the early public internet: absent physical cues in the text-based medium, individuals were free to construct and deconstruct identity as they saw fit. Gender, race, and ability only became a component of social exchange to the degree that individuals chose to introduce it. "We are creating a world that all may enter without privilege or prejudice accorded by race, economic power, military force, or station of birth," optimistically declared another early commentator (Barlow, 1996). Significant amounts of subsequent research energy have been devoted to exploring how computer mediation affects personal identity construction and social interaction (e.g. Donath, 1999; Ellison, Heino, and Gibbs, 2006; Walther, 2007).

A key challenge to such efforts is the fact that the quantity and quality of nonverbal (or nontextual) social cues available to computer-mediated communication (CMC) participants has changed continuously since scholars first began examining them. Where Turkle (1995) explored a low-bandwidth textual social landscape comprised of multi-user dungeons (MUDs) and newsgroups, today's CMC users have options like utilizing voice chat to coordinate raids in the stunningly-rendered World of Warcraft, and participation in asynchronous video discussions via YouTube. Rather than allowing users to experiment and play with their identity, many of today's CMC technologies tie users ever closer to their offline, physical selves. For example, Stefanone and Jang (2007) found that people with large offline social networks adopt blogs with the intent to maintain these relationships. Here, CMC tools were leveraged to reduce the costs associated with maintaining large strong tie social networks. Further, social network sites (SNSs) such as Facebook introduce new types of personal information into social interaction and have become central platforms for interpersonal and group communication. Technological platforms are increasingly likely to mediate interpersonal communication as they diffuse throughout a population, and navigating social environments comprised of mediated identities has become an important communication skill.

In some regards contemporary societies have long been accustomed to interacting with completely "mediated" identities. Although relatively few Americans have had any direct interpersonal interaction with Britney Spears, Brad Pitt, or Kelly Clarkson, many could claim intimate knowledge of these individuals' daily lives. Many people may see photos of Gisele Bündchen and Heidi Klum more frequently than photos of distant friends or family members. Further, celebrity fans are using communication technologies to interact with their idols in many new ways. Manhattanites, for example, are encouraged to plot celebrity "sightings" on the online map available at Gawker Stalker (http://gawker.com/stalker).

Celebrity, of the type enjoyed or endured by actors, models, and athletes, was both a consequence and generator of the mass audience, and resulted in an informational flow that was primarily unidirectional. While the glamour of celebrity was something that mass audiences were encouraged to aspire to (generally by participating in fashionable consumption of advertisers' products) (McCraken, 1989), the world of celebrities was fundamentally removed from the comparatively mundane world of the audience. We argue that the normative and behavioral distinction between the celebrity world and the everyday world is being eroded, and that the dissolution of this boundary is observable in two distinct trends: the development and explosive popularity so-called "reality television," and the concomitant adoption of "Web 2.0" technologies like SNSs that allow individuals to potentially be identified by and communicate with mass-scale audiences.

Reality television has become a dominating component of the contemporary television environment. Reality television focuses on the (purportedly) unscripted interaction of nonprofessional actors who are often framed as "ordinary people" (Reiss and Wiltz, 2004). The transformation of "regular people" into "celebrities" whose every move is worthy of a mass audience's attention was a powerful concept. This media programming model turned out to be hugely popular, immensely profitable, and changed the overall media landscape in significant ways. While the specific components of reality television shows do vary, it is possible to identify broad generic values that generalize to the bulk of content in reality television programming. For example, actors on these shows regularly engage in "confessions," where they ritualistically disclose their private thoughts and feelings to the broadcast audience. This is analogous to the non-directed self disclosure of a specific Web 2.0 application—personal-journal style blogging—as discussed by Stefanone and Jang (2008). Blogs and other easily-accessible communication platforms have enabled an increasingly large population to publish their thoughts, photos, and videos on the Web, and are an example of how new CMC tools are appropriated for social, interpersonal goals (Stefanone and Jang, 2007).

Web 2.0 (O'Reilly, 2005) refers to a changing orientation between online content producers, consumers, and web technologies. While the utility of this particular term remains questionable (e.g., Berners-Lee, 2006) it does serve to highlight the increasing prevalence of user-created and user-focused online content and the development of media sharing sites. People without special technical skills are now able to interact with mass audiences via platforms such as blogs, media sharing sites like YouTube, and SNSs like Facebook.

Taken together, reality television and Web 2.0 set the stage for a major shift in the way individuals perceive their role in the media environment. Rather than simply being the target of mediated messages, they can see themselves as protagonists of mediated narratives, and can integrate themselves into a complex media ecosystem. The media tools and strategies employed by celebrities and their handlers – airbrushed photos, carefully coordinated social interactions, strategic selection and maintenance of the entourage – are now in a sense available to everyone, and are increasingly employed in everyday interpersonal interaction. Today, much CMC is thus marked by an increasing emphasis on existing offline relationships, physical and nonverbal communication cues, and their manipulation.

In this paper, Bandura's social cognitive theory (Bandura and Walters, 1977; Bandura, 1986; 2001) frames an analysis of the relationship between reality television consumption and online behavior with Web 2.0 tools like SNSs. Consistent with social cognitive theory, viewers are operationalized as active processors of television content who learn and model behavior portrayed on television programming. Five broad categories of television viewing are analyzed, and used to predict a range of SNS user behavior. Results suggest that social behaviors commonly associated with mediated celebrity are now being enacted by non-celebrities in an increasingly mediated social environment. Andy Warhol predicted in 1968 that everyone would receive fifteen minutes of fame, and contemporary observers such as David Weinberger (2001) suggest that internet technologies such as weblogs will make everyone famous to fifteen people. Reality television, however, demonstrates to viewers that anyone can become famous to an audience of millions, and Web 2.0 tools and applications put that potential within reach.

LITERATURE REVIEW

Reality Television and Affect TV

The relationship between the content of mass media and cultural attitudes is among the most-examined issues in mass communication research. Previous studies have explored the impacts of mass media upon attitudes towards violence (Dominick, 1984), sex (McGee and Frueh, 1980), and smoking (Shanahan, Scheufele, Yang and Hizi, 2004; Wakefield, Flay, Nichter and Giovino, 2003) among many additional topics. One trend observed in the last two decades is the relative increase of reality-framed television programming. Reality television makes the personal thoughts, behaviors, and interactions of its characters the main focus of audience attention. Bente and Feist (2000) refer to this genre as *affect TV*, which presents viewers with "the most private stories of non-prominent people to a mass audience, crossing traditional borders of privacy and intimacy" (p. 114). As the term is used here, the defining characteristic of reality television is that ordinary people (not professional actors) serve as the main characters (Reiss and Wiltz, 2004), and includes programs such as *Survivor*, *The Bachelor,* and *Blind Date* among many others.

Recently, Ferris, Smith, Greenberg and Smith (2007) conducted a content analysis of reality dating television and found that watching these shows was related to perceptions of dating relationships consistent with those modeled on television. The authors used social cognitive theory to explain the connection between television viewing and subsequent attitudes. The current study, however, differs in that reality television is conceptualized more broadly as described below.

Big Brother and Temptation Island

Calvert (2000) refers to reality television's realignment of the private and the public as "mediated voyeurism," and suggests that this is becoming endemic to the culture at large. This culture of mediated voyeurism may have real impacts on those who are most involved in it, and specific personality traits may also exacerbate these effects. For example, previous research has found links between personality traits and media use, such as aggressiveness with violent media use (Slater, Henry, Swaim and Anderson, 2003) and openness to experience and television consumption (Finn, 1997).

As early as 1922, Lippman suggested that we live in a mediated "pseudo-environment," while a host of later cultural scholars have debated the nature of the "spectacle" and the "simulation" of the media-saturated culture. The present focus of mass media research is "social constructivism" which recognizes that media exert influence upon audiences' structuring of social reality (cf. Shanahan and Morgan, 1999). This influence is moderated by discourse about media, as peer groups negotiate meanings among themselves. While revealing statistically meaningful media effects is difficult, it is reasonable to hypothesize that heavy consumers of specific media genres, such as reality television, may be influenced by its messages (Shrum, Wyer and O'Guinn, 1998).

Previous researchers have suggested that the symbolic world portrayed in the media (particularly television) may differ from the "real world" in important ways – the televised world is more violent (Gerbner et al. 1980a), more youthful (Gerbner et al., 1980b) offers employment that is high-status but requires low effort (Signorelli, 1990b), and over-represents traditional gender roles and stereotypes (Morgan, 1983; Rothschild, 1984). The cultivation perspective of media effects (Gerbner, et al., 1982; Shanahan and Morgan, 1999) suggests that television viewers attempt to align their attitudes and beliefs with those observed in television programming. Cultivation theory has been criticized, however, for being overly broad and unable to account for underlying contextual factors of attitude formation (Rubin, Perse, and Taylor, 1988, Shanahan and Morgan, 1999) and insufficient methodological rigor (Schrum, 2007). Bandura's (1986) social cognitive theory allows for the integration of social contextual factors into the effects model, and provides a useful framework for discussing the effects of celebrity culture and mediated voyeurism.

Web 2.0

Recent studies indicate that many concepts behind Web 2.0 are more than just marketing hype, and that younger people are increasingly engaged with these social technologies. Over half of all internet-using teens, for example, are "content creators" who create websites or blogs, share original media such as photos and videos, or remix content into new creations (Lenhart and Madden, 2005). SNSs such as MySpace (http://myspace.com) and Facebook (http://facebook.com) are often a cornerstone of this information space, with many recent surveys finding that 95% or more of college students have active profiles (e.g., PACS survey, 2007).

There appears to be substantial congruence between Web 2.0's culture of personal self-disclosure and the "reality culture" that has come to dominate some segments of the television market. Recent research on blogging, for example, operationalizes disclosures via personal-journal style blogs as non-directed in nature (Stefanone and Jang, 2007), analogous to behavior typified by the reality television genre wherein characters engage in "confessional" style disclosures to viewers. In the current paper, two social web behaviors that enable individuals to emulate mediated celebrity are discussed: SNS use and digital photo sharing.

Social Network Sites

The explosion in popularity of SNSs represents one of the fastest uptakes of a communication technology since the web was developed in the early 1990s. As of February 2008, three of the top 10 most popular websites worldwide were SNSs: Facebook, Myspace, and Orkut (Alexa Top Sites, http://alexa.com). Academic research on SNSs is growing, and has focused on a range of issues including privacy (Gross, Aquisti, and Heinz, 2005), identity and reputation (boyd and Heer, 2006; Walther et al. 2008), and the role these sites play relationship maintenance and the accumulation of resources like social capital (Choi, 2006; Ellison, Steinfeld, and Lampe, 2007). These sites typically allow an individual to connect their personal profile to the profiles of other users, resulting in a public display of one's entire (online) social network. Adding someone else to one's social network, or "friending," results in a publicly-displayed affiliation between the two individuals. On a technical level, becoming a "friend" requires only a few clicks of the mouse, rather than any investment in conversation or social support. This has resulted in a diversity of approaches and understanding of SNS "friendship." Further, these "friends" cover a wide range of tie strengths, including strong ties which offer social support and increasingly instrumental weak ties (Ellison, Steinfeld, and Lampe, 2007).

Many analyses of social networks focus on groups as a whole, while others focus on abstractions of relationships centered on an individual. Social network structures centered on individuals are known as ego-centered networks and include social relationships of all kinds (Mitchell, 1969). Every individual is the focal actor of an ego-centric network, a social structure which is explicitly reproduced by SNSs like Facebook. In pursuing an ego-centric approach to network analysis, the breadth and intensity of social relationships are typically measured, resulting in a collection of ego networks for a group of participants. The networks of relationships manifest on SNSs are thus readily conceptualized as ego networks.

The "intensity" of SNS use has been operationalized in multiple ways. For example, Ellison, Steinfeld and Lampe (2007) combined number of SNS friends and time spent online to create a "Facebook Intensity Scale." As the current study seeks to measure behaviors that may be influenced by reality television, further refinement of this construct was required. Previous research has found that some SNS users will only link their profile to those of individuals they know in another context, while other users will link relatively indiscriminately (Donath and boyd, 2004). If an individual's SNS network contains "friends" with whom one has no external relationship, then an individual must have motivations for creating the public connection other than affirming friendship.

Photo Sharing

Digital devices have radically transformed the social landscape of photography. Digital cameras began to outsell film cameras in 2004 (Musgrove, 2006), while camera-enabled mobile phones began to outsell digital cameras just a year later (Sharma, 2005). The profusion of digital cameras and camera devices suggest that digital photography, like other new media practices, is an increasingly "banalized" activity (Graham, 2004) that may play a subtle but important role in social relationships. A host of web services such as Flickr (http://flickr.com) and Snapfish (http://snapfish.com) have emerged to support the storage, organization and sharing of digital photos, while general social networking sites such as Facebook and MySpace include photo-sharing as a key functionality.

Photography has a long and social history. As Americans grew increasingly mobile in recent decades, photos helped keep distant family members and friends in close emotional proximity. In many cases, the digitalization of photography just made these sharing processes faster and more convenient. The web has also enabled relatively new forms of interaction through photos. Miller and Edwards (2007) note that two relatively distinct modes of photo sharing can be observed online – both the traditional sharing of photos with an existing social network of friends and family, and an emergent form of public sharing with strangers and online acquaintances. These two groups are perhaps better understood as representing ends of a spectrum of sharing behaviors, as the boundaries of intimacy are increasingly blurred by technological affordances. However, it is reasonable to assume that a primary goal of digital photo sharing, like analog photo sharing, is the development and maintenance of interpersonal relationships.

Social Cognitive Theory

People's adoption and use of Internet-based communication tools and applications have been studied from a wide range of perspectives including the diffusion of innovations (Moore and Benbasat, 1991), the technology acceptance model (Davis, Bagozzi, and Warshaw, 1989; Lee, Cho, Gay, Davidson, and Ingraffea, 2003), and social cognitive theory (Bandura, 1986). Social cognitive theory (formerly social learning theory) attempts to explain how and why people acquire and maintain certain behavioral patterns. Human functioning is explained as the product of the dynamic interaction of personal, behavioral, and environmental influences. Personal influences include cognitive, affective, and biological factors. Environmental factors include social context and the informational environment. Finally, Bandura includes behavior as a component of function because individuals can reflect on the effects of their own behavior. This tripartite construct is thus dynamic and highly contextual.

Social cognitive theory uses the term *modeling* to characterize the process through which an individual observes others, interprets the observed behavior, and adjusts their own behavior in response. Such observational learning may be the intended outcome of a given behavioral process, such as teaching a child to feed itself. However, Bandura (1986) notes that modeling may occur in many other contexts, indeed wherever an individual is able to observe others' behavior. The development of television is viewed by Bandura as an especially important source of behavior models, enabling people to "transcend the bounds of their immediate social life" (1986, p. 55). In comparison to the quantity of information about the world available in daily life, the amount of environmental information provided via media is vast. To the extent that one's images of reality are mediated and vicarious rather than directly experiential and experimental, the greater the impact of the media (Bandura, 1986). Bandura is careful to show that modeling is a more complex process than simple mimicry or imitation, and identifies several specific functions of the process.

The observational learning process requires a model, a learnable attitude or behavior, and a conducive personal/behavioral/ environmental context. In the present study, the characters in reality television programming serve as models and the Web 2.0 environment provides a new context for enacting observed behavior. The wide adoption of platforms such as SNSs among the largest demographic of reality television viewers – young adults (Hill, 2005) – suggest significant potential for interrelated media behavior.

Reality Television: Models, Attention and Retention

As reality television programming can be considered a coherent genre, its characters may serve as symbolic models for behavior. While previous discussions of the genre have included shows such as COPS, where some participants (the "suspects") are ostensibly unwilling, this discussion is limited to socially rewarding reality television with participants who willingly disclose their private selves.

Bandura (1986) identifies four subprocesses of the observational learning process: attentional processes, retention processes, (re)production processes, and motivational processes. The attentional process addresses the models ability to attract the observer's attention, as well as the observer's ability to cognitively attend to the model. People are more likely to pay attention to models that are perceived as similar to the self (Bandura, 1994; Eyal and Rubin, 2003), and a major component of reality television characters' appeal is that they are "ordinary," like the viewing audience (Reiss and Wiltz, 2004).

In order for learning to occur, observed behavior must be retained in some form by the individual. Bandura (1986) argues that the human mind retains abstractions of modeled events, rather than historical mental pictures: "As a result of repeated exposure to modeled events, observers extract distinctive features and form composite, enduring images of behavior patterns" (p. 56). One of reality television's strongest messages regards *non-directed self-disclosure*, where personal revelations are not targeted toward specific, individual others, but rather targeted to an abstract audience. As the personal thoughts of the characters are not (yet) directly accessible to the viewing audience, the narrative structure of many reality television shows requires the characters to transgress traditional boundaries of privacy, a sacrifice they are happy to make. Characters are subject to high levels of surveillance, and are often required to present their motivations and private thoughts to a camera in the format of confessionals. Further, high levels of surveillance and disclosure are presented as necessary and fundamentally normal. Reality television participants are rewarded for this behavior with celebrity status and financial gain in the form of prize money (Andrejevic, 2003, Hill, 2005).

Many participants parley their fame into advertising or entertainment careers. For example, after appearing as a contestant on the second season of *Survivor*, Amber Brkich appeared in a subsequent *Survivor* series, then in two seasons of *The Amazing Race*, then in a short-lived series focused on her and her husband (also a reality television show participant) titled *Rob and Amber: Against the Odds*. The couple is currently developing their own reality series (http://amber-brkick.com). Such developments are chronicled in detail in celebrity magazines, entertainment talk shows, and celebrity and gossip blogs.

The primary argument lies in the final two sub-processes of the observational learning theory: productive and motivational processes. The ability of individuals to participate in the mediated environment as pseudo-celebrities is now entirely feasible via the web.

Web 2.0: Production and Motivation

A production process is the enactment of an observed behavior, while a motivation process refers to the fact that the enactment of any behavior is subject to contextual incentives and disincentives. These two sub-processes will be discussed in tandem as they are related to the emergence of Web 2.0 technologies.

SNSs provide a unique platform for the *re*production of behavior observed in and modeled by reality television programming. For most of its history, "the media" was the domain of those who were, by definition, celebrities. With the wide scale adoption of media sharing, blogging, and SNSs, a much broader range of people now have the capability of creating mediated identities. The creation of a SNS profile allows a user to become a mediated character to others. While probably few SNS users would compare their Facebook profile to a talk-show appearance by a movie star, the SNS platform both enables and encourages activities that have been traditional associated with celebrity, such as the primacy of image and appearance in social interaction.

One behavior which may result from reality television modeling is what we term "promiscuous friending." While many users have articulated SNSs that map closely to their own external social networks, other users have SNS friend networks that contain many people who they have not actually met or have no external relationship with. Promiscuous frienders may be reproducing the fame-seeking behavior that is modeled by reality television characters. Having a large social network on a SNS site can been construed as a sign of popularity (being at the center of a large social network) and conversely as a sign of superficiality (e.g. "whores," [boyd, 2006] who are blatant status-seekers). In either case, a large "friends" list implies a large number of social connections, even if many of those connections have little social value in the traditional sense of friendship. In this scenario, users are actively competing for attention via expansive social networks.

Bandura (1986) recognized that not all observed behaviors are ultimately reproduced, and attributes this to motivational processes. The capacity to enact a behavior is insufficient grounds for most people to do so. Motivational processes like positive anticipated outcomes must be entered into the equation. Once again, reality television provides plentiful motivational input. Within the genre of socially rewarding reality television, participants are rewarded with celebrity and cash prizes for their participation. Consistent with Calvert (2000), we suggest the reality television genre demonstrates a value system which equates celebrity status and fame with social prestige and personal value.

In sum, celebrity culture is a real and significant factor in the contemporary media environment. One component of this culture is the development and increasing popularity of so-called reality television programming. Reality television has introduced a new idea into celebrity culture; namely, that the interactions of everyday people are worthy of the attention of broad audiences, and that anyone can become a public celebrity -- special talents, looks, skills, or wealth not required. Further, the development of powerful, accessible tools for self-expression – the platforms of Web 2.0 – now make it possible for individuals to "mediate" themselves, and reach audiences on the same scale as movie stars and fashion models. Taken together, these trends suggest both motivations and predictable outcomes for online behavior. Social cognitive theory provides a framework for understanding how viewers of reality television may enact specific behaviors online.

The following specific hypotheses are proposed:

1) Reality television consumption is positively related to time spent logged into SNS profiles.
2) Reality television consumption is positively related to the size of users' online social networks.
3) Reality television consumption is positively related to the proportion of users' online social networks who have never been met face to face (F2F).
4) Reality television consumption is positively related to sharing photographs via SNSs.

Because reality television is hypothesized to model a specific set of attitudes and behaviors, viewing other categories of television content should not correlate with these behaviors. Earlier effects theories of television considered all television content as having accumulative effect upon viewers (e.g. Bandura, 1986; Gerbner et al, 1986). Like Ferris et al. (2007), we suspect that heavy consumption of particular forms or genres of television content may be differentially associated with online behavior. Thus, we pose the following research question:

RQ1. How Does The Consumption Of News, Fictional, And Educational Television Content Relate To Online Behavior?

Methods

A total of 452 online surveys were completed by a sample of university students from a large, public, American university. Data from this population is ideal for testing the relationship between reality television viewing and online behavior, as it heavily engaged in both activities. Today's college students, part of Generation Y, remain one of reality television's most lucrative and targeted demographics (Hill, 2005) and are considerably more likely than other demographics to participate in social media, including social networking sites (Jones and Fox, 2009). Approximately 58% of the sample was female; the average age of participants was 20.3 years (SD = 2.6). The majority of participants identified their ethnic background as Caucasian (approximately 62%). About 16% were Asian, 6% were African-American, and 3% were Hispanic. The rest (about 13%) identified with a variety of other ethnicities.

Social Network Site Measures

To measure the length of time spent logged into SNSs, participants were asked, "when you typically log into your SNS account, how many hours do you spend online," and "when you typically log into your SNS account, how many minutes do you spend online." Overall, participants reported spending an average of 47.4 ($SD = 37.7$) minutes per session. Social network sites require users to manage networks in a public manner (Donath and boyd, 2004). Whereas offline social networks may be conceptually amorphous and indistinct (at least until they are examined by curious researchers), SNSs constrain social activity according to their technical design. In general, two SNS users are either "friends," or they have no relationship whatsoever. This particular constraint is articulated by the SNS platform. One question was used to measure how many "friends" participants have connections to via their SNSs, and participants reported an average of 282 friends ($SD = 235$).

Because of the likelihood that these "friend" connections do not accurately reflect the makeup of user's social networks, participants were asked to estimate the number of these friend connections they have not actually met F2F. The average proportion of network contacts not met was 14 percent ($SD = 22$), with the majority of respondents indicating they knew all of their SNS contacts. Finally, participants were asked to indicate the number of photographs they have publicly available on their SNS profiles ($M = 71.6$; $SD = 68.2$).

Television Viewing Measures

Television viewing was measured using a series of questions addressing five categories of content. Participants were asked "how many hours per day" and "how many days per week" they watched reality television, news, fiction, education, and "other" kinds of content. These items are consistent with Salomon and Cohen (1978) who suggest this is an appropriate approach when measuring time spent viewing television. Reality television consumption was prompted with examples like Real world and American Idol. Fiction shows were prompted with examples like The Simpsons, CSI, etc. Educational content was prompted with examples like The History and Discovery Channels. Overall, participants reported viewing approximately 30 hours of television weekly ($SD = 27.8$). On average, participants reported watching about 6 hours of reality television ($SD = 8.5$) and news ($SD = 7.3$) weekly, 9.3 hours of fiction ($SD = 9.7$), and 5.4 hours of educational programming ($SD = 7.3$). The "other" category accounted for 4.7 hours weekly ($SD = 7.1$).

RESULTS

Table 1 below summarizes the relationships between variables used in this study. Age was negatively correlated with time spent logged in SNSs, the number of friends participants report having connected to their profiles, and the number of photographs available online. Younger people clearly are investing the most resources into these tools. Social network size and the proportion of friends not met were strongly correlated to each other, and the number of photographs available correlate with network size, as expected. Also expected was the strong correlation between all of the television viewing variables.

Ordinary least squares (OLS) regression was used to test the influence television viewing had on each of four dependent variables related to SNS use: time logged into SNSs, the size of participants social networks, the proportion of these networks not met, and the number of photographs shared on these sites.

Table 2 below summarizes the results from the four regression models. In these models, age, gender, and aggregate television viewing were entered as independent variables and regressed onto the SNS measures to determine how television viewing affects each. This analysis begins with aggregate television viewing to explore its influence before examining each category of television content. Both age (β = -.163) and aggregate television viewing (β = .111) were significant predictors of time spent logged into SNSs. Not surprisingly, younger people spend more time managing their online profiles. Further, the more television participants watch, the more time they spend online. Overall, this model explained approximately 5 percent of the total variance.

In the next model, the size of participant's social networks was entered as the dependent variable. Although television viewing is not significant, younger people tend to have larger network sizes (β = -.167). Age was the only significant predictor of network size, and this model explained about 3 percent of total variance.

When these variables were regressed onto the proportion of network contacts not met F2F, only television viewing (β = .101) was a significant predictor. As participants spent more time watching television, the likelihood that they "friend" people they haven't actually met also increases. The last model presented in table 2 reveals that television consumption did not have a relationship with photo sharing. However, age (β = -.181) and gender (β = -.187) were both significant.

The next series of OLS regression models presented in table 3 were designed to explore the roles specific content categories of television play in terms of influencing online behavior and address the specific hypothesis outlined in this study. In these models five categories of television content were regressed onto the same three dependent variables constituting user behavior on SNSs in an effort to more clearly understand the role television content has on behavior.

In the first model, the only variable which predicted average time logged in to SNSs was frequency of reality television viewing (β = .169). The model was significant, $F(8, 445)$= 3.41, $p < .001$), and explained approximately 6 percent of the variance. The addition of the television content variables moderated the relationship between age and time spent logged in, which was no longer a significant predictor.

Reality television viewing was also significant (β = .147) in the model predicting network size, as was age (β = -.240). The addition of the content variables to this model strengthened the coefficients for. Younger participants who watch more reality television tend to have larger social networks via SNSs. When the proportion of network contacts not met F2F was designated as dependent variable, only reality television viewing emerged as significant (β = .182). Here, reality television viewing alone explained more variance in the model than aggregate television viewing (table 2).

The last model aims to explain the frequency of photo sharing via SNSs. Again, reality television viewing was significant (β = .107). Consistent with earlier analyses, age (β = -.290) and gender (β = -.236) were also significant.

Table 1. Descriptive statistics and zero-order correlations for variables; means (standard deviations) along the diagonal

	Age	Gender	Time logged	# friends	Prop not met	Photos	Rtv	News	Fiction	Edu	Other	All tv
Age	20.3 (2.7)	0.10*	-0.16**	-0.17**	0.01	-0.18***	-0.06	0.07	0.01	0.11*	0.01	0.03
Gender		-	-0.12*	-0.11*	-0.07	-0.26***	-0.24***	0.08	0.02	0.22***	0.17***	0.06
Time logged in			47.4(37.7)	0.17**	0.20***	0.08	0.22***	0.01	0.05	0.02	0.11*	0.11*
# friends on SNS				282(235)	0.27***	0.47***	0.12*	0.04	-0.02	-0.01	-0.09	-0.01
Proportion not met					.14(.22)	0.18***	0.19***	0.12*	0.05	0.10*	0.05	0.15**
# Photos shared						71.6(68.2)	0.11*	-0.05	-0.02	-0.01	-0.11*	-0.02
Rtv hrs/wk							6.1(8.5)	0.26***	0.49***	0.24***	0.33***	0.70***
News hrs/wk								6.0(7.3)	0.38***	0.38***	0.28***	0.61***
Fiction hrs/wk									9.3(9.7)	0.44***	0.44***	0.82***
Education hrs/wk										5.4(7.3)	0.29***	0.64***
Other hrs/wk											4.7(7.1)	0.65***
All tv hrs/wk												31.28(27.8)

Reality Television and Computer-Mediated Identity

Table 2. Standardized beta coefficients for range of dependent variables, aggregate TV viewing

	Time logged in, session	Size of SNS network	% *not* met F2F	Number of photos shared
Age	-.163***	-.167***	.016	-.181***
Gender	-.101	.071	-.057	-.187***
All TV Viewing	.111*	.061	.101*	-.014
$F(4, 449)$, R^2	6.25***, .049	4.85**, .031	1.77, .006	11.36***, .080

Table 3. Standardized beta coefficients for range of dependent variables, individual categories of TV viewing

	Time logged in, daily	Size of network	% *not* met F2F	Number of photos shared
Age	-.077	-.229***	0.43	-.269***
Gender	-.073	-.061	-.030	-.239***
RTV hrs/week	.169**	.147*	.182**	.110*
News hrs/week	-.021	.059	.076	-.040
Fiction hrs/week	-.097	-.122	-.052	-.065
Education hrs/week	.049	.031	.040	.059
Other hrs/week	.091	-.071	-.029	-.091
$F(8, 445)$, R^2	3.41**, .058	3.01**, .047	2.29*, .024	8.89***, .130

Note: *$p<.05$; **$p<.01$; ***$p<.001$.

Reality television viewing was the only significant television viewing category significant in all four models; none of the other content categories were significant. Together, these results support the hypotheses presented above.

To further highlight the trend in these analyses, Figure 1, above, was created to show the significance of reality television viewing in terms of each dependent variable. Because the scale varies between these variables, the data was first standardized before comparing differences in mean values for each. As Figure 1 shows, there are systematic differences between viewers and non-viewers of reality television in terms of the behavior indices used in these analyses. ANOVA analyses confirm that the between group differences are all statistically significant at greater than $p < .01$.

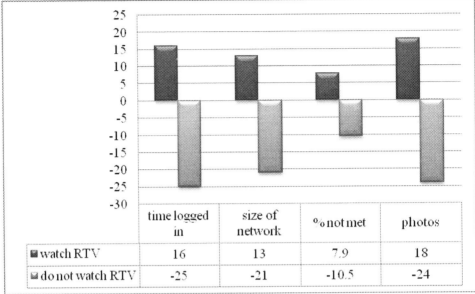

Note: data have been standardized for comparison between variables.

Figure 1. Systematic differences between viewers and non-viewers of reality television.

DISCUSSION

This research is founded on the premise that the confluence of the rising popularity of both reality television and Web2.0 applications has resulted in a fundamental shift regarding people's roles as media content consumers and producers. The purpose of this study was to explore the relationship between the increasing popularity of reality television and people's behavior on SNSs like Facebook. The evidence presented herein suggests that behavior traditionally associated with celebrities is being adopted en mass as people's interpersonal communication becomes increasingly mediated.

Utilizing social cognitive theory as the theoretical foundation, a positive relationship was expected between the amount of reality television young people consume and a range of online behavior in the context of SNSs including time spent logged in, online social network size, the proportion of network contacts not met F2F, and the number of photos shared online. These behaviors are believed to reflect the systematic processing of messages and behavior broadly modeled within the genera of reality television. Recall that the critical change in people's media diets over the past fifteen years lies in a shift from consumption to production. Internet users are faced with low time and financial costs as they enthusiastically contribute to the production of "mass media." This study adds a unique perspective to people's motivations to participate with the social web, and several valuable insights are revealed.

First, aggregate television viewing was used in an attempt to explain the three dependent variables highlighting relevant user behavior on SNSs. The dependent variables used in these analyses are believed to represent a range of generalizable behaviors users regularly engage in when using SNSs, which also correlate with behavior modeled in reality television programming. For example, if people believe that being the object of others attention is positive (as portrayed by socially rewarding reality television) then they should be more likely to engage in promiscuous friending. Concomitantly, viewers of reality television should also be increasingly comfortable with digital images of themselves publicly available via the internet, hence should share more photos via these sites. Reality television viewing should also affect the length of time people spend online managing their profiles and the overall size of their networks.

Overall, respondents indicated they watched approximately 31 hours of television weekly, and exposure to television generally was a significant predictor of the time people spent logged into their SNS accounts. This evidence is consistent with Gerbner et al's (1986) cultivation theory, and partially explains user motivation for spending time online managing social networks.

Although age had a strong negative relationship to the size of people's networks and the number of photos they share, aggregate television viewing did not. Younger people clearly had larger SNS networks, but watching television did not impact this variable. However, when the percentage of network contacts *not met* was considered, results suggest that television viewing was influential. After controlling for the size of people's online networks, there was positive and significant relationship between the amount of television consumed and the likelihood that these network contacts are relative strangers. Extant research shows that people use networking sites to connect to others with whom they share off line connections (Ellison et al., 2007). For example, students typically friend others with whom they have shared a class, lived together with, or met F2F at a social event. While this may be the case most of the time, the results presented herein suggest that television viewing is associated with increased promiscuity in "friending" behavior online.

Aggregate television viewing did not have a significant relationship with photo sharing frequency. Age and gender combined to explain the most variance compared to the other three models and younger female respondents were the most heavily engaged in this practice; these results are clear. It may be surprising that given the limited range in age among participants that this variable was consistently significant in the models. However, this indicates an interesting trend: it seems that as more reality television programming becomes available over time, younger people experience greater levels of exposure to the messages embedded in these programs. As such, they are more likely to engage in the behaviors examined in this study. Future research should focus on this trend because the results presented herein suggest that tomorrow's college freshmen will manifest these behaviors with greater intensity.

Perhaps there are gender differences inherent in this behavior that should be considered in conjunction with media use, as well. For example, French and Raven (1959) differentiate between reward, coercive and legitimate power, and Johnson (1976) discussed gender differences in the utilization of these bases of power suggesting that men are more likely to exhibit reward and coercive power, but *referent* power should be more strongly associated with women due to the interpersonal nature of referent power. Simply, referent power is a function of a person's social attractiveness to others, and *power* derives from association with desirable others. Although the findings in the current research begin to clarify the connection between reality television, the drive for celebrity, and SNSs, many questions remain. If stereotypical gender differences in power persist online, then future research should address in greater detail women's motivations to share photos online.

Although these models generally did not explain a great deal of variance (on average about 5.5 percent), overall the models were significant. If one considers the multitude of stimuli people are exposed to day after day, it is not surprising that television viewing explained relatively small portions of behavior. This is consistent with a wealth of extant research on media effects.

Next, aggregate television viewing was parsed into 5 broad content categories, and these genres were then regressed onto the same four dependent variables used earlier. This was done in an attempt to further delineate the relative influence each content category has on people's behavior in the context of SNSs and to address the research question regarding reality television consumption. Recall that television viewing was measured by prompting respondents to indicate how many hours per day and days per week they viewed reality television, news, fiction, educational and "other" kinds of programming. Each of these variables was then regressed onto the four SNS behavior variables. The results point to a consistent, positive, and significant relationship between reality television consumption and each of the dependent variables. In other words, exposure to reality television programming which models a range of behavior promoting non-directed self-disclosure and positive outcomes associated with celebrity status had a strong and positive relationship with each of the dependent variables used in this study. It is also important to note that the lone reality television consumption variable explains more variance in every model than the aggregate viewing variable used in the first series of analyses.

While many issues and questions remain, the current study shows that motivations for SNS use can be explained in part by traditional mass media consumption. It is clear that Web 2.0 tools allow people to build and maintain extensive social networks and encourage activity traditionally associated with celebrity. These tools reinforce the central position image and appearance hold in social interaction, regardless of the mediated nature of communication today. Foster (2004), in an analysis of the online fan communities of the *Survivor* series, found that the fractious interactions of fans mirrored the competitive nature of the show. In comparison to the fan communities of other "cult" television series, "in reality TV (. . .) the fan isn't drawn into the simulation of a 'fantastic' world as much as offered the chance to participate in one that is meant to be already familiar" (Foster, 2004, p. 274). The reality television industry itself has not been slow to take advantage of these parallels, by setting up discussion forums for particular shows, allowing fans to create detailed personal profiles on these sites, and targeting marketing to social network sites like Facebook and MySpace. The *American Idol* presence on Facebook, for example, includes a fan page (532,803 fans at the time of this writing), a multitude of "widgets" that display *American Idol* branded content on users' profiles, and several message boards where fans discuss and debate the show's current contenders.

Finally, the reality television industry is poised to fully integrate itself with its surrounding online fan culture through direct interaction with fan-created media: the reality series *Iron Brides* encouraged potential participants to audition by uploading YouTube videos for judging – and all audition tapes are of course publicly viewable on YouTube (http://www.youtube.com/ironbrides). With the proliferation of SNSs and the aggregation and documentation of comprehensive "social networks," future research should address how the contemporary definition of "friend" is changing. One way to begin this investigation is to explore the utility and accessibility of resources embedded in SNS-mediated social networks. As the debate about whether Internet-based communication tools are enhancing our social lives or restricting them continues (see McPherson et al., 2006 for recent discussion), additional research is needed to explore people's *motivations* to connect and ultimately whether these connects have instrumental utility for users. Perhaps these tools are simply the latest platform on which people compete for attention.

REFERENCES

Andrejevic, M. (2003). *Reality TV: The work of being watched.* Lanham, MD: Rowman and Littlefield.

Bandura, A. (1986). Social foundations of thought and action: A social cognitive theory. Englewood Cliffs, NJ: Prentice-Hall.

Bandura, A. (1988). Organizational application of social cognitive theory. *Australian Journal of Management, 13*(2), 275-302.

Bandura, A. (2001). Social cognitive theory: An agentic perspective. *Annual Review of Psychology, 52*, 1-26.

Bandura, A., and Walters, R. H. (1977). *Social learning and personality development.* Holt, Rinehart, and Winston: London.

Barlow, J.P. (1996). A declaration of the rights of cyberspace. Retrieved April 23, 2009 from http://homes.eff.org/~barlow/Declaration-Final.html

Bente, G., and Feist, A. (2000). Affect-Talk and its kin. In Zillman, D. and Vorderer, P. (Eds.), *Media entertainment: The psychology of its appeal*, (pp. 113-134). Mahwah, NJ: Lawrence Erlbaum.

boyd, d., and Heer, J. (2006). Profiles as conversation: Networked identity performance on Friendster. Proceedings of the Thirty-Ninth Hawai'i International Conference on System Sciences. Los Alamitos, CA: IEEE Press.

Berners-Lee, T. (2006). developerWorks Podcast [Interview transcript] Retrieved April 23, 2009 from http://www-128.ibm.com/developerworks/podcast/dwi/cm-int082206.txt

Calvert, C. (2000). Voyeur nation: Media, privacy and peering in modern culture. Boulder, CO: Westview Press.

Choi, J. H. (2006). Living in Cyworld: Contextualising cy-ties in South Korea. In A. Bruns and J. Jacobs (Eds.), *Uses of Blogs* (pp. 173-186). New York: Peter Lang.

Davis, F. D., Bagozzi, R. P., and Warshaw, P. R. (1989). User acceptance of computer technology: A comparison of two theoretical models. *Management Science, 35*(8), 982-1003.

Donath, J. and boyd, d. (2004). Public displays of connection. *BT Technology Journal, 22*, 4, 71-82.

Dominick, J. R. (1984). Videogames, television violence, and aggression in teenagers. *Journal of Communication, 34*, 136–147.

Ellison, N. B., Steinfeld, C., and Lampe, C. (2007). The benefits of Facebook "friends:" Social capital and college students' use of online social network sites. *Journal of Computer-Mediated Communication, 12*, 1143-1168.

Eyal, K., and Rubin, A. (2003). Viewer aggression and homophily, Identification, and parasocial relationships with television characters. *Journal of Broadcasting and Electronic Media, 47*, 77-98.

Ferris, A. L., Smith, S. W., Greenberg, B. S., and Smith, S. L. (2007). The content of reality dating shows and viewer perceptions of dating. *Journal of Communication, 57*, 490-510.

Finn, S. (1997). Origins of Media Exposure: Linking personality traits to TV, radio, print and film use. *Communication Research, 24*, 507-529.

Foster, D. (2004). 'Jump in the pool': The competitive culture of *Survivor* fan networks. In S. Holmes and D. Jermyn (Eds.) *Understanding reality television* (pp. 270-289). New York: Routledge.

French, R. P., and Raven, B. (1959). The bases of social power. In D. Cartwright (Ed.), *Studies in social power* (pp.150-167). Ann Arbor: University of Michigan Press.

Gerbner, G., and Gross, L. (1976). Living with television: the violence profile. *Journal of Communication, 26*, 172-194.

Gerbner, G., Gross, L., Morgan, M., and Signorielli, N. (1986). Living with television: The dynamics of the cultivation process. In J. Bryant and D. Zillman (Eds.), *Perspectives on media effects* (pp. 17–40). Hillsdale, NJ: Erlbaum.

Graham, S. (2004). Beyond the `dazzling light': from dreams of transcendence to the `remediation' of urban life. *New Media and Society, 6*, 33-42.

Gross, G., and Acquisti, A., and Heinz, H. J. (2005). Information revelation and privacy in online social networks. In the Proceedings of the 2005 ACM workshop on *Privacy in the Electronic Society*. November, Alexandria, VA.

Hill, A. (2005). *Reality TV: Audiences and popular factual television*. New York: Routledge.

Johnson, P. (1976). Women and power: Toward a theory of effectiveness. *Journal of Social Issues, 32*(3), 99-110.

Jones, S. and Fox, S. (2009) *Generations Online in 2009*. Washington, DC: Per Internet and American Life Project. Retrieved April 23, 2009 from http://www.pewinternet.org/Reports/2009/Generations-Online-in-2009.aspx

Lee, J. S., Cho, H. C., Gay, G., Davidson, B., and Ingraffea, A. (2003). Technology acceptance and social networking in distance learning. *Educational Technology and Society, 6*(2), 50-61.

Lenhart, A. and Madden, M. (2005). *Teen content creators and consumers*. Washington, DC: Pew Internet and American Life Project. Retrieved April 23, 2009 from http://www.pewinternet.org/PPF/r/166/report_display.asp

Lippman, W. (1922). *Public opinion*. New York: The Free Press.

McCraken, G. (1989). Who is the celebrity endorser? Cultural foundations of the endorsement process. *Journal of Consumer Research, 16*,310-321.

McGhee, P. E., and Frueh, T. (1980). Television viewing and the learning of sex-role stereotypes. *Sex Roles, 6*, 179–188.

McPherson, M., Smith-Lovin, L., and Brashears, M. E. (2006). Social isolation in America: Changes in core discussion networks over two decades. *American Sociological Review, 71*, 353-375.

Miller, A. D., and Edwards, W. K. (2007). Give and take: a study of consumer photo-sharing culture and practice. In *Proceedings of the SIGCHI Conference on Human Factors in Computing Systems* (San Jose, California, USA, April 28 - May 03, 2007). CHI '07. ACM, New York, NY, 347-356.

Mitchell, J. C. (1969). The concept and use of social networks. In J. C. Mitchell (Ed.), *Social networks in urban situations: Analyses of personal relationships in Central African towns* (pp. 1-50). Manchester, UK: Manchester University Press.

Moore, G. C., and Benbasat, I. (1991). Development of an instrument to measure the perceptions of adoption an information technology innovation. *Information Systems Research, 2*(3), 192-222.

Morgan, M. (1982). Television and adolescents' sex-role stereotypes: A longitudinal study. *Journal of Personality and Social Psychology, 43*(5), 947-955.

Musgrove, M. (2006). Nikon says it's leaving film-camera business. *Washington Post*, D01. Retrieved April 23, 2009 from http://www.washingtonpost.com/wpdyn content/article/2006/01/11/ AR2006011102323.html

O'Reilly, T. (2005). What is Web 2.0? Design patterns and business models for the next generation of software. Retrieved April 23, 2009 from http://www.oreillynet.com/ pub/a/oreilly/tim/news/ 2005/09/30/what-is-web-20.html

Profile of the American College Student (PACS) Survey (2007). *Profile of the American College Student: University of Missouri-Columbia.* Columbia, MO: Institutional Research, UMC. Retrieved April 23, 2009 from http://ir.missouri.edu/reports-presentations.html

Perse, E. M. (1990). Involvement with local television news: Cognitive and emotional dimensions. *Human Communication Research, 16*, 556-581.

Reiss, J., and Wiltz, S. (2004). Why people watch reality TV. *Media Psychology, 6*, 363-378.

Rothschild, N. (1984). Small group affiliation as a mediating factor in the cultivation process. In G. Melischek, K. E. Rosengren, and J. Stappers (Eds.) *Cultural Indicators: An international symposium* (pp. 377-387). Vienna: Verlag der Osterreichischen Akademie der Wissenschaften.

Rubin, A. M., Perse, E. M., and Taylor, D. S. (1988). A methodological examination of cultivation. *Communication Research, 15*, 107–134.

Sharma, D. C. (2005). Study: Cameraphone market will top digital cameras. *CNET News.com*. Retrieved April 23, 2009 from http://www.news.com/Study-Camera-phone-market-will-top-digital-cameras/2100-1041_3-5827024.html

Shanahan, J., Scheufele, D. A., Yang, F., and Hizi, S. (2004). Cultivation and spiral of silence effects: The case of smoking. *Mass Communication and Society, 7*, 413-428.

Shanahan, J. and Morgan, M. (1999). *Television and its viewers: Cultivation theory and research.* New York: Cambridge University Press.

Shrum, L. J. (2007). The implications of survey method for measuring cultivation effects. *Human Communication Research, 33*, 64-80.

Shrum, L. J., Wyer, R. S., and O'Guinn, T. C. (1998). The effects of television consumption on social perceptions: The use of priming procedures to investigate psychological processes. *Journal of Consumer Research, 24*, 447-458.

Slater, M. D., Henry, K. L., Swaim, R. C., and Anderson, L. L. (2003). Violent media content and aggressiveness in adolescents: A downward spiral model. *Communication Research, 30*, 713-736.

Stefanone, M. A., and Jang, C.Y. (2007). Writing for friends and family: The interpersonal nature of blogs. *Journal of Computer-Mediated Communication, 13*(1), 123-140.

Stefanone, M. A., and Jang, C. Y. (2008). Social exchange online: Public conversations in the blogosphere. In the *Proceedings of the 41st Annual Meeting of the Hawaii International Conference on Systems Science*. Los Alamitos, CA: IEEE Press.

Turkle, S. (1995). *Life on the screen: Identity in the age of the Internet*. New York: Simon and Schuster.

Walther, J. B., Van Der Heide, B., Kim, S. Y., Westerman, D., and Tong, S. Y. (2008). The role of friends' appearance and behavior on evaluations of individuals on Facebook: Are we known by the company we keep? *Human Communication Research, 34*, 28-49.

Wakefield, M., Flay, B. R., Nichter M., and Giovino, G. (2003). Role of media in influencing trajectories of youth smoking. *Addiction,* 98, 79-103.

Weinberger, D. (2001). Weblog Stat Quotations. Joho the Blog. Retrieved April 23, 2009 from http://www.hyperorg.com/blogger/archive/2001_12_01_archive.html#8043585

SECTION II: EUROPE

In: Reality Television-Merging the Global and the Local
Editor: Amir Hetsroni, pp. 47-63

ISBN 978-1-62100-068-6
© 2010 Nova Science Publishers, Inc.

Chapter 3

YOU'LL SEE, YOU'LL WATCH: THE SUCCESS OF *BIG BROTHER* IN POST-COMMUNIST BULGARIA

Maria Raicheva-Stover
Washburn University, USA

Until recently, Bulgarians thought of Big Brother as the embodiment of a totalitarian government capable of subjecting everybody to an uninterrupted surveillance apparatus. Having survived 45 years of communist control, Bulgarians were all too aware of George Orwell's gripping description of the totalitarian state in his *Nineteen-Eighty-Four* novel. Not a lot of people would have heard of the reality show by the same name. The introduction of *Big Brother* as the first Bulgarian reality show, however, marked a sharp transition to a new media reality in this post-communist country. Almost overnight, *Big Brother* became a big media event, attracting not only unparalleled attention from audiences, but dramatically changing the television landscape of the country. According to official data, 2.1 million viewers (62.5 percent of all viewers) were glued to the screen for the first season finale (Antonova and Kandov, 2005).

Big Brother's global success has been the subject of numerous scholarly publications. Studies of the show's reception and significance in the Western world – Great Britain, Australia and the United States – are truly noteworthy (Hill, 2002, Hill, 2005; Kilborn, 2003; Mathijs and Jones, 2004, Bignell, 2005, Andrejevic, 2004; Murray and Oullette, 2009). Yet existing publications are narrowly restricted to Western countries (see Mathijs and Jones, 2004) in spite of the fact that Russia and Poland adopted the format as early as 2001, and Romania, Bulgaria, Croatia, the Czech Republic, Hungary and Serbia created their own versions four years later (Johnson-Woods, 2002). Although the reality television format has fared well in most post-communist countries, its spread to Central and Eastern Europe has failed to generate scholarly research. The pioneer in this respect became a study by Volcic and Andrejevic (2009) examining the nationalistic underpinnings of a Macedonian show modeled after *Big Brother*. Clearly, additional research is needed to illuminate how the reality TV format has fared in other post-communist nations. The purpose of this chapter is to disentangle the formula for success of a global format like *Big Brother* by using post-communist Bulgaria as a case study. In particular, the chapter examines the introduction of *Big Brother* as a hybrid television format, which combines elements of Western and local discourses and mediations. The claim for hybridization, however, needs to be placed within the proper context of Bulgarian media history and recent market trends.

BRIEF THEORETICAL FRAMEWORK

The ability of various television formats to transcend international borders has been well documented (Robertson, 1992; Moran, 1998; Magder, 2004). Yet unlike scholars in the 1950s and 1960s, current analyses examine the transnational television flow as more than just an attempt at cultural homogenization (Waisbord, 2004; Magder, 2004; Bignell, 2005). Instead, they see it as a complex process where globalization of television formats happens in conjunction with, not against, national cultures. Formats are places for negotiation of global and local trends, rather than "Trojan horses" of Western Culture (Moran, 1998; Moran, 2005). In fact, scholars see unscripted formats, and particularly hit reality TV formats, as the perfect instances of glocalization, where "we encounter a program that is abstract and international in type while simultaneously local and concrete in its particular manifestations" (Moran, 2005, p. 299). The glocalization claim can be supported on two levels. On the one hand, reality game shows like *Big Brother* provide the dependability of a tried and tested format that has achieved great commercial success in other markets (Bignell, 2005; Madger, 2004). Add its low production costs, and the format quickly becomes a much-desired global commodity. Madger (2004) observes that formats are more popular outside the United States, "where in general, commercial broadcasters cannot afford the trial-and-error approach that has been the hallmark of developing hit shows in the United States" (p. 147). On the other hand, reality shows like *Big Brother* can be seen as open texts that can be easily adapted to local cultural preferences and idiosyncrasies (Moran, 1998). These shows offer audiences the opportunity to associate and recognize themselves as members of a distinct cultural community, yet they are not tied to any culture. Formats therefore are designed to be adapted locally (Madger, 2004). In the words of Brian Briggs, who is an international line producer at Endemol (the most prolific producer of global TV formats), the success of global formats is due to the fact that:

> ...we take a format that works in one country, strip everything cultural off of it, export it to a new country and then, over time, add cultural aspects of that country to it (as quoted in Madger, 2004, p. 147)

Undoubtedly, such an argument foregrounds the importance of examining the introduction of a global format like *Big Brother* to a post-communist setting.

BIG BROTHER ORIGINS

Of all reality television shows, none has achieved the legendary status and popularity of *Big Brother*. The first episode aired in the Netherlands in the fall of 1999, where it became the most watched show of the year and attracted an average 27.5 percent audience share (Johnson-Woods, 2002). The official description of the show on Endemol's web site, the production company responsible for it, states that the format is centered around four elements:

- The environment in which the contestants live is stripped to basics.
- The knock-out system by which the contestants are voted out of the house by the audience at home.
- The tasks, set by the production team, which the contestants must complete on a weekly basis.
- The diary room, in which the contestants are required to record their feelings, frustrations, thoughts and nominations.

For 100 days, the houseguests live in a completely media-free environment –they have no access to television, radio, clocks, or telephones; even paper, pens, or anything else one can write with, are forbidden. Similar to a totalitarian regime, the houseguests are subordinated to one unseen individual, called "Big Brother," who assigns the tasks and sets the rules. "Big Brother," for example, can evict a contestant or change the rules whenever he chooses. The show's similarity to a game – the last three left in the house win cash prizes – contributes to its characterization as "part social experiment, part real-life soap, and part competition" (as cited in Andrejevic, 2004, p. 72).

Big Brother's claim to authenticity is augmented by the use of surveillance cameras and microphones that follow the actions of the houseguests 24 hours a day, seven days a week. Night-vision cameras film the houseguests during the night or in the bathroom. Although the houseguests are warned that anything they say or do could be broadcast, some of them seem to (consciously) ignore that fact. Unlike other reality shows, where content is first shot and then edited and packaged into weekly half-hour segments, *Big Brother* is aired at least five days a week and is usually packaged with a nonstop real-time feed on the Internet (with slight delays to allow editing for inappropriate content).

According to Endemol's web site, the show became so successful that the company has sold the format to more than 42 countries and has produced various spin-offs since then (Big Brother, n.d.). For example, the British produced *Teen Big Brother*, while the Germans created an entire Big Brother community (Johnson-Woods, 2002; Hill, 2002). The year 2000 brought the *Big Brother* format to Germany, Switzerland, Sweden, Italy, the United Kingdom, Portugal, Spain and Belgium (Hill, 2002; Johnson-Woods, 2002). Denmark, France, Greece, Norway, Turkey, Argentina created their versions in 2001 (Hill, 2002; Johnson-Woods, 2002). The Italian version became so successful that 69 percent of the population said they have watched it (Hill, 2002; Johnson-Woods, 2002). Globally, *Big Brother* has transcended cultural and linguistic borders to appear in Mexico, South Africa, Thailand, Brazil, Israel and Iceland.

BIG BROTHER IN POST-COMMUNIST COUNTRIES

Poland (italics)

Poland became the first post-communist country to create its own version of *Big Brother* in March of 2001. Although Polish audiences had been previously exposed to lifestyle reality shows, *Big Brother*'s resemblance to a game managed to captivate the viewers (Johnson-Woods, 2002; "Big Brother is Coming," 2001). Close to 10,000 candidates appeared for the first casting session, yet in the Polish version, the viewers could select two of the 12 houseguests who were competing for the final prize of 500,000 slotys (or $125,000) ("All Eyes on Big Brother," 2001). In line with the country's grim communist past, *Wielki Brat*, or *Big Brother* in Polish, took militaristic underpinnings. The contestants were subjected to a lie-detector test prior to entering the house and then had to participate in a military training task where they were required to wear gas masks (Johnson-Woods, 2002).

Even before the start of the first season, *Wielki Brat* received lukewarm media coverage and poignant comments from local scholars about the show's low cultural value ("All Eyes on Big Brother," 2001). One Polish sociologist called it a shameful pornography "because it is insincere, filled with ideas that we will learn something about ourselves" ("All Eyes on Big Brother," 2001). In response to the increasingly bold content of realty shows, several Polish directors – including internationally known Andrzej Wajda – wrote an open letter to the National Council for Radio and Television condemning the negative influence of reality television on viewers and asking for a government regulation. A particular point of concern was that "the shows become more and more drastic. They sell more and more intimate spheres of life" ("Reality TV Under Fire," 2001). Possibly as a censorship measure, the National Council for Radio and Television soon after gave *Wielki Brat* a fine equal to $72,200 for showing violent and erotic scenes.

TVN, which was one of Poland's leading commercial broadcasters, originally bought the rights to *Big Brother*. Although the first season attracted four million regular viewers and six million viewers for the final episode (Johnson-Woods, 2002), the show ended after its third season in 2002. On the other hand, TVN did consider airing a ready-made version of the UK *Big Brother* (Holmwood, 2008) and achieved better success with two dancing shows – *Dancing With the Stars* and *You Think You Can Dance* – that still enjoy great popularity with Polish audiences. In 2008 TVN sponsored an event that broke the world record for the largest number of dancing couples (1,635) and the largest number of people dancing the cha-cha together (3,270) ("Dancers Set," 2008).

Russia (italics)

In October of 2001, *Big Brother* came to Russia in a slightly changed format. The show, called *Za Steklom*, or Behind the Glass, became Russia's first reality show. Endemol threatened to initiate legal actions against the Russian copycat, yet the show run for two seasons on Russia TV6, before the station was shut down in what some describe as a political move (O'Flynn, 2001). In the first season of *Za Steklom*, the typical *Big Brother* house was exchanged for a glass apartment in a hotel facing the Red Square where three men and three women had to live for 34 days (Johnson-Woods, 2002). The big prize was a Moscow flat ("Russian Big Brother," 2001). The show's producer, Ivan Usachev, commented that "The purpose of the experiment is to take people back to the times when there was no television. We will see whether today's youth has retained a love of talking" ("Russian Big Brother," 2001). Interestingly, the show quickly became the center of public discussion because of its participants' extroverted behavior, not love of talking ("Russian Big Brother," 2001). The Russian newspaper Izvestia called the show "porn" while another newspaper accused it of having no limits (Iqbal, 2001). One Russian columnist bemoaned the fact that "people who used to go to the museum and theaters are being corrupted by this sort of Western trash" and called the show "vulgar, sick and filthy" (as quoted in Johnson-Woods, 2002, p. 18).

The show and its web site, however, enjoyed phenomenal popularity among the younger audiences. Each day close to 3,000 people would wait on a long line and pay 20 rubles for the opportunity to watch the contestants through the one-way glass of their apartment. Sixty-seven percent of the Russian population claimed to have watched the show at least once (Johnson-Woods, 2002; "Russian Big Brother," 2001). During the show's first season, its web site, zasteklom.tv6.ru (now closed), received a record number of visitors – 3.1 million for the first four weeks (Korkina, 2001).

A number of reality formats have emerged on the Russian television market since *Za Steklom*. By 2005 Russian audiences were exposed to locally produced versions of *Fame Academy*, *Survivor* and *Pop Idol*, while plans were being made to bring *Wife Swap* and *The Apprentice*. And although the Russian channel TNT launched the Endemol version of *Big Brother* in 2005, the show failed to achieve high ratings (Malpas, 2005).

Other Post-Communist Nations

As a true global format, *Big Brother* continued its expansion to other post-communist nations. Its low production costs might explain, at least partially, (Volcic and Andrejevic, 2009) why) , reality shows invaded so many Eastern European television markets. The Croatian channel, RTL Televizija, launched *Big Brother* in September of 2004, and the show is currently in its sixth season. The Slovenian spinoff of a Portuguese show called *Bar*, in which the houseguests work in bar attached to their apartment, became the most watched TV program in the 18-49 age group and reached 54 percent of all viewers (Marko, 2007). In Macedonia a show called *To Sam Ja* (That's Me), which was similar in format but unaffiliated with the *Big Brother* franchise, turned into an experiment at ethnic reconciliation when it combined twelve contestants from six different nationalities (Volcic and Andrejevic, 2009). Promoted as "the first Balkan reality show," *To Sam Ja* became the most watched program in Macedonia after the national news (Volcic and Andrejevic, 2009). Serbia is planning its third *Big Brother* season after a tragic car accident killed three of its houseguests in the 2007 season ("Serbian Big Brother," 2007).

BIG BROTHER BULGARIA

Since its remarkable introduction, *Big Brother* Bulgaria has featured four regular and three VIP seasons and continues to be one of the top rated shows in the country. The first *Big Brother* aired on Nova TV on October 18, 2004, although speculations about the show started circulating a year earlier (Popova, 2003). As a novelty format on the Bulgarian airwaves, *Big Brother* generated unprecedented media coverage. Newspapers immediately started to define the term 'reality television' for their readers as well as speculate about the show's success on local soil (Dimitrova, 2004; Auret, 2005; Leviev-Sawyer, 2004; Daskalov, 2005). Yet by the time the casting call had ended, close to 15,000 candidates tried to secure a position on the first ever Bulgarian reality show. The 12 finalists proved to be a diverse group of six men and six women, selected to represent not only different regions and social classes, but also, as the show's producers admitted, clashing personalities ("Nova TV Thinks," 2004).

The Bulgarian version did not deviate substantially from the established global format. Unlike the Polish or Russian versions, the show kept its original logo – white letters on a blue background—and even its original name. The show's name was spelled and pronounced in English in spite of the fact that English is not one of the official languages of the country. Again in accordance with the global *Big Brother* format, the premise of the show was to live isolated from the rest of the world for 13 weeks. Viewers could evict the houseguests by calling or sending SMS messages. The finalist in the first season was entitled to a prize of 200,000 leva (equal to $147,000 or 700 average salaries in Bulgaria) while the second- and third-place winners won a trip to Tunisia (Peteva, 2005). The best moments of each day in the house were broadcast primetime every night, with an uncensored version aired daily after 11 p.m. that attracted the biggest audience. Nova TV selected one of the cable TV providers, CableTEL, to created a dedicated channel, called Big Brother Channel, to air live footage from the house 24 hours a day for the duration of the show ("SEM Reprimanded Nova TV," 2004).

In light of *Big Brother*'s phenomenal success, the plans for the second season were announced two weeks before the end of the first season (Peteva, 2005). This time the casting call attracted more than 30,000 people, including enthusiasts from the UK, Italy, Russia, Ukraine, Lybia and even an Orthodox monk who passed the casting but was not allowed to participate by the church (Leviev-Sawer, 2005). Among numerous speculations about the success of the second season (Bondokova, 2005; Antonova, 2005), the show's producers announced that they would change the rules to make it even more exciting. Of the 25 people allowed to come to the final casting, only 15 would enter the house, which would be stripped of any amenities. The houseguests had to live in a rural setting, grow their own vegetables and take care of a goat, several hens and a cow. During the first week, the houseguests did not have beds or hot running water. To further reinforce the Spartan atmosphere of the second season, the show's producers gave the logo a Stone Age look.

The second season's colorful combination of participants, offering a double dose of shocking tasks, tears, intrigues, and love triangles proved to be even more successful than the first (Antonova and Kandov, 2005). The ratings for the second season were five percent higher and so were the show's audience shares. In its first two weeks, *Big Brother 2* achieved 13.5 percent higher share than the first season, proving that the reality format has a place on Bulgarian TV. As one prominent Bulgarian newspaper commented: "the [Big Brother] show is the most commented television show" (Antonova, 2005).

Subsequent *Big Brother* seasons capitalized on attention-grabbing approaches to draw viewers in. The third season featured triplets, one of whom became the winner, and increased *Big Brother's* average rating in comparison to its second season by three percent (from 11.9 to 15) (Grancharova, 2007). The second VIP season, which featured the first wedding as part of a Bulgarian TV show, had an average rating of 21.3 percent while its bTV competitor,

Music Idol, achieved a "modest" 15.4 percent rating (Grancharova, 2007). In the 8 p.m.–9:45 p.m. time slot, 43.5 percent of the Bulgarian audience preferred *VIP Big Brother* while only 34.3 percent chose *Music Idol* (Grancharova, 2007). Only the future would tell whether Bulgaria would follow Poland's penchant for talent reality TV shows.

BIG BROTHER'S SUCCESS FORMULA

The cult status of *Big Brother* Bulgaria is indicative of this global format's adaptability. One is compelled to ask, therefore, what are the reasons for *Big Brother's* popularity in this post-communist country? There are several possible explanations.

Big Brother as a Must-Have Format

First and foremost, in the face of *Big Brother*, this global format offered the Bulgarian audiences a dramatically different television experience. For more than four decades, the Bulgarian National Television was the sole source of news and entertainment for the Bulgarian people. As one of two available TV channels during communist times, Channel One functioned as the official carrier of informational programming in the form of communist propaganda. After the fall of the Berlin Wall in 1989, the government continued to hold a firm grip on national broadcasting by turning Channel One into a national public service broadcaster, called BNT, that remained tightly controlled and financed by the government (IREX, 2003). In the decade following the fall of communism, the slow transformation of the television sector was characterized by excessive state interference and the limited programming scheme of the only national broadcaster (Schweitzer, 2003). The mushrooming of private cable operators in bigger cities – some 29 companied were operating by 1994 (Bakardjieva, 1995) – allowed viewers a better assortment of programs, yet the quality and selection varied greatly from city to city. BNT remained the only national broadcaster until 2000, when bTV, an affiliate of Rupert Murdoch's News Corp., won the license to become the first private national broadcaster. In line with its initial promise to create a programming structure that will attract a steady stream of advertising revenues (Popova, 2001), bTV brought such shows as *Everybody Loves Raymond, Friends, Dharma and Greg, Married with Children, Ally McBeal, CSI* and *Sex and the City* to Bulgarian audiences. The approach paid off – within a year bTV became the most popular channel in Bulgaria, dominating the national air as well as advertising revenues (Ibroscheva and Raicheva-Stover, 2007).

Nova TV's positioning as the second private national broadcaster has an interesting history. Nova TV was established in 1994 and became the first private cable channel in Bulgaria to broadcast 24-hour programming. Its reach, however, was limited to the capital city of Sofia. When Nova TV made a bid to become the *first* private national broadcaster, it was disqualified because of suspicions over the transparency of its capital (Ivantcheva, 2000). After Nova TV's failed bid, the channel became the property of the Greek Antenna group, which sold it to the Swedish MTG in 2008. In July of 2003 Nova TV won the license to become the *second* commercial broadcaster, yet at that point it had to deal with bTV's dominance (see Table 1). To challenge bTV's leading position with audiences and advertisers, Nova TV had to resort to innovative programming approaches. Nova TV had already introduced a spin-off of *Who Wants to Be a Millionaire* with the Bulgarian name *Stani Bogat* (Get Rich) in 2001, which according to Nova TV's web site, became the most successful game show in Bulgaria (Nova TV, n.d). Yet, with the introduction of *Big Brother* in 2004, Nova TV was looking for ways to capture audiences with a must-see format. Hoping that *Big Brother* will be "a revolution for TV in Bulgaria," George Zois, executive director of Nova TV, said that "Never before have we had a reality show on Bulgarian television and it's definitely the correct choice to start with Big Brother" (Waller, 2004).

Indeed, *Big Brother* changed everything. The first reality show became the burgeoning format that re-energized the television market and challenged the established status quo. It gave Nova TV the competitive edge it needed by dramatically raising its popularity with Bulgarian viewers (see Table 1).

Table 1. *TV Channels Rate (in percentages) for the Period January 2003-January 2009*

	Jan. 2003[a]	Jan. 2004[b]	Dec. 2004	Jan. 2005	Dec. 2005	Jan. 2007	Jan. 2008	Jan. 2009
bTV	85.8	87.1	83.8	83.9	80.3	80.5	82.2	78.6
NovaTV	23.9	38.5	49.7	50.7	52.4	62.4	65.9	64
BNT	60.8	64	56.2	56.6	59.9	55.7	51.2	40.3

Note. [a] Period before Nova TV became the second private broadcaster
[b] Period before start of *Big Brother*
Data from Alpha Research, http://aresearch.org.

Moreover, it raised Nova TV's profile as a commercial broadcaster and cemented its position as a key player on the national market. It must be pointed out, however, that Nova TV's approach was not without a precedent – reality TV formats have been crucial to gaining large (and primarily young) audiences for fledgling channels in other countries (Bignell, 2005). According to Nova TV's web site, "Nova's market share is constantly growing with demographic samples showing that Nova TV is preferred by the young, socially active and highly educated people" (Nova TV, n.d.)

The popularity of *Big Brother* forced the other key players to quickly adapt to the new media reality by adopting similar formats. By the fall of 2008, Bulgarian audiences could see *Big Brother, Fear Factor* (called *Fear*) and the locally produced *Sing with Me* on Nova TV, *Dancing Stars, Music Idol* and *Survivor* on bTV, and *The Life of Others*, which was specifically created for BNT by the Bulgarian producer of *Dancing Stars* (Antonova and Stoilova, 2008). Copy-cat shows like *Caught Cheating* started appearing on smaller cable broadcasters as well (Antonova and Stoilova, 2008). As the executive producer of *Big Brother* Bulgaria put it: "Reality is the innovative format for the moment. If you want to be relevant …you need to have a reality format" (Mitovski, 2008).

BIG BROTHER AND CELEBRITY (FORMATTING)

Despite accusations of encouraging a pathologic voyeurism (Boytchev, 2005), *Big Brother* became the symbol of a format that provided a viewing experience that was refreshingly different from all that Bulgarian television had to offer. In the wake of the new millennium, fictional entertainment has been described as formulaic, predictable and boring, while television personalities were perceived as annoying (Kilborn, 2003; Andrejevic, 2004; Antonova, 2004). That was the case with Bulgarian television; as a remnant of a communist past, programming was often stiff and boring – same faces, same stories, different night. With the introduction of *Big Brother*, for the first time in Bulgarian television history, viewers could see ordinary people (as opposed to professionals) in the ordinary rhythm of their daily lives – eating, sleeping, cooking and talking. It became a way to expand the medium's democratic potential (Murray and Oullette, 2009).

Moreover, instead of watching celebrities on the screen, ordinary people were given the chance to become celebrities themselves - something Andrejevic (2004) called "de-mystification of celebrity.". Couldry (2002, p.289) argued that "it was precisely the transition from ordinary (nonmedia) person to celebrity (media) persons that was the purpose of [Big Brother]. This was the master-frame without which the game made no sense." The celebrity version, on the other hand, had the opposite effect. While *Big Brother* promised to turn ordinary people into celebrities, the appeal of the VIP version of *Big Brother* was that celebrities were revealed to audiences as ordinary people (Andrejevic, 2004).

Big Brother Bulgaria proved to be refreshingly different in another aspect – it sent the message that it is work that anybody could do (Andrejevic, 2004). The reality TV format had the "lottery-like ability to make a star out of 'nobody'" (ibid., p. 4). No special skill or talent was needed to be selected as a participant; after all, "anyone could perform the work of being watched" on *Big Brother* (ibid, p. 6). In this sense, the houseguests represented the thousands of people who showed up for the casting call thinking that they could, too, be on TV. Undoubtedly, the selection of "ordinary" participants was subject to the "apparatus of celebrity production" (ibid., p. 5) and was limited to those "who make good copy for newspaper and magazine articles and desirable guests on synergistic talk shows and news specials" (Murray and Oullette, 2009, p. 11). Zara, the first season runner-up in the Bulgarian *Big Brother*, was the one who took full advantage of her celebrity status. She appeared in the Bulgarian version of Playboy, became the advertising face of a mobile phone company and a new wafer called Big Brother, and hosted a new reality show (Mihalev, 2005). As a matter of fact, all participants capitalized on their fame, although it was clear that becoming a genuine media star required much more than fame. As one editorial pointed out: "If *Big Brother* was an example of how unknown people managed to focus the interest of the entire society upon them for a short time, then the fame of the show's participants is indicative of the fact that fame is a heavy burden, which not everybody could carry" ("One-time Use," 2005). Yet it seems that in a culture where celebrity status was strictly reserved for the selected few for so long, audiences were eager the see ordinary people "making it" to the top. Audiences could identify with the people on the screen and their tears, fights and revelries, no matter how contrived they were (Mihalev, 2005). *Big Brother* allowed the thousands of people who tuned in every night at 8 p.m. to vicariously experience celebrity status.

BIG BROTHER AND THE DIVERSIFICATION OF TELEVISION CULTURE (FORMATTING)

In the backdrop of rigid cultural traditions, *Big Brother* exerted a democratizing effect in yet another way. It gave unprecedented visibility to taboo topics Bulgarian television had been cautious to handle. As one article commented, "because of inertia, narrow mindedness or unwillingness to take risks, phenomena as homosexuality and bisexuality in a Bulgarian context were lacking from the screen" (Lazarov and Tavanie, 2004). During the first season, one of the favorites, Zara, openly flaunted her bisexuality while another participant divulged his homosexuality. The end result was that,

> ....by a skillful blurring of the public and private spheres, *Big Brother* made people talk not only about politics and soccer, but brought important topics to the public discussion. Topics like aggression, sexuality, human relations, and even mental health. In this sense, we can say the *Big Brother* became the catalyst of a long overdue social transformation (Lazarov and Tavanie, 2004).

In the same vein, the wide mix of *Big Brother* participants contributed to the diversification of television culture. The roster of participants in the first *VIP* season, for example, included an opera singer, a couple of pop and folk singers, a beauty queen, a fashion model, a writer, a gay stylist, and a businessman with a Master's Degree in Management. The fourth *Big Brother* season challenged ethnic, religious and sexual taboos by featuring an all-female cast consisting of women from the Turkish and Roma minorities, a blind woman, a woman who overcame a serious illness as well as a woman with an alternative sexual orientation (Stoilova, 2008). And while the show's producers might have been insrumental in selecting this diverse mix of participants in search of good entertainment and higher ratings, one fact remains – Bulgarians were ready for the experience.

As the success of *Big Brother* Bulgaria indicates, audiences welcomed the opportunity to experience a different type of television. Indeed, the country was divided in two – those who watched *Big Brother*, and those who hated the ones who watched. Supporters of the game called it a unique social experiment and pure entertainment, while opponents called it primitive and voyeuristic (Boytchev, 2005). In spite of this, audience measures unequivocally show that *Big Brother* was watched by people from the entire country, coming from all social groups and ages (see Table 2).

Ironically, the first season was advertised with the catchy phrase "you'll see, you'll watch" and the Bulgarian audiences were ready to watch. Audience research shows that more women than men watched the show; an average 22 percent of the viewers were in the 45-55 age group, with 17 percent of the viewers in the 15-25 age group (Antonova and Kandov, 2005). Close to 27 percent of the viewers were in the service sector while another 24 percent were students (Antonova and Kandov, 2005). While audience ratings and shares do not necessarily guarantee viewing, it is clear that *Big Brother* was something people would tune to.

Table 2. Who Watched the First Season of Big Brother in 2004

		Total viewers above 4 years of age			Place of residence in percentages				Education in percentages			
		Rating in %	Rating in thousands	Share in %	Sofia	Big city	Small city	Village	No primary	Primary	Secon- dary	Post - secondary
Oct.	Big Brother 8 p.m.	10.3	769	25.4	26	28	27	20	17	20	44	20
	Big Brother 11 p.m.	4.1	305	20.6	32	25	26	18	12	20	44	24
Nov.	Big Brother 8 p.m.	12.7	946	31.5	22	28	29	20	18	18	45	19
	Big Brother 11 p.m.	4.5	335	24.6	30	28	29	13	11	18	47	24
Dec.	Big Brother 8 p.m.	14.2	1057	36.3	17	28	31	24	16	22	45	17
	Big Brother 11 p.m.	4.5	333	21.4	22	28	35	15	12	20	46	23
Jan.	Big Brother 8 p.m.	11.8	881	30.1	14	28	33	25	14	23	44	18
	Big Brother 11 p.m.	3.0	224	14.5	20	29	32	19	13	20	45	22

Note. From Antonova and Kandov (2005), compiled with data from TNS/TV Plan.

BIG BROTHER AS A MEDIA EVENT (FORMATTING)

The success of *Big Brother* was also directly linked to its producers' ability to orchestrate the show as a big media event, which, according to Scannell's definition of an event (2002), has the ability to generate much-desired publicity and spin-offs. Ironically, after decades of official party propaganda, Bulgarian audiences were ready to succumb to yet another form of cleverly contrived manipulation. For a start, Nova TV run an aggressive ad campaign, making the casting call, the selection of the participants, the bios of the final 12, and the house itself the center on national attention. Even politicians could not resist commenting on *Big Brother*. A month before the start of the first season, General Boiko Borisov, now Bulgaria's prime minister, visited the house and called it a luxury prison, but praised *Big Brother*'s high-tech approach (Novkov, 2005). Indeed, there was no detail too minor to be spared – Bulgarian viewers knew how much beer, meat, veggies, cheese and condoms were used by the houseguests (Peteva, 2005). As one journalist observed, everything on Nova TV, even the evening news, was somehow related to *Big Brother* (Boytchev, 2005).

Pavel Stanchev, the executive director of Nova TV, called the novel approach of promoting *Big Brother* "total communications" (Antonova, 2005). *Big Brother* was broadcast 24 hours on a specially designated cable channel; the houseguests were featured in a special edition of the popular magazine *Bliasak* (Sparkle), which sold a record 70,000 copies; the biggest national newspaper *24 Chassa* (24 Hours) and the popular radio *Express,* which broadcast updates four times a day, became the show's media partners (Antonova and Kandov, 2004). In addition, the first season's finale was celebrated in Sofia with "unique fireworks" and a music concert, while the towns of the three finalists erected live stages where "beer was in abundance" (Peteva, 2005). Not by coincidence, Nova TV decided to introduce its new logo, new music clips and new programming format on the premiere night of the second season (Antonova, 2005). And just before Christmas, two popular general interest magazines, *Koi* (Who) and *Hai Klub* (High Club) issued special *Big Brother 2* editions (Antonova, 2005).

BIG BROTHER AS AN INTERACTIVE EXPERIENCE (FORMATTING)

As a last ingredient in the formula for success, one should not overlook to examine the interactive component of the show, . Indeed, the popularity of *Big Brother's* web site only reinforces the idea that the show's global success was at least partially due to its clever use of the Internet as an interactive technology (Andrejevic, 2004, Johnson-Woods, 2002; Kilborn, 2003). Discussing the influence of the Internet on the show's popularity, Andrejevic (2004) quoted producer Doug Ross who commented that "Big Brother was generally considered to be more of a success on the Internet than on television" (p. 160). Indeed, being able to actively participate and comment on the experience is one the most attractive features of the interactive format (Kilborn, 2003; Andrejevic, 2004). As Kilborn (2003, p.81) comments:

> It is by exploiting the range of interactive possibilities....and by creating the illusion that audiences are calling all the shots that Big Brother, and shows like it, have been able to attain such landmark status in the history of contemporary television.

The success of the Bulgarian *Big Brother* web site was not unusual. In Germany, the official website generated 90 million hits during the first several weeks; the Dutch web site attracted 52 million visitors during the course of the show, while in the United States the official web site, hosted by AOL, received 9.4 million hits (Andrejevic, 2004).

The official Bulgarian web site, *www.bigbrother.bg*, which is updated each new season, boasts that it provides access to the show even to those without a television set. The lack of data on site traffic is regrettable yet an examination of the web site shows that it is not only technologically advanced, but also extremely popular, possibly because of the updates that appear every 30 minutes during an active season. Users post messages on the chat forums daily and some discussion topics get 80,000 hits. Among the customary participant bios, gallery and news, the attractive web site offers games, wallpaper, screen saver and ring tone downloads, a live chat and a live forum, as well as opportunities for fans to create a fan web page and post their MMS pictures. The most advertised feature, however, is Big Brother Live which allows fans, with the purchase of a special e-card, to "watch the life in the house 24 hours a day, seven says a week." The ardent viewers could also listen to what's happening in the house via a "spyline" or get news updates on their cell phones. The show also had an equally attractive fan web site, www.bigbrotherfans.info (closed in 2006), which offered fewer features, but had detailed site statistics. This is how one could gauge the show's initial success among Bulgarian fans. Since its creation on October 2004, the site had registered 16.5 million hits.

CONCLUSION

The success of *Big Brother* in Bulgaria foregrounds the hybridity of the reality format and its adaptability to a post-communist setting. In its ascent, the spread of the reality TV format is reminiscent of the "global resonance" of such supersoaps as *Dallas* several decades ago (Kilborn, 2003). The start of the new millennium witnessed a phenomenal interest in reality television, to the extent that no media pundit could afford to ignore the fact that reality television is not a passing fad. And whether liked or disliked by media critics, reality shows become tremendous hits with the audiences. The ratings of *Big Brother* across the world indicate that.

It must be pointed out, however, that *Big Brother's* success across the Eastern bloc was not universal. Even with shared memories of Big Brother from the communist past, *Big Brother* did not translate well in all post-communist settings. The show fizzled out in Romania, Slovakia and Hungary after its first few seasons (Auret, 2004) and failed to sustain its hit status in Poland and Russia. Yet, *Big Brother* did become a hit in Bulgaria and Macedonia (see Volcic and Andrejevic, 2009, for an extended discussion of the Macedonian version). The fact that these are two neighboring nations, with similar historic, cultural and religious experiences, might suggest something about the importance of cultural proximity as a factor in the reception of *Big Brother*. Clearly, additional cross-cultural research is needed to examine why *Big Brother* fares well in some post-communist cultures and not in others.

By examining the development and success of *Big Brother* in Bulgaria, this study offered an overview of the dynamics of the reality television format in a post-communist country. The discussion leaves us with the conclusion that *Big Brother's* formula for success represents an amalgamation of carefully orchestrated media events and the introduction of a dramatically different television experience for Bulgarian audiences after 40 years of government-controlled media. It also suggests that *Big Brother's* formula for success, with its emphasis on creating a new type of celebrity and unique opportunities for participation and interactivity, does work in a post-communist setting.

Big Brother not only revived the inert television market in Bulgaria by drawing unprecedented numbers of viewers, but also became responsible for generating a public debate over taboo topics. Moreover, by becoming a programming trail-blazer, *Big Brother* helped Nova TV change the status quo of the market and accelerate the adoption of reality game shows by its competitors. Further research is needed to explain the economic ramifications of such high ratings or the complex web of reasons behind audiences' attraction to this format. Yet the new reality after the introduction of the reality television format in Bulgaria is that, for now, reality rules.

REFERENCES

All eyes on Big Brother (2001, Feb. 11). *The Warsaw Voice.* Retrieved October 26, 2005, from http://www2.warsawvoice.pl/old/v642/News03.html

Andrejevic, M. (2004). *Reality TV: The work of being watched.* Maryland: Rowman and Littlefield Publishers.

Antonova, V. (2004, Dec. 18). The season of Big Brother. *Capital (50).* Retrieved October 16, 2005, from http://www.capital.bg/show.php?storyid=229339

Antonova, V. (2005, Oct. 15). Love Drama=Ratings: The second season of Big Brother is more successful. *Capital (41).* Retrieved October 16, 2005, from http://www.capital.bg/show.php?storyid=233325

Antonova, V. and Kandov, B. (2005, Jan. 25). Big Brother changed the media market. *Capital (03).* Retrieved October 16, 2005, from http://www.capital.bg /show.php?storyid=229777

Antonova, V. and Stoilova, Z. (2008, Sept. 26). The fall starts with songs, dances and intrigue. *Capital (03).* Retrieved April 25, 2009, from http://www.capital.bg/ biznes/media_i_reklama/2008/09/26/555473_naesen_s_pesen_tanci_i_intrigi/

Auret, (2005, Jan. 6). The Big Brother *The Sofia Echo.* Retrieved November 16, 2005, from http://sofiaecho.com/2005/01/06/639135_the-big-bother

Bakardjieva, M. (1995). The new media landscape in Bulgaria. *Canadian Journal of Communication, 20*(1), 67-79.

Big Brother (n.d.). Retrieved August 26, 2009, from http://www.endemol.com/what'big-brother.html

Big Brother is coming (2001, Feb. 25). *The Warsaw Voice.* Retrieved October 26, 2005, from http://www.warsawvoice.pl/archiwum.phtml/1533/

Bignell, J. (2005). *Big Brother: Reality TV in the twenty-first century.* New York: Palgrave Macmillan.

Bondokova, P. (2005, July 20). Reality after the reality show. *Novinite.com.* Retrieved August 4, 2005, from http://www.novinite.com/view_news.php?id=50240

Boytchev, A. (2005). The Big Brother of changes on the Bulgarian TV market.

Media Online. Retrieved October 28, 2005, from http://www.tol.cz.

Couldry, N. (2002). Playing for celebrity: *Big Brother* as ritual event. *Television and New Media, 3*(3), 283-293.

Dancers set double record (2008, Sept. 17). *The Warsaw Voice*. Retrieved August 26, 2009, from http://www.warsawvoice.pl/view/18651/

Daskalov, G. (2005, Aug. 18). Big Brother: 10 things that are not the same, *Dnevnik.com*, Retrieved November 16, 2005, from http://www.dnevnik.bg/analizi/2005/01/18/86558_big_brother_10_neshta_koito_ne_sa_sushtite/

Dimitrova, C. (2004, Oct. 21). Everyone watches Big Brother. *The Sofia Echo*. Retrieved November 16, 2005, from http://sofiaecho.com/2004/10/21/634472_everyone-watches-big-brother

Grancharova, E. (2007, Apr. 9). Reality show rivalry in Bulgaria. *The Sofia Echo*. Retrieved July 27, 2009, from http://sofiaecho.com/2007/04/09/652405_reality-show-rivalry-in-bulgaria

Hill, A. (2002). Big Brother: The real audience, *Television and New Media, 3*(3), 323-340.

Hill, A. (2005). *Reality TV: Audiences and popular factual television*. New York: Routledge.

Holmwood, L. (2008). Big Brother 3 sold to Polish TV network. *The Guardian*. Retrieved July 27, 2009, from http://www.guardian.co.uk/media/2008/apr/08/realitytv.bigbrother

Ibroscheva, E. and Raicheva-Stover, M. (2007). First green is always gold: An examination of the first private national channel in Bulgaria. In I. Blankson and P. Murphy (Eds.), *Negotiating Democracy* (pp. 219-239). New York: SUNY.

IREX, (International Research and Exchange Board) (2003). *Media Sustainability Index 2003: The development of sustainable media in Europe and Eurasia*. Retrieved August 5, 2005, from http://www.irex.org/msi/index.asp

Iqbal, A. (2001). *Row over Russian Big Brother*. CNN/World. Retrieved July 27, 2009, from http://archives.cnn.com/2001/WORLD/europe/11/22/bigbro/index.html

Ivantcheva, A. (2000, March 20). TV commercials: Bulgaria licenses its first private national TV operator. *Central European Review, 2*(11). Retrieved July 27, 2009, from http://www.ce-review.org/00/11/ivantcheva11.html

Johnson-Woods, T. (2002). *Big Brother*. Australia: University of Queensland Press.

Kilborn, R. (2003). *Staging the real: Factual TV programming in the age of Big Brother*. Manchester: Manchester University Press.

Korkina, S. (2001, Dec. 5). 'Za Steklom' web wite makes RuNet history. *The Moscow Times*. Retrieved November 2, 2005, from http://internal.moscowtimes.ru/indexes /2001/12/05 /01.html

Lazarov. A. and Tavanie, I. (2004, Dec. 25). What happened during 2004. *Capital (51)*. Retrieved October 16, 2005, from http://www.capital.bg/show.php?storyid=229407

Leviev-Sawyer, C. (2004, Oct. 7). Legal Alien - Between fantasy and reality. *The Sofia Echo*. Retrieved October 16, 2005, from http://sofiaecho.com/2004/10/07/632314_legal-alien-between-fantasy-and-reality

Leviev-Sawyer, C. (2005, Oct. 3). Legal Alien - Bulgaria's Big Brother and the holding company. *The Sofia Echo*. Retrieved October 16, 2005, from http://sofiaecho.com/2005/10/03/640752_legal-alien-bulgarias-big-brother-and-the-holding-company

Madger, T. (2004). The end of TV101: Reality programs, formats and the new business of television. In S. Murray and L. Oullette (Eds.), *Reality TV: Remaking Television Culture* (pp. 137-156). New York: New York University Press.

Malpas, A. (2005, Aug. 19). Different realities. *The Moscow Times.* Retrieved August 26, 2009, from http://themoscowtimes.com/arts/2005/08/19/182235.htm

Marko, H. (2007, Sept. 5). Big Brother, give us bread and circuses! *Slovenia Times.* Retrieved August 26, 2009, from http://www.sloveniatimes.com/en /inside.cp2?uid= D4623AEA-9CCF-6AE5-7C18-D3648F9BDED6andlinkid=newsandcid=762059D5-F84D-020A-FBA5-2AD66B5F38CB

Mathijs, E. and Jones, J. (2004). Big Brother International: Formats, critics and publics. New York: Wallflower Press.

Mihalev, I. (2005, April 16). O, Brother, where are you. *Capital (15).* Retrieved October 16, 2005, from http://www.capital.bg/show.php?storyid=230900

Mitovski, D. (2008, Sept. 26). Interview with the Big Brother producer, Dimitar Mitovki. *Capital (39).* Retrieved March 25, 2009, from http://www.capital.bg /show.php?storyid=555486

Moran, A. (1998). *Copycat TV: Globalization, program formats, and cultural identity.* Lutton, U.K.: University of Lutton Press.

Moran, A. (2005). Configurations of the new television landscape. In J. Wasko (Ed.), *A companion to television* (pp. 291-307). MA: Blackwell.

Murray, S. and Oullette, L. (2009). *Reality TV: Remaking television culture.* (2nd ed.). New York: New York University Press.

Nova TV (n.d.). Retrieved Aug. 26, 2009, from http://www.novatv.bg/bg/about/en

Nova TV thinks of continuing Big Brother, (2004). *Mediapool.bg.* Retrieved October 16, 2005, from http://www.mediapool.bg/show/?storyid=11203

Novkov, M. (2005, Jan. 31). For the brother of journalism. *Sega.* Retrieved October 16, 2005, from http://www.segabg.bg

O'Flynn, K. (2001, Dec. 4). Dutch firm warns 'Za Steklom.' *The Moscow Times.* Retrieved November 2, 2005, from internal.moscowtimes.ru/indexes/2001/12/04/01.html

Peteva, A. (2005, Jan. 17). Fireworks for the winner of Big Brother. *Standard.* Retrieved October, 26, 2005, from http://paper.standartnews.com/archive /2005/01/17/theday /s4322_19.htm

Popova, V. (2001, Oct. 27). bTV appeared in an official document of News Corp. *Capital (43).* Retrieved October 9, 2005, from http://www.capital.bg/show.php?storyid=211763

Popova, V. (2003, Aug. 30). The Big Brother against the long one. *Capital (34).* Retrieved October 9, 2005, from http://www.capital.bg/show.php?storyid=222794

Reality TV under fire, (2001, July 3). *BBC.com.* Retrieved November 16 from http://news.bbc.co.uk/1/hi/entertainment/tv_and_radio/1420850.stm

Robertson, R. (1992). *Globalization: Social theory and global culture.* London: Sage.

Russian Big Brother (2001, Nov. 12). BBC.com. Retrieved November 4, 2005, from http://news.bbc.co.uk/1/hi/entertainment/tv_and_radio/1651959.stm

Scannell, P. (2002). *Big Brother* as a television event. *Television and New Media*, 3(3), 271-282.

SEM reprimanded Nova TV for illegally showing Big Brother on cable, (2004, Nov. 8). *Mediapool.bg.* Retrieved October 16, 2005, from http://www.mediapool.bg /show/?storyid=11203

Serbian Big Brother cancelled following deaths of evicted contestants, (2007, Dec. 31). *Canadian Broadcasting Corporation.* Retrieved March 25, 2009, from http://www.cbc.ca/arts/tv/story/2007/12/31/serbia-bigbro-deaths.html

Stoilova, Z. (2008, Sept. 23). Big Brother 4: A hit success. *Capital.bg blog.* Retrieved March 25, 2009, from http://www.capital.bg/showblog.php?storyid=553543

Schweitzer, J.C. (2003). Advertising in Bulgaria: Prospects and challenges in this struggling democracy. Paper presented at the Association for Education in Journalism and Mass Communication Annual Convention, Kansas City, MO.

Volcic, Z. and Andrejevic, M. (2009). *That's Me:* Nationalism and identity on Balkan reality TV. *Canadian Journal of Communication, 33*(4), 7-24.

Waisbord, S. (2004). McTV: Understanding the global popularity of television formats. *Television and New Media,* 5, 359-383.

Waller, E. (2004, Sept. 3). Big Brother reaches Bulgaria. Retrieved September 12, 2005, from http://www.c21media.net/

In: Reality Television-Merging the Global and the Local
Editor: Amir Hetsroni, pp. 65-77

ISBN 978-1-62100-068-6
© 2010 Nova Science Publishers, Inc.

Chapter 4

GOK WAN AND THE MAGICAL WARDROBE

Gareth Palmer
University of Salford, UK

Lifestyle Television can be profitably analysed by considering the way in which magical symbolism combine with commercial interests to sell the ideology of consumerism. In this paper I begin by discussing Lifestyle Television's relationship to Reality Television and trends in consumerism in work very much informed by my reading of Adorno. I then follow with a detailed analysis of Gok's Fashion Fix. In this I draw on the mythical studies of Joseph Campbell to show how the Individual-as-hero and their mythic Quest to find the self presents a magical sheen to what is essentially a drive to align the individual with products more suited to their personality-type. I follow this by analysing the role of the Supernatural aid or Magical Helper. The gifts of this individual to befriend and help the hero-questor achieve self realization has to be considered in the light of the advertising that goes in and around the programme which in turn connects to the distinctly unmagical business of commerce. I then conclude by considering 'Entering the Belly of the Whale' - the Transformation – the onstage/catwalk fulfillment which is presented as a self-overcoming and a re-birth but is also a loud and aggressive celebration of what consumerism can do for the individual. In short I want to consider how the rational and the irrational are blended together to create a productive consumer whose self-fulfilment has a fortunate connection with the world of goods and services.

When reality Television first appeared in the 1980s it took a while to appreciate that this was more than just another innovation in documentary. Reality TV was and remains at the intersection of many conflicting and competing interests. In one sense it was allied to public service television inasmuch as the CCTV cameras which captured a large percentage of the footage on which RTV depended surveyed urban spaces on our behalf while attaching the material to security-related messages. Producers were quick to seize upon the commercial applications of RTV. Formats were developed which concentrated on accidental spectacle and titillation which helped us appreciate the value of conducting ourselves with caution less we be spied on by surveillance cameras. Reality TV reflected new trends in law and order as it was both an instrument of policing and in some cases a replacement of traditional policing as the verb 'to police' now became part of the way in which we were encouraged to understand ourselves. Reality TV focused on urban spaces as well as being informed by them. Today, reality TV is used as an ally of the authorities to inculcate messages about policing for example in programs about the police force such as 'Crimewatch UK'. In Gailey's neat formulation 'reality TV is a self-replicating system that works by appropriating pre-existing structures and 'authenticating' them through the conventions of the documentary form.' (Gailey, 2007, p.108)

The last decade has seen the grammar and techniques of reality TV enter into the subgenre of lifestyle television. Lifestyle formats such as 'Wife Swap', 'Ten Years Younger' 'Extreme Makeover 'and 'What Not to Wear' among many others constitute a sort of unofficial unarticulated Finishing School. In O'Sullivans words:

> Lifestyle continues to offer recurrent lessons and advice in ever-more effective stylistic management and transformation of the body, health, fashion, cookery, gardening, house and home, DIY, cars, travel and holidays and property. These updated and proliferating 'how-to-do-it' guides exists uneasily within the increasingly hybridized formats that emerged to achieve dominance in the 1990s. (O'Sullivan, 2005:30/31)

As the material about deportment and manners is available on television these lessons issue forth from the most democratic Finishing School imaginable. But a closer inspection of lifestyle television reveals a focus on class, identity and the body encouraging us to more tightly police our conduct. The realism and intimacy achieved by RTV techniques are deployed here to convey the emotional emotional intensity of key moments. The personal approach is designed to maximise and then transmit the dramatic out-pouring of the individual facing a personal crisis. Lifestyle television is interesting because it utilizes the techniques of reality TV techniques to open out questions of identity in ways that other television formats generally avoid. It is this strategy I will be exploring here in connection to the Fashion Stylist Gok Wan and in particular his series 'Gok's Fashion Fix.'

Lifestyle television's exploration of conduct and identity blend the aspirational drives of consumer culture with the signs, symbols and narratives of magic. Lifestyle presents raw emotional realism to showcase transformations that are both credible and magical. But although choice is a key concept in the genre the individual is guided to fashion a version of the self that is centred on the commodity. Lifestyle media uses the armoury of RTV to exploit the individuals' uncertainties and insecurities and then direct he or she towards a form of consumer therapy. The result is a narrow and limited field in which we are urged to make sense of ourselves. A new and deepened conformity is the happy product we are urged both to consume and to become. People are encouraged to choose a commodity-identity for that appears to be all there is.

A MAGICAL BUSINESS

> By its regression to magic under late capitalism thought is assimilated to late capitalist forms. (Adorno, 1994:129)

Writers from Adorno to Baudrillard have considered the 'pervasive undercurrent of irrationalism' in consumer culture. Astrology, Psychic Columns, Lottery numbers, Crystals and the claims of advertising are all based on the promise of miraculous individual transformation. However irrational the claim may seem to our logical mind-set the hope is perpetuated that the individual can be made special and find themselves the subject of a remarkable change. Thus we live with a mindset succintly described by Bernstein as 'seeing through and obeying'. (Bernstein: 1991:, p.4) Irrational statements may be treated by individuals with a great deal of suspicion or amusement but they still form the inescapable background to consumer culture. The triumphs of X-Factor contestants for example are presented as democratically-based via the vote, the call-to-action etc. But their success transforms the formerly ordinary winners into a magical realm where events seem determined by a different and wondrous system. Similarly advertising is both real and unreal –'a discourse where all physical and social arrangements are held in abeyance' (Jhally 1989, p.217).

Magical images, transitions and tricks are so pervasive and commonplace in advertising that they no longer excite comment. Writers in the Cultural Studies tradition such as Raymond Williams described how the 'alchemy' of advertising turned abstract objects into desirable things to create differences that may not be there (Williams, 1980). Alongside the established use of design and imagery to sell products are those symbolic associations which may be utilised and renegotiated to emphasise difference in lifestyle to demarcate social relationships. In place of class distinctions advertising helps foreground the democratic value of distinguishing ourselves by modes of consumption (Williamson, 1978). What I mean to highlight here is not the specious nature of their claims but the prevalence of this magical imagery and the potential for irrational thinking they help maintain. The 'bursts of enchantment' provided by angels communicating messages through a magazine column sit alongside claims that rearranging the furniture can magically usher money into the living room. Both form the soil in which dreams of personal transformation can be planted.

We might ask why there is such a concentration on magical signs, symbols and narratives in the early twenty-first century in media as wide-ranging as horoscopes, Tarot Hotlines and makeovers. It has been suggested that the less the world is understood the greater is the belief in superstition (Jahoda, 1974). In a culture of estrangement dominated by signs, work and anxiety it may be that social and psychological needs are satisfied in the irrational. An indulgence in the spells of magic may represent a welcome retreat from the entreaties of a culture that suggests that our work on our self should occupy both work and leisure time.

There are of course obvious commercial reasons for developing a faith in magic. It cannot be overlooked that the greatest concentration of magical goods and services are located in media aimed at those under the greatest social and economic pressure – the working class. Psychic hotlines, astrology and magical crystals are integrated into publications that extol the virtues of survival while steadfastly ignoring the conditions that created such suffering in the first place. Magic may represent a way of dealing with anxiety because it suspends reality, re-envisions it without any sort of critical impulse. It offers those prepared to indulge a shared distance by holding reality at arms length while allowing the enchantments that circulate so widely in our culture to take hold. If, as Weber argues capitalism reduces the world to mere materials, magic performs the necessary work of re-enchantment. (Weber, 2007) The sort of magic we witness in Gok and other lifestyle transformation programs is much like the business of advertising because it is an advert – it works on the same dynamic. A threat is created, a solution is offered, the person adopts the solution and normality is restored until the next advertisement arrives to unsettle the already anxious self. In this way the complex dynamic of contemporary life is conjured up - we are self-estranged in a confusing yet ordered world but in order to carry on we must be sold hope – the very soul of advertising. What is special about lifestyle television now is that it has imported the promise and magic of advertising into programming whose claims of genuine transformation depend upon the realistic effects of RTV.

Lifestyle television is driven by the potentials of transformation. The most popular formats are those that demonstrate in detail how transformation can take place – and then prove it through display. The narrative is always based on an individual plucked from obscurity to go on a quest with a magical helper who transforms them into the person they 'really are'. As Jack Bratich has pointed out these are essentially magical narratives - 'the fairy tale conveys a realm of profound modification and makeovers (as long as the protagonist had the right knowledge and skills' (Bratich,2007, p.17). Story-telling strategies such as these extend the message of advertising not only through marketing goods and services highlighted in the programme but also in selling versions of the self that can be transformed simply by the use of the right commodity and adopting the right attitude to the self. But although these magical signs, symbols and narratives are continually recreated by the media industry such extraordinary transformations are the results of dull mechanical labour. The magic is entirely the product of people seeking to maximise their own career options within a competitive and aggressive industry. Such workers are toiling to create formats that will attract audiences and thus keep them in work. Their aim is nothing less than perpetuating content that will 'seize the individual.' Thus while the processes of transformation must appear magical in order to attract increasingly unstable viewers the actual business-end of such work is dull and repetitive. In brief the magic is mechanical but it works because it is inspired by and feeds into a pervasive current of irrationality.

What is brought together in lifestyle television is the combination of self-policing first highlighted in RTV, the emotional realism needed to engender identification and the signs, symbols and narratives of magic. When all three are combined they effect a satisfying transformation that is both believable and entertaining. While the personal effect on the individual is a combination of the redemptive and literally life-changing the result is in every case a new model of the self based on commodities.

At the core of Lifestyle's strategy are some fascinating dichotomies that turn on the real/unreal divide. For example we might consider how the democratic impulse that allows 'everyman/women' to appear on the shows veils the care that goes into the selection of the contributors. We might contrast how the dull and mechanical labour of programme-making is disguised by the magical techniques that efface the vast majority of such effort. In the following I will explore how some of these are worked through by analysing Gok's Fashion Fix.

Gok Wan was born in Leicester in 1977 of English and Chinese parents. Before working on his first series for Channel Four in 2007 he operated as a high profile fashion stylist for many celebrities. Thus his qualifications for helping the individual on their quest have both the celebrity-magic of his past and commercial connections. But in mythical terms Gok represents a classic figure from the mythology – a sexless trickster/helper figure whose role is merely to aid the hero and to erase him or herself in the process. Crucial here is the notion of self-sacrifice. Gok's principal role is to act as a friend and guide on the quest before disappearing into the magic world of television once more.

Democratic Magic – Conflicting Pairs

Lifestyle's subjects, much like all of those living under the sway of consumer culture, are asked to find meaning and significance not merely in acquiring the right commodities but by becoming commodities. What is significant here is how we are invited to order our experience of ourselves. Beneath the admonition to be happy are messages of conformity and adaptation that silence reflection.

(I)The Adventure of the Hero./ Re-Making the Commodity Self

> "the awakening of the self…whether small or great and no matter what the stage or grade of life, the call rings up the curtain, always, on a mystery of transfiguration – a rite, or moment, of spiritual passage, which, when completed, amounts to a dying and a birth' (Campbell,1975: 52/53)

The ordinary heroes that feature in Lifestyle programs are those willing and able to undergo a degree of tele-treatment for the good both of themselves and the programme. The people represent commodities to the producers and they are more likely to be chosen if they can deliver the most emotional responses. The role of the host is to help the chosen individual re-build him or usually her self. The individual's emotional frailty is a commodity which viewers can identify with and which remains central to the narrative's development. Those chosen for tele-treatment in Gok's Fashion Fix are 'real' women – over thirty and more than a size 14 but not obese. These are women with unexceptional bodies who are presented as having become accustomed, even resigned to themselves and therefore approach any change with a degree of trepidation. They are defined as 'lacking' and this lack is rooted in their low self-esteem. Those chosen are upper working-class or lower middle-class women. It is assumed that this group are the ones most need of sartorial assistance. These are the lives presented as ones in need of rejuvenation by the new cultural intermediaries of fashion and television (Palmer, 2008). Bourdieu has identified this group of workers as members of the new rising fraction of the petit-bourgeoisie. They are the taste-makers in the new media marketplace, acting as cultural entrepreneurs in their own right in seeking to legitimate the intellectualization of new areas of expertise such as popular music, fashion, design, holidays, sport, popular culture. (Bourdieu in Featherstone, 1991, p.91)

The individuals selected for tele-treatment can be taught fashion because it is assumed they have no critical or artistic intelligence guiding their choices in the way it does in the lives of the middle class. This class are often the targets of lifestyle producers because they have aspiration without skills and as such they embody Bourdieu's dictum 'Taste classifies the classifier'. As lower middle-class or upper working class women they betray the signs of their class and in our aspirational culture some origins have to be hidden. The quest is to help the individual find him or her self which on every occasion has meant finding a self of a higher social class – precisely that class who wear their style 'effortlessly' as if it were a simple expression of their personality rather than being the product of a different habitus. The quest then has as its centre something essentially fake – adaptation to the tastes of another. But such fakery is permitted, even encouraged because deceiving others about the real costs of fashion (prices of course not conditions) is an important element of the show's finale. This makes it all comfortable and not too threatening.

The first crucial stage has the individual discussing their wardrobe. The host points out how they have made many mistakes in their choices and that what they really need to do is to focus on their attributes - what they already have - and not to imagine they are someone else or have some other shape. The quest then is clear – to find a true self that embraces the 'already existing' as defining. Rather than stepping beyond the bounds of the real the individual is asked to re-embrace conformity as the one real authority. The work to be done is all on the surface where one can playfully immerse oneself into the maelstrom of fashion with an attitude that is both ironic and accepting.

But for the quest to work as television it must involve a little pain and unhappiness. It involves the individual having to confront themselves via magical devices such as a mirror. Crucially, the individual is not to look for the answers 'inside' as all the answers one needs are to be found in the 'High Street.' The 'true self' is not hidden inside but is waiting to emerge by being correctly attired. The solution offered by this extended advertisement for the saving value of fashion is to reinscribe the subject in fashion. The individuals' new and eager submission to consumer culture 'references out' the ability to countenance any way of thinking outside its frames. The individuality here is one of acceptable differences that can be understood as merely matters of taste. Consumers are 're-aligned' with an idea of themselves that seems personal – the personal has been the impetus of the show – but which rests in advertising's solution of consumerism where any kind of individuality can be bought.

In Goks' last series 'How to Look Good Naked' individuals had their self-perception challenged, corrected, and then directed into a new sense of self-worth. The quest then as now was to accept the self despite it not fitting into the usual thin stereotypes of fashion. The quest means enduring self-torments and self-doubt but ends with a public display (of discreet nudity/the catwalk) from which maximum self-affirming value was extracted. The message here is that the true self is not only one the individual is proud to display but is defined by display. The quest is triggered by the individual feeling a loss or lack of authenticity. As Philips has written of a similar process in Trinny and Susannah (another successful British makeover show): "In rescuing their subjects from fashion disaster, the fashion tastemakers are restoring harmony and redeeming their subjects from personal neglect" (Philips, 2008, p.118)

The magical process of selection, treatment and transformation are laced with magical signs and symbols, but the result is the individual getting in touch with what is supposedly 'the real me'. This is produced partly as what reality TV does – uncover the real – but what is significant here is how this particular brand of authenticity is tied to a self rooted in commodities. Authenticity is to be found on the flat surfaces of commodities and is not connected to any sort of interior. This is a revamped version of the 'democritisation of beauty' that has been the staple element in makeovers since the 1930s (Fraser, 2007, p.178).

It is of course possible that the Gok viewer will decry the shallow futility of an identity rooted only in surfaces. Such a critical reading may be bolstered by the suggestion that putting people into market-driven body-types may not be that liberating. But the emotional realism put in service of the quest underlined by reality TV techniques dissuades alternative readings because they seem to deny or qualify the validity of the individual's responses. The primacy of emotion carries the viewer along with the elan or sparkle of an extended advertisement sprinkled with its trusty magic dust.

(II)Gok – Supernatural Aid/the Astute Businessman

> One only has to know and trust, and the ageless guardians will appear. (Campbell, 1975: 67)

Gok, like many others from the world of fashion as envisioned by lifestyle television is presented as a camp other-worldly creature who has been 'finished' by fashion and no longer share the concerns of the ordinary mortals they have been sent to help. He dresses in an outlandish manner that visually separates him from ordinary mortals. At no point can the contestant or watching viewer forget that Gok is from the magical world of television. The fact that he has already achieved real world successes with individuals re-asserts that while he is rooted in a magical place his successes make him very real and thus of considerable value. In short Gok has charisma. In Weber's terms this is:

> A certain quality of the individual personality by virtue of which he is set apart from ordinary man and treated as endowed with supernatural, superhuman or at least superficially exceptional qualities' (Weber in Dyer, 1978. p.35)

In Gok's Fashion Fix the first magical moment is in the hosts' appearance. On first glimpsing this other-worldly creature the subject/contestant will know that (i) they are in the company of a celebrity who (ii) will befriend them, (iii) who has an impressive track record of transforming individuals and, above all (iv) they are about to go on a quest with him as guide. Gok's out of the ordinary appearance is crucial for contrasting with the ordinary and mundane quality of the chosen subject.

The first illustration of the trickster's power is to have the subjects clothes laid out in a large yard or on poles positioned across a beach etc. In this simple yet dramatic way the ordinary is made to seem extraordinary and the individual is shocked by the signs of the self they have been inadvertently placing in the world. A limited selection of clothes or perhaps an overtly garish one suggests a complacent or ill-thought-out attitude to the self that has to be changed into something more discrete (ie more middle class). When seeing this excess of clothes the individual is shocked into realising something about themselves via Gok – 'you've been wearing the same outfit for twenty years'. This bizarre image of excess can be read in several ways – one of which might be that consumer culture offers so many choices that this leads to waste and excess. Why not donate all the clothes to charity shops and resolve to shop only there in the future? But quickly such alternatives are closed down. The magical helper is there to carefully direct the individual into the bright corridors of the mall. This quest may have magical elements – 'look how that blouse transforms you girlfriend' – but its destination is entirely commercial. The individual has been transported by the host/trickster and given the ability to see their world anew – given a fresh perspective that will enable them to see themselves in a new light.

Aside from his past in fashion what has qualified Gok to take on this role? In the first place a combination of recognisable, identifiable 'problems' with his weight. The story of Gok's own transformation from nineteen stone 'fatty' to thin man has been extensively covered in other media and in a documentary film 'Gok Wan: Too Fat; Too Young' (2007). It is reinforced by Gok's involvement in size and bullying related charities. But merely sharing a history with those-to-be-transformed while ideal for identification is not enough to guarantee his value. Gok's power is derived from his magical location in the world of fashion. This background is expressed in his camp manner which in turn licenses a heightened degree of performativity that enriches the text. The magic of his performance serves to shake contestants up. Gok adopts the excited tones and exaggerated claims of advertising as if he were doing a voice over for a commercial which, in many senses, he is. By using these tones Gok enables us to view the show as a form of advertising but at the same time to help us see that he is playing with the material. He offers viewers a 'knowing wink' that flatters their ability to see beyond the surface while also suggesting that the only way to salvation of the self comes through commodities. (Goldman, 1998).

Gok is at the same time both other and intimate. He is 'other' because of his celebrity status, because he is an ethnic minority, a tv persona, 'a transformer' of the self and others, and a camp man. He is intimate inasmuch as he will become very close with the individual in a way few could. The concentration of time needed by television means that Gok has to demonstrate his empathy quickly. When we witness its effects in moving dialogues with the subject it only enhances the magic of Gok's charisma. He is a wonderful companion – giving without taking, dedicating his time and effort to the neglected, in short a perfect sexless non-threatening friend. Thus while his otherness suggests a magical location – the power from which he derives his power to transform - his intimacy is legitimated by the fact that he has been through the same problems, faced his own struggles and has to live in a culture sometimes antagonistic to camp folk. It is by tying these two together that lifestyle operates most efficiently– at the intersection of felt democratic need and magical effect.

It is important to recall that Gok does represent a brand with a clothing line of his own. While regulation prevents this from ever being directly marketed in the show it cannot be overlooked that his promoting identities based on commodities cannot be separated from his location in the world of fashion. Gok sells his own line of lingerie through Dorothy Perkins and 'SimplyYou.co.uk. These, in turn, link to his website 'simplygokwan.co.uk' which is also of course a promotional aid for his services as a 'fashion consultant, author and tv presenter'. The Facebook site features items chosen by Gok from the program accompanied by friendly advice. But in the logic of consumer culture it has to be this way. While Gok may be working on a semi-public service station his message is a commercial one – that the self can be built and then rebuilt from commodities. For lifestyle television the expertise with the highest value is that which has been tested and proven in the market place. Commercial success and value as a person are united and they share the same values. The helpful advice we gladly receive from such an expert is part of the advertising business that sometimes may instil negative feelings about social acceptability and the self before offering a commercial magical solution.

(III) The Belly of the Whale/the Commodified Self

> the passage into a temple and the hero-dive through the jaws of the whale are identical adventures, both denoting, in picture language, the life-centring, life-renewing act. (Campbell, 1975:81)

The body is the absent-presence in Gok's Fashion Fix. We have seen how Gok has won praise and affection for his strategy of not mentioning the body beyond adopting a language style –'banger boosters', 'boobyliscious basques' etc which celebrates the individual's already existent state. His approach is a long way from recommending surgery and the 'disciplinary technologies and the unequal power relations (it) produces and supports' (Doyle and Karl, 2008). However, it is difficult not to interpret Goks' approach as a convenient marketing strategy which some might link to his own line of clothing which also celebrates larger bodies and drives consumers to the mall. Gok's approach punishes the wallet instead of the waistline. He has recreated consumerism's productive anxiety in which one's stability can only be achieved via products. But this works for Gok (and related businesses) because his response to the body is that of a friend and as such is defined in opposition to those authorities who are always going on about '5-a-day' and assorted bullying messages.

As we consume material on or by Gok we come to read him as a star and develop forms of attachment that cannot but help to sell products. His assistants or competitors have the status of character actors to whom we respond perhaps as the star does. They may later rise in status much in the way Gok did. But the participant finds a parallel in the extra 'whose anonymity is expressed very precisely through his anonymity within the framework of the star system.' The token individual is there to have her anonymity banished for a tele-moment before being invited to understand that 'her uniqueness is best celebrated through conformity' (Heyes cited in Karl and Doyle, 2008). This conformity he or she appreciates is now something to be prized rather than taken for granted. 'No one should think themselves better than they are. The viewer is persuaded of the merit of his own averageness' (Adorno, 1991, p.59)

Psy-Thinking

Mass culture expressly claims to be close to reality only to betray this claim immediately by redirecting it to conflicts in the sphere of consumption where all psychology belongs today from the social point of view (Adorno, 1991, pp. 57-58)

Lifestyle's celebration of transformation has to be articulated with those psychological models that also offer quick-fix solutions to the problems of identity. As Sommers and Satel have pointed out 'Freudian psycho-analysis is too time-consuming and expensive an option for any but the affluent' (Sommers and Satel, 2005, p.37) And so in its place comes the far more simplistic Person-Centred Therapy or Cognitive Behavioural Therapy (CBT) with its theoretical grounding in Neuro-Linguistic Programming (NLP). These approaches to the problems of identity are perfectly designed for consumer culture. Successful titles such as 'The Seven Habits of Highly Effective People' or 'Change Your Life in Seven Days' suggest that our problems are simply those of inefficient planning. It is inevitable that transformational programmes in lifestyle Television sometimes offer such therapies as part of the life-changing package that encourages the user to take a very practical /mechanical view of the self. At the heart of these models if their 'do-ability'. The messages come coded in what are essentially psy-manuals offering measurable models of change that fit perfectly within a speedy media culture. No challenges are offered that cannot be overcome by a determined self. No depths are there to be explored. It can be readily seen how this ties in to the commodity-self. In place of reflection is display. But Lifestyle media can sell the individual this very comfortable achievable idea of the self with a clear conscience because it is the same model other authorities subscribe to.

The more invidious effects of the sorts of magical transformations operated by Gok and others is to re-integrate individuals into the system of commodities. In this sense our host perfectly defines the role of the new cultural intermediaries 'to explain both the use value and exchange value of the new commodities' (2008: 45 Negus in Bonner) The problems of self-esteem or self-worth which may lie behind a tired wardrobe are perhaps rooted in fundamental issues which lifestyle television cannot explore. Its main business is in fostering then exploiting the frames of mind targeted by advertising. In place of muted introspection are stories of hope. The ideal here is not just to belong but to make the achievement of belonging something worth straining for. But what is inadvertedly revealed here is the frailty of people whose insecurities are turned into commodities which in turn sell the programme and its ethos. Rather than offering the individual a perspective on themselves the show (and others of its kind) encourages an outward-facing perspective: one is to take ones value from others – indeed the climax of such shows has to do with the validation given the transformation in the eyes of others. Ideally one has become a hero to this group and thus properly a hero to the self. The new dependency is for consumer culture in the glittering guise of fashion. Just like in advertising, there is no space here for problems which cannot be solved. But unlike the advertising that wears it commercial intentions on its sleeves, lifestyle television makes claims to empower while offering only alternatives that lead to the market.

A more practical reason for televisions' investment in formats that blend the magical and the commercial has to be connected to the medium's own concern for its survival. As online entertainments and social networks multiply television has to invest in formats that foreground its unique selling proposition. Transformation and the deployment of the magical lexicon illustrate in moving and captivating ways what television can do best. As I have illustrated elsewhere (Palmer, 2007) formats such as Extreme Makeover: Home Edition are very much focused on the magic of transformation and even enlist the magical kingdom of Disneyland to help suture the drama of the show into a commercial/mythological context. Furthermore as a recent OFCOM study has suggested with children moving away from television to online as their primary media of choice it is clear that magic may be one of the only ways to hold on to this important audience. As well as deploying the lexicon of magic to attract younger audiences the survival of television depends increasingly on using space to present itself as an honest broker of marketplaces. Television now defines itself as an open space whose legitimacy derives principally from the brand values forged in a time when the medium had both greater power and status.

DEFINING AUTHENTICITY

Reality Television' first excursions into the real seemed to capture the authenticity of the moment. It had a raw untutored quality that made it an ideal instrument for producing stories about the real. But as we have seen in the last decades the techniques and methods of reality TV are so woven into the grammar of television that it is increasingly difficult for us to trust in the authenticity of moments captured as the rhetorical devices seem so worn. As Annette Hill's research has suggested:

> The ability for audiences to see through reality TV and by that I mean critique as well as watch stories in reality programmes, is fundamental to our understanding of the reality genre. In this sense most viewers come to reality TV in a default critical position (Hill, 2005, p.185).

The significance of lifestyle formats is that they use the techniques of reality television to create spaces for displays that reinforce the commodity-self and as such extend the work of identity-formation first begun by reality TV. The irrationalism that underscores such stories – discovery, help-mates, struggle, quest, revelation and transformation – are older than the media but they work because they connect both to other stories and the powerful trend of irrationalism that is so prevalent in our culture. The end-result of such programming defines authenticity in ways that reinforces consumerism and the pursuit of an individuality defined by products. While there is of course considerable value in the 'critical default' position taken by many audience researchers we have to consider the overwhelming dominance of a culture more swamped by the ideologies of consumerism than ever before.

> Lifestyle celebrates importing the work ethic into leisure so that the distinction between the two is erased in the name of efficiency.(Adorno, 1994:25)

The self developed by lifestyle TV is one perfectly in line with those of the authorities in both public and commercial sectors. Conformity is key here in a self that is too self-absorbed to be anything but politically mute. The new psy-models sponsored by therapists such as Carl Rogers offer valuable back up to such a mind-set. According to Rogers:

> 'Your ideal self was already there within you... Personal redemption would come to those who had the courage to discover their true feelings and genuine attitudes – and to accept what they found'. (Rogers cited in Sommers and Satel, 2005, p.75)

It is perhaps an indication of the triumph of consumer culture that while reality TV continues to reproduce scenes of urban life and showcases its potential value as a means of building communities it is now eagerly integrated into selling narratives of the self. Britain' Channel Four is ideally suited for this sort of work. The Channel is part public and part private – it offers programming that has a public service dimension while also offering programs that act as a portal or gateway to a variety of products. Commercialism is never far away here.

CONCLUDING COMMENTS

Lifestyle media encourage us to play the game of self-defining. This invitation is taken up because the traditional means of defining the self via job, faith, and family, are either less stable or in decline. Authority of all kinds is open to question. Thus identity is now a busy fluid game but which becomes more interesting and enriching the more time and money one has. It is in the interests of consumer culture to continue offering modes of belonging that require investment.

Reality television's intervention was to both record and instigate changes in the use of public space. The significance of lifestyle television is that it has borrowed the realism effects of reality TV and used it to sell commodification to individuals. As such the world that might have been opened up the new genre has been pulled into the business of selling commodities. Can it be any wonder that little joy is found in the empty prize of a prescribed happiness?

REFERENCES

Adorno, T. (1994). The stars down to earth. London: Routledge.

Adorno, T. (1991). The culture industry. London: Routledge.

Baudrillard, J. (1988). Simulacra and simulation. Chicago: Stanford University Press

Bernstein, J. M. (1991). Introduction. In T. Adorno (Au.) The culture industry. (pp. 1-29). London: Routledge.

Bonner, F. (2005). Whose Lifestyle is it anyway. In D. Bell, D and J. Hollows (Eds.), Ordinary lifestyle. (pp. 35-47). Berkshire, UK: Open University Press.

Bratich, J. (2007). Programming reality: Control societies, new subjects and the powers of transformation'. In D. Heller (Ed.), Makeover television: Realities remodelled. (pp. 6-23). New York: I.B.Tauris.

Campbell, J. (1975). The hero with a thousand faces. London: Abacus.

Doyle, J., and Karl, I. (2008). Shame on you: Cosmetic surgery and class transformation in 'Ten Years Younger'. In G. Palmer (Ed.), Exposing lifestyle television. (pp. 83-101) London: Ashgate.

Dyer, R. (1978). Stars. London: British Film Institute.

Featherstone, M. (1991).Consumer culture and postmodernism. London, UK: Sage.

Fraser, K. (2007). Now I am ready to tell how bodies are changed into different bodies. In D. Heller (Ed.), Makeover television: Realities remodelled. (pp. 177-193). New York: I.B.Tauris.

Gailey, E. (2007). Self-Made Women. Cosmetic Surgery Show and the Construction of Female Psychopathology. In D. Heller (Ed). Makeover television: Realities remodelled. (pp. 107-119). New York: I.B.Tauris.

Goldman, R. (1992). Reading ads socially. London: Routledge.

Hill, A. (2005). Reality TV. Audiences and popular factual television. London: Routledge.

Jahoda, G. (1974). The Psychology of superstition. (pp. 1-158). London: Pelican

Jhally, S. (1989). Advertising as Religion: the Dialectic of Technology and Magic. In I. Angus and S. Jhally (Eds.), Cultural politics in contemporary America. (pp. 217-230). London: Routledge.

OFCOM. (2007). The future of children's television programming. Retrieved on 1.10.2009 from http://www.ofcom.org.uk/consult/condocs/kidstv/

O'Sullivan, T. (2005). From television lifestyle to Lifestyle Television. In D. Bell and J. Hollows, Ordinary lifestyles: Popular media, consumption and taste. (pp. 21-35). Maidenhead, UK: Open University Press.

Palmer, G. (2007). Extreme makeover - Home edition: An American fairytale. In D. Heller (Ed.), Makeover television: Realities remodelled. (pp. 165-177). New York: I.B.Tauris.

Palmer, G. (2008). The Habit of Scrutiny. In G. Palmer (Ed.), Exposing lifestyle television. (pp. 1-15). London: Ashgate.

Philips, D. (2008). What Not to Buy: Consumption and Anxiety in the Television Makeover. In G. Palmer (Ed.), Exposing lifestyle television. (pp. 117-129). London: Ashgate.

Sommers, C.H., and Satel, S. (2005). One nation under therapy. New York: St Martins Press.

Weber, M. (2003). The protestant ethic and the spirit of capitalism. London: Routledge.

Williams, R (1980). Advertising: the magic system). In R. Williams (Ed.), Problems in materialism and culture. (pp. 170-195. London, UK: Verso.

Williamson, J. (1978). Decoding Advertisements: Ideology and Meaning in Advertising. London: Marion Boyars.

In: Reality Television-Merging the Global and the Local
Editor: Amir Hetsroni, pp. 79-93

ISBN 978-1-62100-068-6
© 2010 Nova Science Publishers, Inc.

Chapter 5

SLOVENE REALITY TELEVISION: THE COMMERCIAL RE-INSCRIPTION OF THE NATIONAL

Zala Volcic (1) and Mark Andrejevic(2)
1. University of Queensland, Australia
2. University of Queensland, Australia and University of Iowa, USA

In the fall of 2007 Slovenia found its most popular reality format to date, a show called simply The Farm, which tapped directly into the country's self-identification with its rural, agricultural history. The Farm easily outperformed all competitors, including locally produced familiar global formats like The Bachelor, PopStars, Big Brother, and Who Wants to be a Millionaire, earning high ratings among the coveted 18-49 demographic, and breaking ratings records with its finale. On average, the show's 2007 season regularly drew almost half of the viewers watching TV during its primetime slot (45%) in a county of 2 million people (POP TV, 2007). This success was even more striking, given the fact that Slovenia is a nation with relatively high penetration of cable television (57%) and a wide selection of regional and international channels. In short, The Farm was a national phenomenon and, as we will argue, a nationalist one, insofar as it tapped into a deep vein of rural nostalgia for Slovene folk culture, complete with traditional costumes, accordian-centred folk music, and a celebration of the country's agricultural way of life.

One of the paradoxes of the show is that, in the name of creating a distinctive sense of Slovene national identity, it drew heavily on the imagery of a past in which Slovenia had never, until 1991, been an independent entity. Thus, the show inevitably draws upon a sense of identity formed under the influence of the Austro-Hungarian Empire in order to portray a historically unique image of Sloveneness. In this regard, the show demonstrates the way in which nation building requires a reappropriation of history, and in this case, crucial for our account, a commercial reappropriation of Slovenia's feudal, agrarian history. Indeed, the goal of this chapter is to suggest that from the perspective of media studies in the current conjuncture it is important to consider the source of nationalist representations and the ideologies they encode. It might, in other words, make a diffference whether the portrayal of national priorities and characteristics is structured by the private sector for commercial purposes or by various state and public actors for political and cultural reasons. Thus, this chapter argues for the importance of highlighting the distinctive character of those forms of nationalism whose purpose is to mobilize a brand community around markers of national identity in order to drive ratings and sell products. In an era of economic globalization, one of the strategies mobilized by transnational media conglomerates is, paradoxically, the mass customization of nationalism, targeted news and entertainment products directed toward national audiences. This is the formation that we identify in our consideration of the success of The Farm – a transnational format that, suggestively, capitalizes on the portrayal of authentic Sloveneness.

Politically, Slovenia today is still involved in the project of national identity building, and broadcasting outlets attempt to focus on consolidating both a sense of distinctive national identity and a national brand identity (Volcic, 2007). As many local scholars have shown (Basic, Kucic and Petkovic, 2004; Splichal, 1992), television in Central Europe remains a central stage on which national identity is displayed and reconfigured, although the landscape is being dramatically transformed by the advent of commercial broadcasting and the presence of global economic players. Amidst the different ways in which cultures connect and overlap in the Central-Eastern European region, national culture has emerged as a preeminent frame of reference. In this regard, the questions of how reality TV formats may contribute to the nation-building process remain vital. We argue that reality television, as produced by commercial broadcasting, can be seen as a site in which national identity is (commercially) represented, shared, invented, and dramatized. Of interest to this chapter is the way the reality shows, caught up in the shifting currents of post-socialist nationalism, channel these into what might be described as an emergent form of commercial nationalism, which represents important transformations in the reproduction of the concept of a nation. Specifically, we will suggest, commercial nationalism refers to the way in which nationalist appeals migrate from the realm of political propaganda to commercial appeal: that is, into the mobilization of a commercialized version of nationalism as a means of increasing ratings, popularity, and sales.

This chapter will first offer some historical frames for understanding the current Slovene media landscape. It will then trace the outlines of an emergent form of commercial nationalism on reality TV, drawing on textual analysis of the selected shows and their reception by viewers, as expressed in chat rooms and blogs.

BRIEF HISTORICAL AND POLITICAL CONTEXT

Slovenia gained its independence and became a parliamentary democracy in 1991 after the disintegration of the former Yugoslavia. The Slovene national identity was, due to 19th and 20th century historical circumstances, a rather weak one. Before its independence in 1991, Slovenia was a part of Yugoslavia, a diverse mix of ethnicities, nations, religions and cultures. Before 1918, when Yugoslavia was established, it was for centuries a part of the Austro-Hungarian Empire with a separate, but weaker Slovene national identity being slowly crafted in the latter half of the 19th century. The Slovenes in the Austro-Hungarian Empire had neither a national church nor a national economic policy. Slovenes also did not possess an aristocracy or a sizable middle class, as traditionally found in 'historic' nations (Zajc, 1997).

Not only was the Slovene national identity weak, it was also a conservative one framed in terms of language, the distinctive Slovene landscape, rural, peasant life, and to some extent Catholicism. Stankovic (2005) writes about the traditional construct of Sloveneness as something deeply related to nature and an idealised rural life. Slovene national identity was from its start defined in opposition to the perception of Germans as 'cold' and 'efficient' – Slovenes perceived themselves as warm, sincere, and close to the nature, sometimes literally and sometimes mediated through the notion of a simple and authentic rural life.

This construction of national identity can be interpreted as the result of a long history of rural peasant life dating at least to the dissolution of the Austro-Hungarian Empire in 1918. Stankovic however also recounts how, at the end of 1960s, a new type of more urban-oriented Slovene national identity was being promoted by communist elites as part of the project of communist modernization. Political independence in 1991 was followed by numerous economic and political reforms, such as the introduction of a market economy, the denationalization of public and state-owned property and the introduction of parliamentary democracy. These changes in the political and economic system are associated with changes in the social system that created a large underclass, where around 13.6% of the population is at risk of poverty (Volcic, 2007).

In Slovenia, the "triumph" of nationalism and (neo)liberalism in the 1990s has been accompanied by the gospel of individualism and a focus on commercialization. However, it was in 2004, when Slovenia entered EU, and when the country's neo-liberal right-wing government initiated radical transformations and changes that the recent trend toward (neo)liberal governance and the triumph of individualism became dominant. Its main goal was to support a total withdrawal of the state from the economy. The impact of the post-socialist condition, accompanied by the rise of capitalist democracy, could be illustrated by increasing uncertainties about the economy and the fate of media reforms.

SLOVENE MEDIA LANDSCAPE: THE TRIUMPH OF COMMERCIAL BROADCASTING

In a small country such as Slovenia (whose population is less than 2 million people), with a unique language, television has crucial importance in public debates about national culture. Especially confrontational are the debates that juxtapose the role of public television with that of commercial, private broadcasters in terms of their role in building national culture. Historically, it was public television that claimed a political role in the nation-building process. As Volcic (2005) argues, national public television's self-proclaimed goals are to represent a cultural and political institution that would help to create, maintain, and reinforce Slovene identity. Volcic's (2005) interviews reveal how the editors and journalists of national public television understand their role as contributing to the process of national integration. However, commercial broadcasters also claim to participate in catering to a sense of what it means to "be a national". But they understands this mission differently, positioning and consequently packaging national identity as a brand. According to one producer for commercial television, "our aim is to make profit, whatever it takes. If that means focusing on local affairs, so be it. If it means to emphasize Slovene national culture, and its customs, folk traditions...that's what we will do. We don't pretend we have some higher goal..." (personal interview, 2008). The image of Sloveneness becomes a powerful marketing tool with which commercial stations can realize profits. It is crucial then to address a question of what happens to national identity when it no longer necessarily links to political and state apparatuses for its construction, but comes to rely extensively on commercial forces for its (re)production. Foster (1999) points to the rising importance of commercial technologies in nation making, but claims these are not new – commercial rituals of nation building, he suggests, should be carefully analysed across the space and time. He shows how Coke consumption "implies participation in a distinctively American way of living – a way of living characterized by its material modernity and abundance" (266). Foster particularly looks at the role of Coke advertising in marking Papua New Guinea as a nation and argues that specific commercial advertising campaigns "achieved the identification of nation and consumer (278). He writes (274) that Coke and other multicorporations are "complicit in the construction of PNG as a multicultural nation in which 'diversity' means benign stylistic variation in dance routines and ceremonial costume". Coke becomes a kind of a floating signifer, attempting to constitute the nation as a totality – either USA or PNG. Macdonald (1995) describes how "becoming Australian" during the 1950s and 1960s was promoted through consumption practices, such as shopping and owning a barbecue. And Kemper (1993) looks at the government Development Lottery activities in Sri Lanka, and analyses how gambling represents a national act, since it sponsors development projects that benefit all citizens. All the authors argue that nations should be conceptualized also as "imagined communities of consumption", and observe that consumption practices may operate as powerful tools for materializing nationality.

While the branding of places, mainly for the purposes of tourism industries, has existed as a practice for a while, it has only recently begun to be explicitly connected to nationhood. The majority of academic studies on the branding of nations are conducted by marketing scholars and tend to have a strictly applied orientation (Kotler and Gertner, 2002). Not surprisingly, the majority of current literature on national branding is produced by advertising practitioners who have been or still are engaged in the practice of consulting for governments of various countries. Thus, the character of the literature is not highly theoretical but, rather, presents various examples of how nation branding has been used by different nations and makes some recommendations about how such branding practices can be implemented. These authors suggest that it is precisely through branding that national governments can make their nations attractive to transnational capital, subordinating national sovereignty to the logic of capital flows. Anholt, a 'branding expert' claims that planning a national branding strategy requires the cooperation of the political organizations of a nation-state and the mobilization of support by the general public, as well as "the personal backing of the 'chief executive' of the country, whoever he or she may be" (2003: 16). This view is consistent with the marketing model of a nation-state as a corporation that can be run according to business logic (Kotler et al. 1997). For the purposes of our analysis of commercial nationalism we would distinguish between longstanding attempts of nations to brand themselves, and, on the other hand, the use of nationalism by commercial entities (often transnational ones) to sell media products, including music, news, and entertainment programming.

In terms of the media industries, broadcasting trends in Slovenia are similar to those in the rest of Central and Eastern Europe (CEE). On one hand, we witness a complete deregulation of the print media field, but on the other hand, governments attempt to retain control over the broadcast media. However, there has been a significant influx of private capital and with that, foreign investors, into the CEE countries. Private media investors have expanded at a fast pace in all countries, with the American Central European Media Enterprises (CME) leading the wave. CME operates stations in Slovenia, Croatia and Bosnia. It owns the most successful and popular Slovene television station POP TV as well.

The beginnings of commercial television in Slovenia can be traced back to 1989 when the first independent television in former Yugoslavia, Kanal A, was established. But the real development of commercial television occurred only in the mid 1990s, with television stations such as TV3, and POP TV mushrooming in the country. POP TV started to broadcast in 1995. CME invested some US$ 16 million and acquired a 58% share in the production company Pro Plus which is responsible for the management, production, technical operations and finances of POP TV and consequently, another commercial station, TV Gajba.

POP TV was organized according to 'a network principle', similar to those existing in the US - the goal being to create a large enough distribution system which would attract advertisers nationally. Through POP TV, a familiar logic was imported to Slovenia: that of encouraging large foreign investment, combined with local capital to attract enough advertising money to finance media production at the national level. In Slovenia, commercial broadcasters continue to devote few resources to in-house production, relying largely on entertainment programming and cheap US programming. They are wary about the ability of investment in local programming to generate adequate profit margins. However, Gorazd Slak, Pro Plus's programming director, claimed in 2008 that " While the likes of *Ugly Betty* and *Desperate Housewives* are nice and fun to watch, most viewers want local actors and stars, whether they're scripted, non-scripted shows, or game shows and quizzes" (in Akyuz, 2008).

For our purposes it is worth emphasizing the rampant commercialization of the broadcast sector. The national public TV station retained a lead in the evening news for some time – one of the last holdout time slots -- but by 2009 was surpassed in popularity by commercial broadcasting. According to Skrinjar (2009), both commercial broadcasters, POP TV and Kanal A have now achieved higher ratings for their entertainment and information programmes. Moreover, economic deregulation led to greater disrespect for broadcasting laws, including caps on advertising time and the percentage of locally produced content aired.

However, the fears that the foreign capital would come to dominate the Slovene media scene were not realized. Basic and Milosavljevic (2008) write that a small number of national media owners with shares in diverse companies (such as the iconic beer company Lasko) control the majority of the Slovene media market. The state does not have the mechanisms to successfully regulate the new commercial media industries. Economic deregulation, as Basic and Milosavljevic suggest, has not provided for a wider range of political content, in large part because of the convergence of interests between political and economic elites familiar in other predominantly commercial media systems. However, economic logics do not fully align with political ones – nor are they accountable in the same way. It is in this difference that the distinction between commercial forms of nationalism and the traditional forms of state supervised nationalism associated with national broadcasting in the communist era emerges. As Billig (1995) points out, national identity gets reproduced not only through top-down, elite discourses, but importantly, through mundane and banal details of every-day life. Commercial television, we argue, adopts national rhetorics that are different from those employed by public television. Commercial television then provides a site in which banal forms of Sloveneness are reproduced. Reality television's 'flagging' of Sloveneness works at different levels, from the use of particular settings to explicit themes in the shows' narratives (Billig, 1995).

POP TV: "THE ONLY SLOVENE TV THAT COUNTS"

POP TV was the first "truly" Slovene commercial broadcaster that offered popular programming and television hits from the US, Great Britain and other Western European countries. In most of the time slots, the ratings of POP TV soon exceeded that of the TV Slovenia's Channel One – the main state, public broadcaster. To maximise its audiences, POP TV relies on cheap programming typically comprised of light entertainment programmes, game shows, and cheap drama. Repeats and US imports continue to be particularly attractive, since to produce programmes can be more costly than to buy them in packages. Reality TV shows offer an alternative: the importation or adaptation of foreign formats (at least in some cases) that rely upon locally supplied content and cast members. Such shows are desirable, not just because of their low production costs, but because they count as "domestic" production, and are able to attract high ratings.

As in the rest of Europe, the main trend in television production in Central and Eastern Europe is commercialisation, with entertainment colonizing all formats. In recent years reality TV has been the fastest growing genre with local production of international formats topping the ratings chart. Like other CEE countries, Slovenia produces numerous reality shows. For example, between 2003 and 2007, seven local versions of reality formats were produced. The first one was *Popstars* (2002), local version of American Idol, produced and broadcast on Kanal A. On POP TV alone, from 2004 onwards, there were five reality television shows, some for three seasons: *The Bachelor [Sanjski moški]* (2004, 2005); *Bachelorette [Sanjska ženska]* (2005, 2006); *Bar (2005, 2006); Big Brother* (2007, 2008); *Kmetija [Farm]* (2007, 2008, 2009). As one indication of the success of the genre, every year, the number of would-be reality stars auditioning for the shows increases: in 2006, there were 425 candidates to take part in *Bar*; a year later, 1370.

Thanks to the local production of reality formats, the shows evoke a sense of national identity, often in the guise of authenticity, within a distinctly commercial context as a means of mobilizing loyalty, gaining audience support, and distinguishing between candidates on the show. Of interest is not just the way in which such shows reinforce forms of national identity, but also, in the post-socialist context, how commercialism frames and reframes national identity in accordance with economic imperatives: in this case, the need to attract viewers and gain ratings. Take, for example, the *Big Brother* format, which has a built-in sense of national or regional appeal. As a glocalizable format, what distinguishes it from transnational fictional formats (and places it in a shared category with other glocalizable formats) is the way in which it incorporates the local, by drawing on cast members from a specific nation or region. When a show like *Sex in the City* or *the Sopranos* circulates in international markets, it carries with it a broad array of markers of its nation of origin. This allows for national identification and comparison, as exempifed by the ways in which shows might come to represent for foreign viewers aspects of an American sensibility or "way of life" or to serve for Americans as a point of national or cultural identification. By contrast, a show like *Big Brother*, even though it may originate in a particular cultural context (the Netherlands, in this instance), can serve as the basis for local forms of identification, perhaps even in contrast to other national or regional instantiations of the show. Critical discussion of such shows have explored the specificity of shows like *Big Brother: Africa* or the differences between the character and success of the format in the UK and the United States (Jacobs, 2007). Each version of the show then, might be conceived as a format container that can be filled with local content, and in this respect invites the attribution of national or regional specificity to the characters, the audience response, and the contest's outcome.

A prominent Slovene commentator, Hrastar (2007: 34) has argued, for example, that the *Big Brother* show offers an insight into what it means to be a Slovene. "If you have missed the show, you have missed the televized essence of Sloveneleness... the characters fulfilled all the essential stereotypes that Slovenes have about Slovenes. The show did not disappoint in showing the most important attribute: how Slovenes are really boring and ordinary. And how they desperately try to be 'happy', 'nice', and 'joyful'."

In the case of the Slovene version of *Big Brother*, audiences responded to the perceived authenticity of those characters who seemed to best exemplify characteristic attributes of Sloveneness. The winner of the first season of *Big Brother*, interestingly, was a member of the Slovene diaspora, who was five years old when his parents left Slovenia for Australia. The fact of his Australian upbringing – he even earned the nickname of "little kangaroo"– did not, paradoxically, detract from his perceived Sloveneness. Rather, it gave him a "prodigal-son" appeal, and more specifically, it allowed him to assert a self-reflexive form of national identity. As someone who had returned to the Slovene homeland after his parents departed, but who still spoke a very classic, traditional version of the Slovene language, who played the accordian and clearly identified strongly with Slovene culture, Andrej came across as a character who, given the choice, had picked Sloveneness over and above other possible choices. Not only did he comport himself on the show as someone who adhered strongly to traditional Slovene values, he made it clear that he wanted to marry a Slovene woman and to stay in Slovenia. He had preserved a sense of Sloveneness as a member of the diaspora (in particular cultivating the language, a historically central marker of Slovene identity), and had made a point of his return to the nation as a form of homecoming. After he won the grand prize, however, he renounced his ambitions to stay in Slovenia and returned to Australia. This final move highlighted the performative character of national identity – the way he had been able to mobilize it as a means of both attracting ratings for the show and winning a commerical competition, only to shrug it off when he had achieved his objective.

The popular TV show *Bar* also helped highlight certain aspects of national self-identity as a means of creating a branded form of commercial synergy. *Bar* featured selected members of the public chosen to live together and work in a bar in the capital city of Ljubljana. The goal was to capitalize on the interactive appeal of the show by anchoring it in a downtown commerical location where the public could interact with the cast members. In keeping with this premise, *Bar* was a heavily branded show, not only did it create a popular nightspot (for which customers often queued for half an hour or more for the chance to buy a drink and perhaps appear on the show), but it also incorporated a wide range of product placements, from beer companies to hairdressers. It also generated an array of branded products, from T-shirts to lighters, all featuring the show's logo. Since the show was based around an audience selection format, one in which viewers could vote to evict cast members from the show, *Bar* prompted reflections on the characteristics that Slovenes looked for in evaluating and indentifying with cast members (Zupancic, 2007). Although the neighborhood pub is a staple of Slovene culture and an important site of social life, the spectacle of an excessive nightlife runs counter to traditional sensibilities of an historically agricultural/rural society. Thus cast members found themselves caught in a contradiction: on the one hand, their task was to participate in the creation of an entertaining spectacle devoted to the city's nightlife; on the other, they found themselves subject to the standards set by a public that frowned on late-night partying, brawling, and drinking, and the consequent effect on ideals of discipline, reliability, and the norms of rural life. After the public voted off one of the harder partying more dynamic cast members rather than a calmer low key one, a third cast member lamented the apparent audience preference: "It looks like we're all going to have to go to sleep at nine, just to satisfy all the Slovenes... And wake up four to five hours before we usually do to drink our coffee... that's what people want, and we have to give it to them..." (in Leskovec, 2007). As one media critic noted in the popular newsweekly *Mladina*, "The members and viewers alike tended to reward punctuality, orderliness, cleanliness... and [people who express themselves in] terms of warmness, sensibility...caring" (Stefancic, 2008).

In this regard, it's probably also worth noting that in the second season of *Big Brother* one of the favorites to win the show was voted off after getting drunk and, oddly, confessing to pulling the nose of a woman he did not know in a bar one night. Again, it was apparently the spectacle of paryting and inappropriate behaviour that turned the audience against him. The first season of the Slovene version of *The Bachelor* followed the same pattern. The coveted bachelor was the co-owner of an import company, an avid sportsman, and someone who came from a traditional family background. The show's "winner" - that is, the woman chosen by the bachelor – was the cast member who portrayed the role of the stereotypyical Slovene and presented herself as hardworking, beautiful, rural, honest, obedient, caring, kind, compassionate, unselfish, witty, simple and good-natured. Interestingly, most of the participants of *the Bar, Big Brother,* and *the Bachelor* alike emphasize their interest in the cash prize and in the fame that can bring them further financial gains. A female *Bar* contestant stated: "I wanted to become famous. This is a small country and it is easier to become a celebrity here. I took my part in the reality show as a business... it should help me to get a job as an actress..." (in Sloveniacom, 2008). A male contestant from the same show said that he won because "I didn't drink and party. I was not vulgar, and I worked hard... a lot of religious people voted for me, because they saw I am a spiritual and honest guy... I got a car as an award, and want to promote myself more as a fashion-model." (in Sloveniacom, 2008). And one of the participants, while being evicted from the *Bar* stated: "I'm not sorry...why would I be sorry...It's about one more experience...that allowed me to grow, you know. And I plan to make the most of my 5 minutes of fame... " (in Zarki, 2008: 77).

"THE FARM": A RETURN TO SLOVENE ROOTS

The success of *The Farm,*which outstripped other popular formats like *Big Brother* and *POP Stars*, has lead at least one cultural observer to write that "the participants felt really 'at home' on the farm, as their grand-fathers did... the farm has proven to be their natural habitat. This was the first authentic Slovene reality show..." (Klemencic, 2007). This wasn't strictly true in the sense of being an original Slovene format, however. *The Farm* was created by the Swedish production company Strix in 2001 and has been a successful and popular format for the company, which has been able to sell it to more than 40 countries. Despite its northern European provenance, the format neatly taps into a deep sense of rural identity that, in terms of mass media representation, can be traced back at least to the first Slovene partisan film *Na Svoji Zemlji* (On Our Own Land, 1948), which, in a stereotypical way, positions Sloveneness in terms of the image of a small, authentic village community, linked to the beautiful surrounding nature (Stankovic, 2005). The film portrays the wily resourcefulness of the simple, unshophisticated villagers in the face of the external threat posed by Italian and German occupiers – and in this sense frames the relationship between East and West. The Westerners are more sophisticated and urbane but, at the same time, corrupt. In this regard, *The Farm* fits neatly with the assertion of Slovene identity in the face of assimilation by the West and reinvigorates the historical identification with rural life against the background of the recent social changes associated with both the project of communist industrialization and post-communist globalization. The use of landscape and a rural setting of a farm in the show similarly points to how the ruralness of a farm is essentialy connected to the reproduction of national identity.

Despite its international provenance, *The Farm* adjusted to the Slovene national market by incorporating cultural markers of Sloveneness into both the show's narrative and its structural elements including the elimination competitions and the role of the show's "master", who supervised the cast members. One of the show's producers emphasized the role that national identity played in casting the show: "We look for real Slovenes, able to live as we did in the past." The show was framed as part social experiment, part game show, and part national history lesson. As a producer put it "The show attempts to teach Slovenes how life was 100 years ago... so that they can easily understand rural life, and their cultural heritage" (personal interview, 2008). From the beginning, the show triggered passionate public debates about the rural origins of Slovenes. Some commentators celebrated the fact that the show's "participants are put in really traditional farm, not a modern one... where it's hard to survive... so no machines... they have to work hard!" (Burja, 2008). Others expressed how they "just like to watch Slovene rural life... I appreciate the topic and want to learn more about how my grandparents lived..." (chef, 2008).

The viewers frequently judged the authenticity and merit of cast members by how well they lived up to the Slovene stereotype of being hard-working and even-handed. The characters who were seen to shirk their duties or avoid work were vociferously criticized in the forums and on the blogs. Typical of this response was one blog post that said, "I root for goran and alex on the show. These two are the only ones who work, the others are lazy. Spela has some blisters from cutting the grass – but she behaves as if she lost 6 fingers! It's only blisters! spela,davor,sergej are so lazy – don't they realize they have to work in order to survive?" (Kalinic19, 2008). Another viewer on his blog posed the sarcastic question "are the participants on the reality show picked out because they're peasants? Or is the Farm just a show that fits Slovenes extremely well [so that anyone in Slovenia who was picked would "fit"]? (Hadblog, 2008).

In terms of format, *The Farm* was a hybrid of live and pre-taped footage. Every episode summarized the week's happenings and included interviews with cast members and detailed extracts from their video diaries. The first nine weeks of the show are taped in advance, and cast members are eliminated through regular challenges and competitions with one another. Each week, one of the cast members is selected to be the "master" of the farm. As leader he or she is responsible for organizing the rest of the cast to meet the challenges posed by producers such as getting the cows milked, cut firewood, plow the field, and so on. The leader is also responsible for selecting the two cast members who will be the servants for the week. These two – one male and one female – are given the traditional titles of *hlapec* and *dekla*, meaning roughly, groom and maid. The structure of house master and servants directly invokes the structure of Slovene life under the Austro-Hungarian empire, when Slovenes were traditionally the farm workers and household servants for members of the Austrian aristocracy. At the end of the week, the rest of the participants vote to decide who did the worst job, the *hlapec* or the *dekla*. The one who is chosen has to participate in an elimination competition and is given the choice of competitor of the same gender. The master for the week, however, is exempt and cannot be chosen as a competitor for the elimination round. Whoever loses the duel is eliminated from the show and decides the master for next week.

The elimination round features three types of competition, all of which invoke a sense of national identity rooted in the traditions of farming life. In Billig's terms (1995), we might describe them as "flags" -- a way of signalling a sense of national identity without necessarily reflecting upon it. The least popular competition is a knowledge quiz based on information gleaned from a traditional handbook, called *Pratika* which serves as a guide to the tasks and traditions of rural life. *Pratika* is a book which most Slovene families own and serves as a compendium of traditional recipes, proverbs, and advice for how best to complete the myriad tasks of family and productive life on a farm. Once a well-used encyclopedia of farm life, *Pratika* is no longer read as a matter of course by young people, but it is accorded the respect of tradition, and is often turned to later in life for advice on gardening and cooking, as well as for information about traditional holidays and other interesting tidbits of Slovene tradition. Contestants were able to study this book before each elimination round to increase their chances of winning. However, this was the least popular challenge and was not chosen during the first two seasons.

The second type of challenge is called "endurance", sometimes also referred to in terms of "technique", "skills" or "agility". The two competitors sit facing each other on two parallel horizontal logs, each trying to saw through the other's log. The loser tumbles down when the winner succeeds in cutting through his or her perch. The competition draws on one of the traditional occupations of Slovene rural life - that of the wood cutter.

The final category is called "strength" and features two contestants attached to one another at either ends of a rope that masses through a post between them. The goal of the game is to pull towards the outer side of the ring in which the competition takes place to grab horseshoes arranged around the perimeter and throw them into a box. The first one to collect five horseshoes wins the game. Both competitions took place in the setting of a traditional Slovene hay barn. Horseshoes in Slovenia are traditional signs of good luck and most houses still have a horseshoe over the door to bring good luck. The final week of the show is live in the studio and features an interactive element: members of the audience are given the opportunity to challenge cast members to the same competitions that have been the basis for elimination during the first nine weeks of the show.[1] These elements of the show represent the

---

[1] Reality TV triggered controversy in many countries including Australia,Germany, Malawi, Turkey, and especially France. In Arab rcountries they were able to mobilize protests, street riots, high-level political resignations and contributed to clerics issuing hostile fatwas, and mobilize transnational media wars (Kraidy, 2008). In Slovenia, the Farm generated hot public debates because of possible violation of animal rights, when the participants on the show were killing a pig. A Slovene human rights ombudsman was called to follow closely Big Brother and the Farm. She found that Big Brother contained violence and degrading behaviour

adaption of a global format to the Slovene market, taking elements that may have different cultural resonances elsewhere (like woodcutting) and linking them directly to the traditions of Slovene rural life in ways that are readily recognizable to Slovene audiences. In this regard, the show works to craft a sense of authentic Slovenenes to which the participants are held accountable. The "interactive" element of the show in which the public votes for the most popular cast member (who receives the prize of a new car) – invites viewers to participate in the accountability process.

CONCLUSION: SELLING THE NATION AND MARKETING THE SELF

We want to use the example of the Farm to trace the outlines of several elements of what we are describing as commercial nationalism – the phenomenon whereby commercial media intstitutions take on an increasingly important role in framing the nation in the era of globalization, neo-liberalization and, in central Europe (as elsewhere) the galloping commercialization of the media industries. We also might describe the selling of nationalism – which has become a common theme in populist commercial outlets globally in a number of countries – as a form of national identity building: the choice to consume a particular version of national identification as marked by consumption practices. We might, for example, describe in terms of commercial nationalism the choice to watch a particular newscast or to purchase the products of a specific artist: country singer Toby Keith in the United States (whose song, "Courtesy of the Red, White, and Blue" caters to nationalist revenge fantasies, with lyrics like "You'll be sorry that you messed with the US of A, 'Cuz we'll put a boot in your ass, it's the American way") or nationalist turbofolk star Ceca in Serbia. This is not to say that nationalist identification is the only reason for making such a choice, or that it can fully explain the choice, but rather that that such identification can be interpreted as an element of it. Our central interest, however, is in the way that commercial nationalism imports commercial imperatives into the version of nationalism it purveys – and that these imperatives may not fit comfortably with prior conceptions of state-oriented nationalism. We therefore propose the following elements of commercial nationalism as a starting point for thinking about shifting constellations of national identity in a globalized era:

1) Perhaps most obviously, commercial nationalism is a means to a very specific end: sales. Shows like *The Farm* are not necessarily produced with the direct goal of promoting a sense of national idenity as was the case in the era of state ownership of the media industries. Rather, the show mobilizes themes of national identity and authentic "Sloveneness" to win ratings and advertising revenues. If state-oriented forms of nationalism served to foster public identification with the imperatives of political leaders, to the extent that such identification is an element of commerical nationalism it is subordinated to a "higher" end – one that transcends loyalty to particular political formations, political representatives, government policies, and even, paradoxically loyalty to nation. Commercial nationalism is not, primarily, a political principle" (as quoted in Billig, 1995: 19). Nor does it necessarily rely on the

---

with and among the participants, which violates the dignity and other human rights of the house tenants (more in Kovacic, 2008).

congruency of the political and the national – indeed, as a form of brand idenity, commercial nationalism can target "authentic" national subgroups within the nation state. There is no clear political agenda attached to commercial nationalism – it is, as Zizek (1989) points out – an empty signifier, but in its commerical form, it serves not solely as a point of national identification, but of brand loyalty. The allegiance is only secondary to a sense of national community or identity and primarily to a product and its authenticity. Just as it is not necessary to provide determinate content to the notion of "The Real Thing" to sell Coke, it is not necessary to provide political content to "The Real Slovene" to promote a reality TV show.

2) In an era characterized by what Hearn (2009) has described as the importance of personal branding as a means of managing the risks and exploiting the opportunities associated with neo-liberal regimes of governance, commerical nationalism provides a resource for the construction of a nationally branded self. To put it somewhat differently, if, in an era marked by the replacement of the social safety net and other forms of security associated with the welfare state by increased flexibility and its associated risks, nationalism becomes a resource for the entrpreneurial self. This leads not only to the emergence of a relatively new type of celebrity – the ubiquitous (commercial) nationalist commentator – but, more generally, makes nationalism as brand identity available as a personal marketing resource. Thus, for example, the cast members on *the Farm* saw their identification with autnentic Sloveneness as a means to an end: their own self-marketing as reality TV personalities in the afterlife of their appearance on the show. For example, two participants in season 2 of *the Farm* openly claim that they are "real Slovene rural girls" but they did not hesitate to confess their plans to capitalize on their participation in the show. One of them said: "I am a real rural girl also in real life. If I would not live on a farm, I would be bored. Because I decided to be a part of a reality show *The Farm*, I postponed my studies of medicine. But I think it is a good decision: I think I became famous and people recognize me now... so it will be easy for me to get a fancy job for me now." (in Skala, 2008). In the neo-liberal era, nationalist identification is more (or other) than a political statement; it is a commercial asset.

3) From the perspective of the state, commercial nationalism serves as a resource for public relations and propaganda, but, given its higher (or lower) allegiance to the profit motive, it can be a problematic or ambiguous resource. The tendency of commercial nationalism is to read the political through the lens of marketing – not just to view politics as one more forms of salemanship, but to offload forms of national identification onto the private sector. Indeed we might describe the advent of commercial nationalism as a kind of reflection of imagined community in an era characterized by anxiety over its fate in the face of fragmentation, mobility, and globalization. If, for example, the national newspaper helped create a sense of shared community traveling through time, commerical nationalism provides a point of identification in the face of the fragmentation of the public sphere in the era of the internet and associated forms of mass customization. The recent tendency of political figures to appear on reality shows as cast members – and of cast members to see the shows as potential stepping stones for their political careers – marks a further permutation of the relationship between politics and celebrity mediated by the notion of authenticity as a key component of national identification. Consider, for example,

in the United States, former House Majority Leader Tom Delay's decision to compete on *Dancing with the Stars*, and the attempt by disgraced former Illinois Governor Rod Blagojevic to join the cast of a reality show *I'm a Celebrity. Get Me Out of Here*. Shows like *American Candidate* in the United States and related formats in Australia and the United Kingdom encourage feature cast members with political ambitions in political contests and debates. The conjunction of reality TV celebrity and politics is perhaps not surprising in an era when politicians's personal lives are leveraged by their campaigns as symbols of their authenticity. Thus, the notion that behind-the-scenes access to political celebrities provides evidence of their personal authenticity parallels the promise that one's authentic national character might prove a useful resource in the political sphere.

The Farm is not unique in its mobilization of forms of banal, commercial nationalism as a means of fostering viewer loyalty and creating a sense of brand identity. We see similar processes at work in formats that are well known by viewers to be global franchises – such as the *Idol* music competitions where cast members come to reflect the distinctively national version of a transnational format. We do think that reality TV shows, precisely because of their global/local mix and their emphasis on access to behind-the-scenes reality are sites where the relationship between national identity and personal authenticity becomes a marketing strategy that aligns itself with the more general phenomenon of commercial nationalism. It is a phenomenon that bears close examination in a media environment which is becoming increasingly ubiquitous, accessible, interactive, perpetually accessible and subordinated to commerical imperatives.

REFERENCES

Akyuz, G. (2008, January 14). Broadcaster profile – Pro Plus – Slovenia. *Central European Media Enterprise*. Accessible at http://www.cetv-net.com/en/press-center/media/49.shtml.

Andrejevic, M. (2004). *Reality TV: The work of being watched*. Lanham, MD: Rowman and Littlefield.

Anholt, S. (2003). *Brand new justice: The upside of global branding*. Oxford: Butterworth-Heinemann.

Basic Hrvatin, S., Kucic L. J., and Petkovic, B. (2004). *Media ownership. Impact on media independence and pluralism in Slovenia and other post-socialist European countries*. Ljubljana: Peace Institute.

Basic Hrvatin, S., and Milosavljevic, M. (2008). *Media policy in Slovenia in the 1990s. Eurozine*, *2*, 1-35.

Billig, M. (1995). *Banal nationalism*. London: Sage.

Burja, B. (2008). *Kmetija!* (The Farm! - Slovenian) Accessible at http://www.mavricni-forum.net/index.php?showtopic=3976

Chef A. (2008). *POP TV klavnica show*. (POP TV butchery show - Slovenian). Accessible at http://simonarebolj.blog.siol.net/2007/12/01/POP-tv-klavnica-sov-ena-mozakarsko-lovska-in-polhasta-tradicija/.

Foster, R. J. (1999). The commercial construction of 'New Nations'. *Journal of Material Culture*, 4(3), 263-282.

Hadblog A. (2008). Kmetija na POP TV (The Farm on POP TV - Slovenian). Accessible at: http://www.had.si/blog/2007/10/02/resnicnostni-sov-v-naravi-kmetija-na-POP-tv/

Hearn, A. (2009). Variations of the branded self. In D. Hesmondhalgh and J. Toynbee (Eds.), *Media and social theory* (pp. 194-210). London: Sage.

Hrastar, M. (2007). Zivaloljubci (Animal-lovers - Slovenian). *Mladina 47*(4). Accessible at http://www.mladina.si/tednik/200747/clanek/nar-femina_sceptica--mateja_hrastar/.

Hrastar, M. (2007). Skepticarkin zagovor (Defence of a sceptic - Slovenian). *Mladina 23*(4). Accessible at http://www.mladina.si/tednik/200723/clanek/skepticarka23/

Jacobs, S. (2007). Big Brother, Africa is watching. *Media, Culture and Society, 29*(6), 851-868.

Kalinic19 (2008). *Diskusija kmetija* (Discussion about the Farm). Accessible at http://www.velenje.com/DISKUSIJEsporocila.php?stev=532576.

Kemper, S. (1993). The nation consumed: Buying and believing in Sri Lanka. *Public Culture 5*(1), 377-93.

Klemencic, I. (2007). The Farm on POP Tv. *Indirect.* Accesible at http://www.indirekt.si/scena/ocene/117651

Kotler, P., Jatusripitak, S., and Maesincee, S. (1997). *The Marketing of nations: A strategic approach to building national wealth*. NY: The Free Press.

Kotler P. and Gertner, D. (2002) "Country as brand, product, and beyond: A place marketing and brand management perspective." *The Journal of Brand Management*, 9(4/5), 249-261.

Kovacic, V. (2008) The Show Must go on. *The Slovenia Times*, 16.5.2008. Accessible at http://www.sloveniatimes.com/en/inside.cp2?uid=BEDFD6BB-3724-3D1C-E511-9CF87ABDB102andlinkid=newsandcid=95BA81FE-636A-933D-C350-7EE6FECF5929.

Kraidy, M. (2009). *Reality television and Arab politics: Contention in public life.* Cambridge: Cambridge University Press

Leskovec, B. (2007). *Resnicnost resnicnostnih shovov* (The reality of reality shows - Slovenian). Ljubljana: FDV.

Macdonald, R. (1995). Selling a dream«. *The Australian Magazine*, 4(5), 25-28.

Nahtigal, Z. (2008). *Resnicnostni sovi: Slovenski nastopajoci in njihove vloge* (Reality shows: Slovene participants and their roles - Slovenian). Ljubljana: FDV.

POP TV (2007), *The farm and its ratings*. Accesible at http://24ur.com/ekskluziv/tuja-scena/internetna-kmetija-odslej-brezplacno.html?ar=

Skala, N. (2008). *Intervju: Veronika, Anka, in Simona s Kmetije* (The Interview: Veronika, Anka, and Simona from the Farm - Slovenian). Accessible at http://zadovoljna.si/clanek/trend_report/nedokoncano-kmetice-s-studentskim-statusom.html

Skrinjar, K. (2009, August 18). Kanal A pred nacionalno TV (Kanal A before national TV - Slovenian). *Delo.* Accessible at http://www.delo.si/clanek/86505.

Slovenijacom (2008) "Slovenci smo Voajerji!" (Slovenes are Voajers). Accessible at http://slowwenia.enaa.com/prikaziCL.asp?ClID=16813.

Slovenska Pratika (2009) *Pratika*. Ljubljana: Mladinska Knjiga.

Splichal, S. (1992). Izgubljene utopije?(Lost Utopias? - Slovenian) Ljubljana: Znanstveno in publicisticno sredisce.

Stankovic, P. (2005). Rdeci trakovi Reprezentacija v slovenskem partizanskem filmu (The representation in Slovene Partisan Film - Slovenian) Ljubljana: Fakulteta za druzbene vede.

Stefancic, M. (2008). Opustite vsakršno upanje, vi, ki vstopate!" (Give up all your hope, you, who enter... - Slovenian). *Mladina 15*, 5.

Volcic, Z. (2005). The machine that creates Slovenes: The role of Slovene public broadcasting in re-affirming the Slovene national identity. *National Identities Journal, 7*(3), 287-308.

Volcic, Z. (2007). Yugo-nostalgia: Cultural Memory and Media in the former Yugoslavia. *Critical Studies of Mass Communication, 24*(1), 21-38.

Zajc, D. (1997). The changing political system. In Drago Zajc (Ed.), *Making a new nation: The formation of Slovenia* (pp. 156-172). Aldershot: Dartmouth.

Zarki, M. (2007). *Resnicnostni sov v Sloveniji* (Reality shows in Slovenia - Slovenian). Ljubljana: FDV.

Zizek, S. (1989). *The sublime object of ideology*. London: Verso.

Zupancic, T. (2007). Temno srce Big Brotherja (Dark Heart of Big Brother – Slovenian). *Sobotna Priloga, 1*, 22-23.

In: Reality Television-Merging the Global and the Local
Editor: Amir Hetsroni, pp. 95-111

ISBN 978-1-62100-068-6
© 2010 Nova Science Publishers, Inc.

Chapter 6

TALKING ABOUT BIG BROTHER: INTERPERSONAL COMMUNICATION ABOUT A CONTROVERSIAL TELEVISION FORMAT

Helena Bilandzic and Matthias R. Hastall
Zeppelin University Friedrichshafen, Germany

At the beginning of the 21st century, episodes of the reality television show *Big Brother* were watched by millions of viewers worldwide and became the subject of countless media and interpersonal debates (Bignell, 2005). *Big Brother* was "in many ways a watershed for our understanding of media audiences" (Ross and Nightingale, 2003, p. 3), as it provoked unprecedented levels of audience ratings and audience involvement. This chapter explores the relationship between a media spectacle like *Big Brother* and interpersonal communications about such events by viewers and non-viewers. We follow Hartley's (1999) understanding of interpersonal communication as a face-to-face communication from one individual to another, in which personal characteristics, social roles and social relationships of the communicating individuals are reflected by form and content of the communication. Our perspective is not restricted to family communication (e.g., Larson, 1993), but encompasses all situations and locations in which interpersonal communication about television programs occurs.

The significance of interpersonal communication for the selection and effects of mass media offerings has been acknowledged decades ago (Lazarsfeld, Berelson, and Gaudet, 1944). Several scientific attempts to combine mass and interpersonal communication have been made since then, and it has been argued that "many of the richest approaches to inquiry about mass communication effects acknowledge a role for interpersonal communication in some way" (Southwell and Torres, 2006, p. 335). Interpersonal communication processes play a role in theoretical approaches like Agenda Setting (Yang and Stone, 2003), the Two-Step Flow hypothesis (Lazarsfeld, et al., 1944), and Diffusion research (Rogers, 1962). Provoking interpersonal talks is also a frequently employed strategy to boost the effectiveness of communication campaigns (e.g., Hafstad and Aaro, 1997). In the majority of communication research, however, interpersonal communication remained a rather neglected topic. In the case of *Big Brother*, interpersonal communication processes deserve a particularly thorough scientific consideration: "Viewers watch [Big Brother] for many reasons—it's something new, you can vote people you don't like off the show—but perhaps the most striking reasons for watching [Big Brother] are that everybody else is watching and talking about it" (Hill, 2002, p. 324). After briefly exploring media consumption motives related to interpersonal communication, we will examine motives for conversations about media content in greater detail.

MEDIA CHOICES FOR INTERPERSONAL COMMUNICATION

Functional media choice approaches assume that media content differs in its ability to satisfy the needs of media users. The hypothesis that mass media offerings are sought and used as a means for subsequent interpersonal communication has been expressed repeatedly (e.g., Chaffee, 1986; Lull, 1980) and can also be directly derived from the Uses and Gratifications Approach (Blumler and Katz, 1974) and from the original Informational Utility Model (Atkin, 1973, 1985). Uses and Gratifications research acknowledges media use as a convenient way to overcome feelings of loneliness, as an activity that may involve interpersonal communication with other people during media use, and as a means for information acquisition for anticipated interpersonal communication (e.g., Wenner, 1985). The Informational Utility Model considers *communicatory uncertainty*, defined as "a cognitive state of incomplete familiarity with a potential conversation topic" (Atkin, 1973, p. 217), as a determinant of media choices. Both approaches share the assumption that interpersonal communication purposes constitute an important motive for media choices, among others. Although the functional logic of these theoretical approach received severe criticisms (McQuail, 1984; Carey and Kreiling, 1974), this interpersonal communication motive appears regularly among the most important self-reported reasons for media choices (Hastall, 2009).

MOTIVES FOR INTERPERSONAL COMMUNICATION

Interpersonal communication can occur before (pre-communicative phase), during (communicative phase), or after (post-communicative phase) exposure to media (Levy and Windahl, 1984). Conversation topics are likely to vary greatly depending on the point of time that the conversations occur: Interpretative and evaluative elements are likely to constitute the biggest share of audience comments in the communicative and post-communicative phase, while expectations about upcoming developments appear more likely in the pre-communicative phase. The diversity of potential topics reflects the range of possible motives to start or sustain conversations.

Although it is widely accepted that interpersonal communication can serve different needs at the same time, little has been done to theoretically elaborate the question why people start interpersonal communication (see Rubin, Perse, and Barbato, 1988, for an overview). Schutz's (1966) Fundamental Interpersonal Relations Orientation theory suggests the existence of three central interpersonal needs: Inclusion (need to belong to others), control (need to exert power), and affection (need to love or be loved). Burgoon and Hale (1984) distinguish seven dimensions of relational communication: Control, intimacy, emotional arousal, composure (self-control), similarity, formality, and task-social orientation. Based on a thorough literature review, Rubin et al. (1988) identify 18 possible motives for initiating interpersonal communication. One third of these dimensions (pleasure, affection, inclusion, escape, relaxation, and control) is empirically validated and included in the Interpersonal Communication Motives (ISM) scale (Rubin, et al., 1988).

How are these motives linked to reality TV programs like *Big Brother*? Given the highly entertaining nature of this show, motives like *entertainment* (pleasure) and *arousal-seeking* appear fairly obvious. The same holds true for *escapism*, the desire to avoid unpleasant thoughts and feelings by seeking interpersonal communication (see also Katz and Foulkes, 1962), as well as for the motives *relaxation*, *convenience*, and *pastime*. Considering the numerous violations of behavioral norms featured in *Big Brother* (e.g., Pawlowski, 2005), interpersonal communication can be initiated to relieve anger or frustrations about the program (*emotional expression*). Furthermore, individuals are likely to be aware of friends and family members watching the program as well; thus, *information-sharing* and *information-receiving* motives may initiate interpersonal conversations. The *Big Brother* motto "You decide!", reflecting the participatory character of this program (Holmes, 2004), can be linked to the *control* motive in the classification cited above. Feelings of *self-esteem* may play an important role too, either to the extent that the own voting decision is in line with the final voting decision, or through social comparison processes that will be discussed later.

MORAL CONSIDERATIONS AND INTERPERSONAL COMMUNICATION

The classification of interpersonal communication motives by Rubin et al. (1988) contains *social norms* for situations in which conversations are required by societal rules. In the case of *Big Brother*, we find it crucial to consider these norms for another reason as well: The main themes of reality television is "to portray subjects engaging in behaviors that tend to violate social norms" (Pawlowski, 2005, p. 1245). Social norms refer to relationships of group members to each other and can be distinguished from procedural and task norms, and also from formally established group rules (e.g., Adler and Rodman, 2006). Deviations from formal rules and informal group norms are likely to instigate emotional discussions, as these conventions constitute the grounds for social relationships. Such discussions about show elements in the mass media and in interpersonal conversations have been repeatedly reported for *Big Brother* (Bignell, 2005). A major ethical concern was that the contestants had to live under a constant surveillance of dozens of TV cameras and microphones, without any contact to the outside world. The shows' title *Big Brother* explicitly refers to the Orwellian nightmare of a society under constant surveillance, and the norm deviation was made obvious with further visual elements like depicting the CBS logo with an open eye (Kellner, 2003).

A SOCIAL COMPARISON PERSPECTIVE ON BIG BROTHER CONVERSATIONS

Selective exposure research suggests that people choose media to acquire valuable information about themselves (Knobloch-Westerwick and Hastall, 2009). Individuals have a desire to evaluate their abilities and opinions, and people depicted in the media are likely to be a useful source for such *social comparisons* (Festinger, 1954). Two main directions of comparison processes can be distinguished, which both may lead to viewers' intensified feelings of self-enhancement or self-esteem (Wood, 1989; Wills, 1981): First, individuals can compare themselves with media personae who are in a less fortunate situation (social downward comparison) to feel better about themselves and their current situation. Second, a social upward comparison can be performed with media personae in a superior situation, in order to learn from them. Social Identity Theory (Tajfel and Turner, 1986) suggests that both positive information about the own group (in-group) and negative information about out-groups can bolster feelings of self-esteem. Consequently, watching *Big Brother* and talking about the show, the contestants and its voyeuristic audience can have a self-esteem bolstering function for members of the same social group – and likely a similar effect for members of higher social groups.

MEDIA EVENTS AND GOSSIP

The *Big Brother* producers employed an extremely successful cross-media strategy that included television, print media, and internet coverage, as well as music spin-offs. *Big Brother* information belongs to the few examples of online media content, apart from pornography and financial information, for which consumers were willing to pay (Freedman, 2006). This cross-media strategy worked well in terms of public attention, which is a precondition of interpersonal communication. Another important function of the show was the potential for *gossip*, which has been labeled "an intrinsic feature of *Big Brother*" (Scannell, 2002, p. 271). Gossip, "a kind of small talk that concerns people who are not present" (From, 2006, p. 231), constitutes an important element of human communication (Thornborrow and Morris, 2004). Instead of perceiving it as a low-status or worthless form of communication, sociolinguists nowadays consider gossip as an important activity for social relationships, identification of group membership, social status, reassurance of social norms, and entertainment (Thornborrow and Morris, 2004). *Big Brother* offered countless possibilities to watch the tenants' gossip, and also allowed viewers to gossip about the contestants. The simplicity of the show and the high levels of media coverage made this possible for non-viewers as well. The *voyeuristic* nature of the program offered viewers many private insights in the contestants' lives, which further fuelled interpersonal discussions as well as passionate criticism (Rayner, Wall, and Kruger, 2004).

This brief and selective review of theoretical approaches for interpersonal communication about reality television programs illustrates the diversity of the field. Although it appears obvious that interpersonal communication can serve a variety of needs at the same time, little is known about motives that instigate conversations about a controversial reality TV show like *Big Brother*. We know *that* people talk about it a lot, but why they do so is less clear. To what extent are reality television shows sought and watched with the intention to talk about them, either during or after exposure? What types of communication about the show can be distinguished? The current investigation explores these questions in more detail and provides an empirical description of the type and the content of *Big Brother* conversations, as well as charts connections to judgments about and exposure to the show, motives for exposure and conversation.

METHOD

Sample

A representative telephone survey with German adults and adolescents was conducted in winter 2000/2001 when the second season of Big Brother was broadcast on the German television stations RTL and RTL 2. Two thousand three hundred and fifty two valid telephone numbers were randomly drawn from electronic telephone directories. Each number was contacted up to five times. Nine hundred and fifty seven interviews were completed (response rate: 41%). Of these, 12 had to be eliminated due to missing data. This left 945 respondents for the analysis. The structure of the sample roughly corresponds to the general population (over 14 years) regarding sex (with a slight overrepresentation of women; 59% female in the sample vs. 51% in the population) and age (15 to 24 years: 18% (population: 11%); 25 to 44 years: 41% (30%), 45 to 64 years: 26% (25%) and 65+ years: 15% (17%); see Statistisches Bundesamt, 2009). However, a bias occurred with regards to the appropriate representation of different education levels: elementary/secondary school: 40% in the sample vs. 68% in the population; high school/college: 55% in the sample vs. 22% in the population. Thus, we check for differences in educational groups across all analyses and report differences whenever they occurred. The interviews were conducted by 50 trained student interviewers, who completed course requirements.

Measures

Spontaneous judgments about the show. The first seasons of Big Brother were accompanied by heated public controversy about the new format. Controversial topics, especially those which are morally disputable, lend themselves readily for interpersonal conversation. To capture a spontaneous evaluation of the show and have an indicator for the tendency as well as extremity of judgments, we asked respondents for their spontaneous reaction when they hear "Big Brother". This question was open-ended, and responses to it were noted verbatim by the interviewers. Later, the responses were coded into categories by two coders. The coders discussed cases of disagreement and agreed on one coding option.

Exposure to Big Brother was measured with the question "How often do you watch the daily one-hour summary show about Big Brother?" Respondents answered in an open-ended fashion, and interviewers coded the answer from 1 (never) to 6 (almost every day, 4-7 times a week). The mean was 2.77 with a standard deviation of 1.75.

Motives for exposure to Big Brother were measured with six items. Respondents could agree (1) or disagree (0). This dichotomous scale was chosen to keep the questionnaire as simple as possible in the telephone situation and to avoid that respondents refuse to continue the survey. As the survey duration was limited to 15 minutes for the same reason, each of the items on this list of motives represents a different dimension and will not be collapsed into a scale or index. The items were: I watch *Big Brother*, because... "I like the fact that I can influence the course of the game as a viewer" (agree: 11%); "it's a good topic for gossip" (55%), "it represents an interesting social experiment" (44%), "my curiosity was evoked by media reports" (46%), "the show is unconventional and controversial" (34%), "I can sort of see into the living room of other people" (28%).

Channels for information about Big Brother. Respondents indicated which media they use as source of information about *Big Brother* (scale: 1=do not use at all; 4= use it very often): newspapers ($M = 2.45$; $SD = 1.09$), magazines ($M = 2.12$; $SD = 1.10$), television ($M = 2.99$; $SD = 1.10$), radio ($M = 1.80$; $SD = .95$), and internet ($M = 1.46$; $SD = .91$).

Last conversation about Big Brother. Three questions concerned the last conversation about the show that took place within the previous two days. If respondents indicated that they had talked about the show within the last two days, they were asked how many people participated in the conversation and were given an open-ended question about the content of this conversation. The open-ended question was again coded by two coders; differences in coding were negotiated until one solution was agreed upon.

Conversations about Big Brother in general. A set of questions concerned informal conversations that respondents had about the show. First, they were asked in which social group they talk about the show (family, friends, people from a school or work context, casual acquaintances). Second, they indicated whether they talked about *Big Brother* during watching the show or independently of the show. Third, they were asked to estimate whether their frequency of talking about *Big Brother* had increased, decreased or remained the same compared to the first season. Then, a set of items assessed the content of these conversations (scale: agree: 1; disagree: 0). Three items measured *involved communication*, where respondents talk about the candidates, their conduct and the events in the *Big Brother* house from an involved perspective. These items ("I like to discuss possible nominations and evictions from the *Big Brother* house", "I like to talk about the conflicts and intrigues inside the *Big Brother* house.", "I enjoy talking about the relationships and kiss-and-tell stories inside the *Big Brother* house.") were combined into a mean score (Cronbach's $\alpha = .76$; $M = .29$; $SD = .37$).

The other items measured *reflective communication*, a type of conversation in which participants take on a distanced perspective or even assume the role of a media critic and talk about the show concept and the effects that the show may have. These items were too heterogeneous to be combined; thus, we used the single items for analysis (moral considerations: "I often discuss whether it is morally correct to keep individuals under surveillance 24/7", $M = .40$, $SD = .49$; psychological damage: "I like to discuss the effects that *Big Brother* may have on the life of the candidates, $M = .43$, $SD = .49$; success: "I often talk about the success of *Big Brother* and its reasons", $M = .53$, $SD = .50$).

Finally, we asked the respondents for their motives to talk about the show. Seven single-item measures were used to tap different aspects of the specific conversational motivation (items: I talk to other people about *Big Brother*...: "to get information about the show", $M = .22$, $SD = .41$; "because it is an easy way to start a conversation", $M = .25$, $SD = .43$; "because it is an good topic for small talk", $M = .40$, $SD = .49$; "to gossip about the candidates"; $M = .39$, $SD = .49$; "to show that I am up-to-date", $M = .14$, $SD = .34$; "because people impose conversations on me", $M = .41$, $SD = .49$; "because I am a true *Big Brother* fan" $M = .07$, $SD = .26$; scale: agree: 1; disagree: 0).

RESULTS

Talking about Big Brother

In our sample, 80% of the respondents (n= 738) had at least one conversation with other people about *Big Brother*. Talking about *Big Brother* was also a widespread phenomenon among respondents who never watch this show on television: 57% of those who never watch it have talked about it at least once (see Figure 1). Thus, *Big Brother* provided conversational topics for viewers and non-viewers alike – and, in a sense, started a conversation within society as a whole. Of course, an increase in exposure frequency also brings about more conversations: If people watch *Big Brother* several times a week, more than 90% talked about the show at least once. Even those who rarely watch it (less than once a month or just once) have talked about it in more than 80% of the cases.

Demographics. No differences in the frequency of conversations can be found between men and women (see Table 1; $\chi^2 = .11$, $df = 1$, n.s.), or between people with different education attainment (see Table 1; $\chi^2 = 3.40$, $df = 1$, n.s.). However, there is a pronounced tendency for younger people to talk about *Big Brother*: Those who had at least one conversation were on average 40 years old, while those who never had a conversation about *Big Brother* averaged at 55 years. This difference is significant ($T = 10.21$; $p < .001$).

Parameters of conversations about Big Brother. Most respondents stated that they talk about *Big Brother* outside of the viewing situation (64%); only 16% indicated that talking happened during viewing (multiple responses were possible). Thus, conversations about the show did not merely accompany viewing, but kept people busy even after they watch the show. The conversations most often went on among friends (50%), among family (32%), and work or school colleagues (39%); conversations among strangers rarely happened (6%).

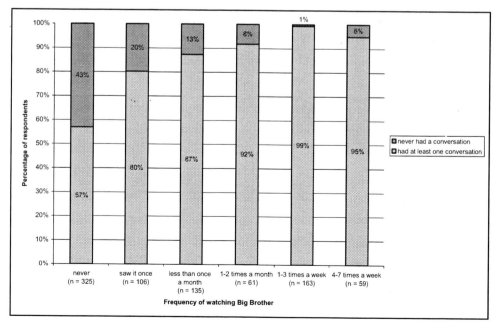

Figure 1. Frequency of Watching Big Brother and Conversations about Big Brother.

Table 1. Demographics and Conversations about Big Brother

| | Conversations about *Big Brother* | | |
	had at least one conversation (%)	never had a conversation (%)	Total n
Sex			
Male	80	20	390
Female	81	19	552
Education			
Elementary/secondary school	78	22	374
High school/college	83	17	523
Total	80	20	945

Note. Values represent percentages.

Compared to the frequency of conversations during the first *Big Brother* season, 38% of the respondents said they talked less about the show during the second season, only 9% talked more during the second season, and 14% thought that the level remained about the same. Apparently, the novelty effect wore off to some extent after the first season; however, this did not eliminate conversations completely.

Last conversation about Big Brother. To obtain a clearer picture of the nature of the conversations, we asked respondents to recall the last conversation they had about the show within the past two days. 17% (n=161) of the respondents had such a conversation. Most respondents (93%) indicated that they had talked with five people or less; a two-person constellation was the most common group size (42%). Then, we asked respondents to sketch the content of the conversation for us. Of those who reported their last conversation, 67% indicated some form of involved communication (e.g., about the behavior of candidates, guesses who may win, evaluations of who is a good or a bad candidate, about events), 20% reported reflective communication (e.g., how the channel is making money with the show, moral evaluations of show, other media coverage of the show) and 13% detailed other topics (e.g., discussion about allowing a minor to watch the show, about *Big Brother* merchandise, about being relieved not to be in the *Big Brother* house).

Conversations and spontaneous judgments about the show. To capture spontaneous evaluations we prompted respondents to indicate what comes to their minds when they hear "Big Brother". It is remarkable that negative evaluations prevailed in this very first comment (see Table 2): 19% of the 920 respondents with valid responses gave a definition that contained a negative evaluation (e.g., boring, not interesting, impossible, waste of my time, don't like it); another 25% used very strong negative evaluations or even swearwords (e.g., absolute nonsense, imbecility, pathetic, primitive, human zoo, garbage, dumbing down of society, exhibitionism, or voyeurism). Approximately a third of respondents used neutral definitions or named candidates, 5% spontaneously gave a positive evaluation and only 4% gave a reflective comment such as "psychologically interesting", "social phenomenon", "receives a lot of media attention", or "brilliant marketing idea" as their first thought about the show.

Table 2. Types of Communication about Big Brother and Spontaneous Reactions (open-ended)

	Conversations about *Big Brother*		
	had at least one conversation (%)	never had a conversation (%)	Total
Definition and neutral evaluation, names candidates	31	19	29
Definition and positive evaluation	6	1	5
Definition and negative evaluation	18	24	19
Definition and strongly negative evaluation, swearword	22	36	25
Reflection	4	4	4
No spontaneous thoughts	15	13	14
Other	4	4	4
Total %	100	100	100
Total n	n = 738	n = 182	n = 920

Note. Values represent percentages.

Talking about the show with other people increased positive and neutral evaluations, and decreased negative and extremely negative judgments (see Table 2): While 31% of those who had a conversation about the show at least once reacted in a neutral way, it was only 19% of those who never had a conversation. In a similar vein, only 22% of respondents who talk about the show express an extremely negative evaluation, whereas it is 36% of those who do not talk about it. Thus, conversations may serve as outlets for positive opinions that have been formed during watching the show; or, talking with other people may also actually improve evaluations of *Big Brother*.

Content of the Conversations

Overall, reflective communication seems to be the more common content of *Big Brother* conversations: More respondents agreed to at least one of the reflective communication items (75% of 651 respondents who talk about the show) compared to the group that agreed to at least one of the involved communication items (44%).

Demographics and conversational content. We cannot find any significant differences between men and women except for the reflective communication item "Success" (see Table 3; $T = 2.01$; $p < .05$). There are, however, some differences regarding education: People with lower education tend to have involved communication more often ($T = 2.81$; $p < .01$), talk about morals less often ($T = -2.11$; $p < .05$), and talk about success less often compared to more highly educated people ($T = -2.11$; $p < .05$). This result suggests that conversations about *Big Brother* may serve as a vehicle for downward social comparison for the more highly educated and provide the grounds for starting a dialogue about ethics in television.

Table 3. Demographics and Content of conversations about Big Brother

	Involved communication	Reflective communication		
		Moral considerations	Psychological damage	Success
Sex				
Male	.27	.37	.46	.57
Female	.31	.42	.40	.49
Education				
Elementary/secondary school	.35	.34	.45	.47
High school/college	.26	.43	.39	.56
Total	.29	.40	.42	.53

Note. Values are means. Range for involved communication: 0-3; reflective communication: 0-1.

Involved communications are not necessarily affirmative of the events in the *Big Brother* house; they too can involve moral issues centered on the candidates' behaviors and statements. But as far as people think about television's role in society and what commercial programs should be allowed to show, it is the more highly educated people who choose to talk about this.

Age was also related to the content of the conversations that people had about *Big Brother*. The older people are, the less they engage in involved conversations ($r = -.32$, $p < .001$) and the more they talk about moral considerations ($r = .15$, $p < .001$); talking about success is not related to age ($r = -.04$, *n.s.*), neither is talking about the possible psychological damage of candidates ($r = -.08$, $p < .05$).

Frequency of watching Big Brother. Next, we computed partial correlations between the frequency of watching *Big Brother* and types of communication, controlling for age, sex and education (see Table 4). The more people watch *Big Brother*, the more frequently they talk about it in an involved way ($r = .54$, $p < .001$). Similarly, the more they talk about possible psychological consequences for the candidates, the more often they watch ($r = .17$, $p < .001$). Conversely, if people indicate to talk about moral considerations regarding the show, they watch the show less often ($r = -.13$, $p < .01$). Talking about success is not at all related to watching the show.

Motives for exposure. As watching *Big Brother* is positively related to only two conversational contents (involved communication and psychological damage), *motives* for watching the show also tend to correlate with these two contents only (again, partial correlations were computed, controlling for age, sex and education; see Table 4): The strongest correlates with involved communication are gossip ($r = .35$, $p < .001$) and voyeurism ($r = .39$, $p < .001$), followed by considering the show as an interesting social experiment ($r = .33$, $p < .001$), appreciating the option to participate as an audience member ($r = .24$, $p < .001$), seeing the show as controversial ($r = .26$, $p < .001$) and curiosity evoked by the media ($r = .12$, $p < .05$). Talking about the psychological damage that candidates may suffer from participating is related to audience participation ($r = .11$, $p < .05$), considering the show as an interesting experiment ($r = .15$, $p < .01$), and to curiosity ($r = .12$, $p < .05$).

Table 4. Partial Correlations of Types of Communication about Big Brother, Exposure and Motives for Exposure

	Involved communication	Reflective communication		
		Moral considerations	Psychological damage	Success
Watching *Big Brother* on TV	.54***	-.13**	.17***	.07
Motives for exposure :				
Audience participation	.24***	.07	.11*	.12*
Gossip	.35***	.02	.12*	.05
Interesting social experiment	.33***	-.01	.15**	.13*
Curiosity evoked by media	.12*	.10	.12*	.16**
Controversial show	.26***	-.00	.11*	.09
Voyeurism	.39***	-.10	.08	.02

Note. n = 428-449; partial correlation coefficients (Pearson's r) controlling for age, sex, and education. *$p < .05$ **$p < .01$; *** $p < .001$.

Curiosity evoked by media reports also motivates people to watch the show when they have the tendency to talk about the success of the show ($r = .16, p < .01$). Talking about moral concerns is not related positively to any of the motives.

Motives for conversation. Respondents who talk about *Big Brother* in an involved way seem to be active parts in the conversation and willing to talk about it – they do not feel that conversations are imposed on them ($r = -.17, p < .001$) – very much in contrast to people who talk about moral concerns and success who have the only positive correlations with this motive (all partial correlations: see Table 5). Similar to motives for exposure, gossip is the strongest motive for conversation for involved talkers ($r = .42, p < .001$). This is followed by conversational motives (small talk: $r = .33, p < .001$; and easy way to start a conversation: $r = .31, p < .001$). Finally, involved talkers also commit to being fans ($r = .32, p < .001$) and wanting to get information about the show ($r = .27, p < .001$). People who talk about psychological damage and success tend to agree to talk about the show because it's an easy way to start a conversation ($r = .17, p < .001$ and $r = .15, p < .01$, respectively), because it's a good topic for small talk ($r = .18, p < .001$ and $r = .09, p < .05$, respectively) and to show that they are up-to date ($r = .10, p < .05$ and $r = .12, p < .01$, respectively).

Channels to learn about Big Brother. Television seems to be the most important channel to learn about the show ($M = 2.99; SD = 1.10$), followed by newspapers ($M = 2.45; SD = 1.09$) and magazines ($M = 2.12; SD = 1.10$). Radio and internet are by far less important ($M = 1.80; SD = .95$ and $M = 1.46; SD = .91$, respectively). Internet use, of course, can be expected to be much more important today than it was during the first seasons eight years ago. Differences between the means are significant (overall ANOVA for repeated measures: $F(df=4) = 374.69; p < .001$; all contrasts between the adjacent means are also significant: internet vs. radio: $F(df=1) = 72.89; p < .001$; radio vs. magazines: $F(df=1) = 61.30; p < .001$; magazines vs. newspapers: $F(df=1) = 49,33; p < .001$; newspapers vs. television: $F(df=1) = 136.66; p < .001$; as contrasts were computed with adjacent means, significances mean that all means are different from each other).

Helena Bilandzic and Matthias R. Hastall

Table 5. Partial correlations of Types of Communication about Big Brother and Motives for Conversations

Motives for conversations:	Involved communi-cation	Reflective communication		
		Moral considerations	Psychological damage	Success
I talk about *Big Brother*...				
to get information about the show	.27***	-.05	.07*	.07
because it's an easy way to start a conversation	.31***	-.04	.17***	.15**
because it's a good topic for small talk	.33***	-.06	.18***	.09*
to gossip about the candidates	.42***	-.08	.06	.00
to show that I'm up-to-date	.17***	.06	.10*	.12**
because people impose conversations on me	-.17***	.11*	.03	.14**
because I am a true *Big Brother* fan	.32***	-.01	.08	.01

Note. $n = 619\text{-}644$; partial correlation coefficients (Pearson's r) controlling for age, sex, and education. *$p < .05$ **$p < .01$; *** $p < .001$.

Table 6. Partial Correlations of Types of Communication about Big Brother and Sources of Information about the Show

	Involved communication	Reflective communication		
		Moral considerations	Psychological damage	Success
Newspaper	-.03	.03	.05	.03
Television	.25***	-.01	.16***	.13**
Radio	.10*	.05	.05	.08
Internet	.15***	-.05	.02	.03
Magazines	.07	.06	.06	.10*

Note. $n = 630\text{-}651$; partial correlation coefficients (Pearson's r) controlling for age, sex, and education. *$p < .05$ **$p < .01$; *** $p < .001$.

However, not all important channels also relate to conversations about the show. All types of conversation except moral concerns correlate positively with television use (see Table 6: significant partial r's from .13 to .25. Internet use only correlates positively with involved talk ($r = .15$; $p < .001$) while newspaper use is not related to any type of conversation. Magazine use is only related to talking about success talk ($r = .10$; $p < .05$).

DISCUSSION

People love to talk about morally disputable issues – especially when issues are easily accessible through media and coupled with public attention and money. The first seasons of *Big Brother* provided ideal grounds for interpersonal dialogue among all parts of society, men and women, persons with high and low education, viewers and non-viewers alike. This chapter explored interpersonal communication that took place during the first seasons of the reality television show *Big Brother* in Germany – at a time when the format was new and highly controversial. Media content is known to permeate into interpersonal communication to a certain degree (e.g., Keppler, 1994). However, it is rare indeed that a large portion of the population chooses the same show to talk about. In our survey, we found that four fifths of the respondents had talked at least once about Big Brother. The show even reached people who did not watch it: Non-viewers were not only confronted with the show through the extensive media coverage; the show also haunted them outside the viewing situation, in their personal contexts – in conversations with their friends, relatives, along with work and school colleagues. In this way, the show expanded its reach considerably, even to those who refused to be exposed to it in the first place.

Judgments about the show are not as favorable as one would expect, considering its popularity on television and as a conversational topic shared by a wide portion of the population. The very first spontaneous reaction to the show was negative in almost half of the cases in our sample. What is more, the majority of the negative reactions consisted of extreme judgments, even swearwords. This may be a reflection of moral panic on an individual level: When *Big Brother* was first broadcast, a flood of media reports and commentaries severely criticized the new format for its numerous transgressions of social norms. This "public anxiety about key social and moral issues, characterized by spiraling debate" (Biltereyst, 2004, p. 91) is often referred to as "moral panic" – or "media panic", if the event is media-generated (Drotner, 1992). Finding negative reactions in private conversations is an expression and a catalyst of this public discourse and shows that the moral indignation was not reserved to persons with a public voice (journalists, politicians, etc.), but also occupied the minds of regular people. Biltereyst (2004) claims that the staging of moral panic has become an integral part of the format of reality TV; indeed, we may conclude from our findings that the audience similarly considers controversy as an integral part of the viewing experience: Negative views on the show do not preclude the audience from watching it; exposure feeds into conversations and conversations in turn fuel more exposure. One of the main motives for exposure is being able to gossip about the show. To some extent, a satisfactory use of *Big Brother* may be only present when audience members complete their understanding of the show through interpersonal conversations. The portion of negative and extremely negative evaluations decreased when people had at least once talked about *Big Brother*. The causality of this relationship, however, is not clear – people may talk about the show because they like it, or they may like the show because they (can) talk about it.

One of the reasons of the show's popularity as a conversational topic may be that it offers material for conversations on several levels: *Involved communication* deals with the world within *Big Brother*, the actions and events, and the candidates' emotions, while *reflective communication* considers *Big Brother* as a cultural and commercial artifact and deliberates moral issues. As the concept of the show is as simple as it was scandalous at its first introduction, people do not need much information about the show to form an opinion and discuss it. This may be an explanation why we found that more conversations deal with reflective rather than involved aspects – for involved communication, detailed information is necessary and actual exposure to the show indispensable, while people are able to carry a reflective conversation with no more than just the concept in mind.

The two kinds of conversations, involved and reflective, are associated with different groups of people and contexts. People who enjoy involved communication are young and less educated, and tend to watch *Big Brother* on a regular basis. The most prominent motives for them to watch the show are gossip, voyeurism and considering the show as an interesting social experiment. Gossip is also the strongest motivator for involved talkers to engage in interpersonal communication. This group appreciates *Big Brother* as a convenient topic for small talk and wants to get more information about the show by talking to others. Involved talkers constitute the group that makes most use of the internet to get information about the show.

People who engage in reflective conversation have quite a different profile: The more they communicate about moral issues, the better educated and the older they are, and the less they watch the show. Reflective talkers tend to watch Big Brother out of curiosity evoked by the media, and because they perceive the show as an interesting social experiment.

Overall, *Big Brother* represents an interesting example of media content that engaged very different kinds of people in a common dialogue. Certainly, people did not agree in their evaluations of the show, but its provocations started a societal conversation (or argument) that many parts of the population shared. In this sense, there is a discrepancy between the conscious, often harsh judgments about the show, and the way in which this show may be functional for society. Providing a common topic for social discourse serves an integrative function: People may disagree about evaluations and enter heated discussions about *Big Brother*. However, in order to talk about it, they still need to agree that *Big Brother* is something that needs discussion, something that is relevant to society even though in the sense of being a threat. Characteristics of people who engage in reflective communication (they are older, have higher education, and watch less of the show) suggest that social comparison may be one the functions that the show readily serves – to increase the viewer's (and talker's) self-esteem. It is not just a discrimination that happens in this type of social comparison. Cultural Studies scholar Robert Pfaller, in a recent newspaper article, analyzed how sexuality is banned from public life, removed from normality, and quarantined into in television talk shows and reality formats as a repulsive caricature. These formats serve as entertainment for viewers, but also as "a dangerous threat: If you don't pull yourselves together, it will be you in the container tomorrow" (Pfaller, 2009). Indeed, such processes surfaced in our survey as well – one of the respondents said that his last *Big Brother* conversation was about him being relieved *not* to be in the container himself. This goes beyond social comparison. In this sense, *Big Brother* may serve to shape one's attitudes about social conduct and moral understanding (see Krijnen and Tan, 2009). Beyond direct exposure to the show, interpersonal communication may be a central vehicle and catalyst for this process – and may multiply the show's effect by reaching non-viewers as well.

REFERENCES

Adler, R. B., and Rodman, G. (2006). *Understanding human communication*. Oxford, UK: Oxford University Press.

Atkin, C. K. (1973). Instrumental utilities and information seeking. In P. Clarke (Ed.), *New models for mass communication research* (pp. 205-242). Beverly Hills, CA: Sage.

Atkin, C. K. (1985). Informational utility and selective exposure to entertainment media. In D. Zillmann and J. Bryant (Eds.), *Selective exposure in communication* (pp. 63-91). Hillsdale, NJ: Lawrence Erlbaum.

Bignell, J. (2005). *Big Brother: Reality TV in the twenty-first century*. New York, NY: Palgrave Macmillan.

Biltereyst, D. (2004). Reality TV, troublesome pictures and panics: Reappraising the public controversy around reality TV in Europe. In S. Holmes and D. Jermyn (Eds.), *Understanding reality television* (pp. 91-110). London: Routledge.

Blumler, J. G., and Katz, E. (Eds.). (1974). *The uses of mass communications: Current perspectives on gratifications research*. Beverly Hills, CA: Sage.

Burgoon, J. K., and Hale, J. L. (1984). The fundamental topoi of relational communication. *Communication Monographs, 51*(3), 193-214.

Carey, J. W., and Kreiling, A. L. (1974). Popular culture and uses and gratifications: Notes toward an accommodation. In J. G. Blumler and E. Katz (Eds.), *The uses of mass communications: Current perspectives on gratifications research* (pp. 225-248). Beverly Hills, CA: Sage.

Chaffee, S. H. (1986). Mass media and interpersonal channels: Competitive, convergent, or complementary? In G. Gumpert and R. Cathcart (Eds.), *Inter/media: Interpersonal communication in a media world* (pp. 62-80). New York, NY: Oxford University Press.

Drotner, K. (1992). Modernity and media panics. In M. Skovmand and K. C. Schrøder (Eds.), *Media cultures: Reappraising transnational media*. London: Routledge.

Festinger, L. (1954). A theory of social comparison processes. *Human Relations, 7*(2), 117-140.

Freedman, D. (2006). Internet transformations: 'Old' media resilience in the 'new media' revolution. In J. Curran and D. Morley (Eds.), *Media and Cultural Theory* (pp. 275-290). London, UK: Routledge.

From, U. (2006). Everyday talk and the conversational patterns of the soap opera. *Nordicom Review, 27*(2), 227-242.

Hafstad, A., and Aaro, L. E. (1997). Activating interpersonal influence through provocative appeals: Evaluation of a mass media-based antismoking campaign targeting adolescents. *Health Communication, 9*(3), 253.

Hartley, P. (1999). *Interpersonal communication*. London, UK: Routledge.

Hastall, M. R. (2009). Informational utility as determinant of media choices. In T. Hartmann (Ed.), *Media choice: A theoretical and empirical overview* (pp. 149-166). New York: Routledge.

Hill, A. (2002). Big Brother: The real audience. *Television New Media, 3*(3), 323-340.

Holmes, S. (2004). 'But this time you choose!': Approaching the 'interactive' audience in reality TV. *International Journal of Cultural Studies, 7*(2), 213-231.

Katz, E., and Foulkes, D. (1962). On the use of the mass media as "escape": Clarification of a concept. *Public Opinion Quarterly, 26*(3), 377-388.

Kellner, D. (2003). *Media spectacle*. London, UK: Routledge.

Knobloch-Westerwick, S., and Hastall, M. (2009). *Please your self: Selective exposure to news about in- and out-groups and its effect on self-esteem*. Paper presented at the Annual Meeting of the International Communication Association, Chicago, IL.

Krijnen, T., and Tan, E. (2009). Reality TV as a moral laboratory: A dramaturgical analysis of The Golden Cage. *Communications: The European Journal of Communication Research, 34*(4), 449-472.

Larson, M. S. (1993). Family communication on prime-time television. *Journal of Broadcasting and Electronic Media, 37*(3), 349.

Lazarsfeld, P. F., Berelson, B., and Gaudet, H. (1944). *The people's choice*. New York: Duell, Sloan and Pearce.

Levy, M. R., and Windahl, S. (1984). Audience activity and gratifications: A conceptual clarification and exploration *Communication Research, 11*(1), 51-78.

Lull, J. (1980). The social uses of television. *Human Communication Research, 6*(3), 197-209.

McQuail, D. (1984). With the benefit of hindsight: Reflections on uses and gratifications research. *Critical Studies in Mass Communication, 1*(2), 177-193.

Pawlowski, C. (2005). Reality television. In J. K. Roth (Ed.), *Ethics* (pp. 1244-1246). Pasadena, CA: Salem Press.

Pfaller, R. (2009, Aug. 8). Lust und Prüderie: Vom Sex in der Medienmoderne [Lust and prudery: About sex in the media modernity]. *Süddeutsche Zeitung*. Retrieved 08-14-2009, from http://www.sueddeutsche.de/u5E38T/2999427/Lust-und-Pruederie.html.

Rayner, P., Wall, P., and Kruger, S. (2004). *Media studies: The essential resource*. London, UK: Routledge.

Rogers, E. M. (1962). *Diffusions of innovations*. New York, NY: Free Press.

Ross, K., and Nightingale, V. (2003). *Media and audiences: New perspectives*. Berkshire, UK: Open University Press.

Rubin, R. B., Perse, E. M., and Barbato, C. A. (1988). Conceptualization and measurement of interpersonal communication motives. *Human Communication Research, 14*(4), 602-628.

Scannell, P. (2002). Big Brother as a television event. *Television New Media, 3*(3), 271-282.

Schutz, W. C. (1966). *The interpersonal underworld*. Palo Alto, CA: Science and Behavior Books.

Southwell, B. G., and Torres, A. (2006). Connecting interpersonal and mass communication: Science news exposure, perceived ability to understand science, and conversation. *Communication Monographs, 73*(3), 334-350.

Statistisches Bundesamt (2009). Genesis Online Datenbank [Genesis online database]. Retrieved 08-14-2009, from https://www-genesis.destatis.de/genesis/online.

Tajfel, H., and Turner, J. C. (1986). The social identity theory of intergroup behavior. In S. Worchel and W. G. Austin (Eds.), *Psychology of intergroup relations* (pp. 7-24). Chicago, IL: Nelson-Hall.

Thornborrow, J., and Morris, D. (2004). Gossip as strategy: The management of talk about others on reality TV show 'Big Brother'. *Journal of Sociolinguistics, 8*(2), 246-271.

Wenner, L. A. (1985). The nature of news gratifications. In K. E. Rosengren, L. A. Wenner and P. Palmgreen (Eds.), *Media gratifications research: Current perspectives* (pp. 171-193). Beverly Hills, CA: Sage.

Wills, T. A. (1981). Downward comparison principles in social psychology. *Psychological Bulletin, 90*(2), 245-271.

Wood, J. V. (1989). Theory and research concerning social comparisons of personal attributes. *Psychological Bulletin, 106*(2), 231-248.

Yang, J., and Stone, G. (2003). The powerful role of interpersonal communication in agenda setting. *Mass Communication and Society, 6*(1), 57-74.

SECTION III: THE MIDDLE EAST

In: Reality Television-Merging the Global and the Local
Editor: Amir Hetsroni, pp. 115-122

ISBN 978-1-62100-068-6
© 2010 Nova Science Publishers, Inc.

Chapter 7

REALITY VS. REALITY TV:
NEWS COVERAGE IN ISRAELI MEDIA AT THE TIME
OF REALITY TV

Dror Abend-David
Ohalo College, Israel

On Monday, April 20, 2009, the Israeli press highlighted an important announcement by the Governor of Israel's National Bank, Professor Stanley Fischer: One or two major companies might find themselves in bankruptcy by the end of the week (Avriel, 2009). The press was immediately buzzing with the names of possible Israeli companies and individuals who might be candidates for bankruptcy. However, by the end of the week, the press was already preoccupied with the hostilities between Fischer and Bank Hapoalim owner Shari Arison, and the identity of the Israeli mogul to be banished from the circles of economic powers was never revealed.

Fischer's success at securing the attention of the national media was achieved through methods not very different from the ones used in reality TV. He employed an attractive promotion that kept audiences glued to the screens for an important end-of-the-week announcement. And the proposed banishment or outing of a participant in a circle of national celebrities echoes rituals of periodical elimination of contestants from chart-topping reality programs such as *Survivor* and *Big Brother*.

This article will examine some of the ways in which reality programming has been influencing news coverage in Israel, and consequently, some of the different ways in which events have been treated, and perhaps even shaped by a new reality perception of real events.

News and reality television are usually grasped in Israeli society as contradictory venues. A blog by Oz Rahamim that was first published anonymously on October 20, 2008, divides Israeli society into those who are preoccupied with reality TV, and the *Big Brother* series in particular, and those who are concerned about reality, and particularly about the fate of hostage Israeli soldier, Gilad Shalit. Oz compares the voluntary imprisonment of the actors in the series, and the involuntary and rather dangerous incarceration of a soldier held in captivity. This initially anonymous blog was copied and forwarded enthusiastically, and was finally published with full credit on the web site of a major radio station (103 fm radio) on

November 11, 2008 and reported on by the established press. The article, which received a great deal of support,[1] echoes a popular notion that reality audiences are shallow people who do not follow national events and serious issues on the news.

However, what I wish to demonstrate in this article is that the gap between reality and reality television is narrowing, as the news – not less a form of entertainment than other programming – is adopting some of the techniques and styles of reality programs. An important catalyzing force in this process was the coverage of the Israeli invasion of the Gaza Strip in December 2008. Particularly as a traditional coverage involving a consistent narrative proved less efficient in terms of echoing the sentiments of the viewing public, methods for sustaining interest, manufacturing pick points and drawing an emotional appeal were adopted as part of the daily news. These remained an integral part of news coverage after the Gaza invasion as well.

We all seem to have a strong sense of what reality television is about. But a clear definition of this genre is by far more slippery and elusive than we might imagine at first. One authoritative and clear definition is provided by J. Dovey in *The Television Genre Book*:

> 'Reality TV' is...any factual programme based on... direct, unmediated account of events, often associated with the use of video and surveillance-imaging technologies. The different kinds of programme described as 'reality TV' are unified by the attempt to package particular aspects of everyday life as entrainment. (Dovey 2001, p.135)

Dovey's definition is mostly formal, and begs a question about the differences between reality television and news-casting: Are breaking news a form of reality television? What about the live broadcast of anchor news? Or even the unrehearsed chat between news announcers, their guests, or even the weather man/woman, who might turn into a reality mini-series in its own right, as in Blayne Weaver's 1994 film, *Weather Girl*. Particularly on morning shows, between the live cooking segment in which the cook might cut his finger on the air, and a live footage of a woman bleeding to death on the sidewalk in Iran, which is news and which is reality television? Both are direct, unmediated accounts of events, but our gut feeling is that they are not one and same.

One possible factor that Dovey provides, and which does not have to do with format, is related to scope, or what Dovey refers to as aspects of everyday life. The subject matter of the program, consequently, might better serve to shape our definition. Accordingly, a live report on gardening one's own back-yard can be classified as reality, while the British Queen's visit to Parliament can be seen as news. Arguably, the latter affects a vast portion of the viewing public, while the former affects only a few of them. But how should one classify a live report on the Queen as she decides to do some gardening of her own? And what should we make out of an appearance of the President of the United States on *Saturday Night Live*? Is this news or entertainment?

On what ground do we find ourselves when, as recently happened in Israel, a well known entertainer has been accused of violent crimes, crossing over between game shows and news from the courthouse? During the summer of 2009, Israeli entertainer Dudu Topaz was the

---

[1] As one of many examples for the public support of this blog, the website of the Israeli Scouts movement has reprinted it on December 12, 2008. The website keeps a record of the number of people who had read the blog since it was posted (3200 to date). The supportive reactions of readers are listed there as well. See: http://www.zofim.org.il/magazin_item.asp?item_id=206584794256. In Hebrew.

subject of headlines very different than the ones he had usually received. He was accused and confessed to having hired hit men to assault some of his rivals in the entertainment business. On August 20, 2009, Topaz committed suicide in his jail cell (Lappin, Y. August 20, 2009). This sad affair had brought a great many people familiar to the public from the world of guest shows, quizzes, and televised games to suddenly appear in news segments either as witnesses or people somehow related to the offender, the victim, or to the Israeli entertainment world in general. This was certainly "news," but was it also an entertainment?

These, starting with the British Queen and ending with cultural celebrities, are only a few out of a growing number of items that transcend the boundary between the private and the public.

These queries are not a part of the viewing experience of most of the public. Most of us have a strong sense of the type of programming that we are watching, and we contextualize the contents of various programs accordingly. In a research conducted by Hill, Weibull and Nilsson (2007), British and Swedish audience's attitudes towards news and reality programming were studied. The survey used in this research included ten types of fact-based programming, including nature programs, documentaries and lifestyle programs. But even the mere presentation of this list betrays the difficulty of defining such sub-genres, and the extent to which they collapse into each other. For example, is a recently advertised Turkish television program that features Christian, Moslem, Jewish and Buddhist missionaries trying to convert a group of atheists (Tait, 2009) a documentary, a lifestyle experiment program, or a reality game show? Given the current volatile religious situation in Turkey, this could very easily be considered a current-affairs program or a political discussion panel as well.

But Hill and her colleagues focus their interest on audience's attitude towards reality programs, and on that they present clear data: Both British and Swedish viewers attribute a great deal of value and quality to traditional news programs such as newscasts, current affairs, investigative journalism and political discussion programs. Ironically, while viewers believe that such programs are more important than reality programs such as lifestyle experiments and reality game shows, they also find such programs less entertaining, and prefer to watch reality television: "Popular factual functions the other way round, they are watched more than they are regarded as important" (Hill et. al., 2007, p.39).

The best case scenario, of course, would have been if viewers were entertained by programs that they consider publicly and educationally valuable. Hill and her colleagues recommend, accordingly, that news programming will adopt the methods of reality television in order to avoid losing the next generation of viewers (Hill et. al., 2007). My argument, of course, is that news programs in Israel have already taken this step and narrowed considerably the gap between news and reality.

But in order to examine carefully the relations between news and reality television, one still needs to develop a clear definition of the differences between the two. A possible method for doing so has to do with context, development and a certain consciousness of the existence and qualities of the medium one uses. Such consciousness is better known as the suspension of disbelief, and is what Miller (2000) sees as the main reason for the differences between early reality programs such as *American Family*, and more recent productions such as *The Real World* and *Survivor*. According to Miller, the nineteen-seventies production of *American Family* was able to address important social and political issues because it featured ordinary participants facing cameras, rather than ordinary participants pretending not to face cameras, and performing the appearance of ordinary reality. In other words - the tendency to present a complete reality, devoid of mediation, has ironically created a greater extent of directing and editing of authentic action:

> Given the construction and tendency of the media industries, any attempt at realness is all but doomed to fail being accurate. Instead of offering diversity, these shows [*The Real World* and *Survivor*] will put forward ideal types as everyday people. Moreover, whereas contestants or participants may not be actors (and hence always acting as if they were their real true and always authentic selves), editing will turn interactions and confessions into montaged performances, showing how hours of dull footage can be edited into dramatic sequences. In these sequences, not all aspects of the story are included, nor will all viewers find themselves in the characters on screen. (Miller, 2000, p.15)

Context and development are also described as an important element in a recent sub-genre of TV comedy that is often referred to as *Comedy Verite* (Thompson, 2007). Shows such as *Curb your Enthusiasm, Arrested Development* and *The Office* use the technique of observational documentary to create the sense of loosely edited footage based on unrehearsed reality. However, despite the overwhelming number of features of reality programming that are included in *Comedy Verite*, these comedies are never mistaken for reality programming. The reason is that programs such as *Curb your Enthusiasm, Arrested Development* and *The Office* retain the Aristotelian structure of comedy, which has undergone only few changes over some twenty-five thousand years.[2] The Shakespearean tradition of weaving a number of parallel storylines throughout the dramatic text, which has been adopted successfully in television comedy still glues together the action in *Comedy Verite*.

The elements of context, development and suspension of disbelief are, therefore, dramatic terms, appropriate to dramatic texts and television drama, whether comedy, historical drama, mystery, suspense etc. Their relevance to news programming and reality television has to do with the social and historical function of storytelling in general, and of eighteenth- and nineteenth-century novels in particular. This link is pointed out by Charles McGrath in his article, "The Triumph of the Prime~Time Novel," first published in *The New York Times Magazine* in 1995 and since reprinted and translated in various anthologies. McGrath looked at television dramas of the nineteen-nineties such as *E.R., Chicago Hope, Homicide: Life in the Streets, N.Y.P.D. Blue, Law and Order, Picket Fences* and *My So-Called Life* and demonstrated how they took over the social and political roles of eighteenth and nineteenth-century novels in accompanying new social trends, commenting and encouraging social change, and reshaping social class and hierarchy (Foucault 1977, 35; Holt 1990; Mitchell, 1992).

But the social role of television drama is only one part of the equation. Since the beginning of time, narrative did not only serve the function of social commentary and the recording of events. Ancient tribes have used structured mechanisms of rhyme and rhythm, beginning middle and end, context and the suspension of disbelief to make sense out of their environment, provide explanations through fiction to various phenomena, and, most importantly, attribute significance to various events. If, as Freud argues in *Civilization and Its Discontents*, the development of an entire society can be compared to the psychological process observed within the psyche of a single individual (Freud 1989, 104-9), then the

---

[2] It might also be worthwhile to examine shows like *The Hills* which consist of reality features yet adhere to more conventional dramatic structures.

function of narrative can be compared with that of the individual dream: it is a therapeutic, critical method of coming to terms with one's surrounding by assigning value, order and meaning to the otherwise incomprehensible mess of human life.

In other words: novels, television drama, and traditional news programming tell a story. This is an important observation when taking into account that all three are the product of a Modernist tradition that privileges structure and presumes a clear and unchangeable hierarchy. Accordingly, within the traditional newscast, a presumed collective memory plays an important part in contextualizing news. The news are also categorized by predetermined hierarchy, and delivered according to fixed order: political news, social affairs, sports and weather (Harrison 2001, 114).

But, within the structural collapse of a Post-Modern reading, there is no certainty that one news item is more significant than another; categorization if fluid; and context is lost on the viewer as all pre-suppositions regarding the common defining texts and myths that audiences ascribe to, the categorization of news items, and the functions and limits of the medium are open to discussion. Television drama, by definition, must retain a Modernist structure, while reality television thrives on the transcendence of boundaries and the pastiche of possibilities. Part game-show, part documentary, and part investigative journalism, there is never an item too trivial, unauthentic, or irrelevant to reality television. In fairness, who is to say that our daily lives are less significant than the empty speech of politicians or misadventures of large business firms that dominate traditional news casting?

For the purpose of this article, the recognition of these differences serves as a better tool for defining the relationship between news and reality television than any formal definition having to do with broadcasting methods and the scope of transmitted content. This is so first, and foremost, because the added polar of television drama provides a necessary perspective to the present discussion. If one places news programming between the two extremes of reality and drama, it is finally possible to compare two opposing traditions: structured and unstructured. In the simplest terms possible: is current news programming more like television drama or more like reality television?

In Israel, the tendency is certainly towards the latter. The extent towards which contexts is often lost on contemporary viewers and the boundaries of traditional news-casting and reality television are transcended, is demonstrated in a rather unique 2008 blog article attributed to Oz Rahamim.[3] This blog article is an interesting communicative phenomenon in its own right. Initially anonymous, this blog was published in a private forum, but had caught the imagination of its readers to the extent that it was rapidly forwarded and read by a large number of people. Finally, it was formally published, citing the author's name on the official website of the Israeli 103 FM radio station. The article highlights in an ironic fashion the extent to which news events can be trivialized and made to compete with similar developments on reality television when taken out of context. Rahamim chooses to compare the lengthy incarceration of Israeli soldier Gilead Shalit, who is being held hostage by Palestinian *Hamas* forces, and the confinement of contestants in the Israeli version of *Big Brother*. As Rahamim demonstrates, both Shalit and the contestants in the *Big Brother* show are held prisoners, kept away from their family, and must adapt to new surroundings. Both the captive soldier and *Big Brother* contestants undergo a difficult psychological process, and, as far as viewers are concerned, both the contestants and the soldier receive a great deal of

---

[3] Available at: http://www.103.fm/blogs/chapter.aspx?pun26r4Vq=GKF&Zrzor4Vq=LJJK. In Hebrew.

public attention. In fact, as Rahamim writes, Shalit, the captive soldier, is not even the viewers' favorite prisoner, as he does not receive most of their sympathy. Most of the attention and admiration is directed at another prisoner, chosen among the better exposed participants in the reality show.

Rahamim finally does present the context that renders this comparison invalid: Shalit is incarcerated against his will, while the prisoners of *Big Brother* participate willingly in the program. *Big Brother* participants must adapt to living in a large house complete with a back yard, swimming pool and other amnesties, while the conditions of Shalit's incarceration are unknown. And finally, *Big Brother* participants wish to stay incarcerated as long as possible, and stand to win large sums of money when they leave. Shalit stands to win nothing, and might never leave his prison alive.

In its conclusion, this blog article repeats the views expressed by the viewers who have been surveyed by Hill et al. (2007): That there is a substantial difference between news and reality television, as the former is socially significant and worthy of attention and the latter is entertaining while inconsequential. However, Rahamim's article also demonstrates the extent to which the two can be conflated, as the differences between reality and reality television are lost on jaded audiences, for whom any content appearing on the screen is taking place in a televised Never-Never Land. Moreover, Rahamim certainly does not partake in Hill's recommendation that news programs will adapt some of the entertaining qualities of reality television. On the contrary: he calls on his readers to abandon the trivial contents of reality television and focus on the serious events reported in news programming.

But can news programs sustain a clear definition of serious and trivial events? Such distinction would be based on a subjective notion, and might become increasingly difficult to sustain within the current reality of Israeli media. During the week of April 19 to 24 (2009), the entire Israeli news media, including newspapers, television, radio and online news sources, occupied itself with the exciting brainteaser (with which I began this discussion) offered by Stanley Fischer, the Governor of Israel's National Bank: That one or two major companies were to find themselves in bankruptcy by the end of the week. Economists and annalists, newscasters, personal columnists and bloggers competed in hinting coyly to viewers and readers that they might know which companies Fischer is referring to, but none of them actually provided the information. The guessing game was more exciting than the actual information, and it has kept readers and audiences in suspense for an implied announcement anticipated by the end of the week. This was an efficient promotion, keeping viewers tuned and not a little blood-thirsty for the banishment of one of the members in an exclusive and well known group of Israeli moguls. In that, it was not much different from the anticipated banishment of participants in programs like *Survivor* or *Big Brother*, as the characters of successful Israeli businessmen appearing on the news enjoy the same celebrity status as that of participants in reality shows.

Coming out of the blue and disappearing into thin air, the excited discussion around a potential dethroned tycoon never came to fruition. Development and contexts proved immaterial in this case, as the upcoming story of the banishment of the moguls cleared the stage at the end of the week to Fischer's publicized feud with Bank Hapoalim owner Shari Arison. No one bothered to find out which company, if any, was bankrupt, as the tension between the Governor of the National Bank and the owner of one of the largest banks in the country received equal attention to that of the occasional hostilities between reality characters, trapped in a sealed house or on a tropical island.

But this is not the only way in which context has been lost here. As Eytan Avriel comments in his newspaper article from the same week (Avriel, 2008), the increased viewership and attention was only one advantage of the mogul banishment scheme. The "added bonus," as Avriel writes, is that it served to divert attention from other issues related to Fischer's work, which are of no lesser significance. Such issues include a public debate over the defense budget, and the wages of government employees, particularly at the National Bank. What Avriel is suggesting is that popular and gossipy headlines do not only attract wider viewership (that being rather obvious), but that they can actually be used to subvert and avoid public scrutiny and controversial issue that might provoke public anger and retaliation if the public was not preoccupied with who might be the next mogul to go bankrupt (and in the meantime public money is being spent for the benefit of various moguls and public officials).

It seems, then, that the "reality" quality of news programming makes the work of news editors (and public figures) easier: In the same way that reality programming can present characters without the laborious process of background and development, news programming is freed from the fastidious need to accompany items with lengthy expositions that might challenge viewers' attention span and intellectual capacity. This, as I mention above, was particularly visible during the war in Gaza between December 2008 and January 2009. It is impossible to discuss here the entire televised coverage of the war, but as Lisa Glodman write in *The Seventh Eye* (Goldman, 2009), war coverage on television consisted of very little reporting from the battlefield. Instead, Israeli channels devised "open studios" which broadcasted mostly personal stories of individuals affected by the war. As Li-Or Averbuch writes in *NRG Maariv* (Averbuch, 2009), most Israelis were quite pleased with this type of coverage, particularly in comparison with previous wars in which more facts and images were reported. While Israeli television did not tell viewers much about the events of the war, it did provide the emotional support and nationalist pride that Israeli reporters have been often blamed of withholding. For example, a mother worried about her two sons fighting in Gaza was invited to the studio on her birthday. Her two sons, located somewhere on the battlefield, sang "happy birthday" to her over the phone, with the camera zooming on her live and unavoidable tears. This was reality television at its best, putting both *Big Brother* and *Survival* to shame, and utterly innocent of informative content. But this was of little consequence to Israeli viewers who were generally pleased with the reports they have been receiving. This type of war coverage comforted them; made them fill better; and provided contents that they could easily handle both intellectually and emotionally. Most of us do not expect television programming to do more.

Media shifts such as the one occurring in Israeli news programming are, for better or worse, unavoidable. The collective memory of a society is to a great extent a product of news reports in the media, and through the process that currently takes place in Israeli news-casting, the extent of the collective memory that can be expected from viewers is gradually diminishing. Perspective and context are equally eroded as personal human stories move steadily from the margins of news programming to its very core. There is yet another outcome to the development of reality-like news programming: Because exposition and causality play a secondary role to viewers' attention and repeated viewing, the news tends to highlight what most attract viewers' attention. To this extent, reality television does not only influence the style of the news, but also its focus and its contents. The advantage, of course, is that news programming does become more entertaining as a result (as suggested by Hill et al., 2007). At

REFERENCES

Averbuch, L. (2009, January 1). Reruns: The Televisied coverage of the campaign in the south. NRG Maariv. Retrieved from: http://www.nrg.co.il/online/1/ART1/833/866.html. In Hebrew.

Avriel, E. (2009, April 24,). Which Mogul is Fisher Thinking About? The Marker.

Dovey, J. (2001). Reality TV. In Creeber, G. et. al. Eds., The Television Genre Book. London : British Film Institute. 135-138.

Foucault, M. (1977). Discipline and ϱunishment - The ϶irth of the ϱrison. Tr. Alan Sheridan. New York: Vintage Books.

Freud, S. (1989). Civilization and its discontents. Tr. James Strachey. NY: W. W. Norton and Company.

Goldman, L. (June 7, 2009). What Israelis Wanted to Know about the War. The Seventh Eye [published by the Israeli Institute for Democracy]. Retreived from http://www.the7eye.org.il/articles/pages/040609_covering_gaza_from_israel.aspx. In Hebrew.

Harrison, J. (2001). Constructing News Values. In Creeber, G. et. al. Eds., The Television Genre Book. London : British Film Institute. 114-116.

Hill, A., Weibull, W., & Nilsson, Å. (2007). Public and popular: British and Swedish audience trends in factual and reality television. Cultural Trends, 16(1). 17-41.

Holt, S. W. (April 1990), Rhetorical analysis of three feminist themes found in the novels of Toni Morrison, Alice Walker, and Gloria Naylor, Dissertation Abstracts International, 50(10). 3224A.

Lappin, Y. (2009, August 20). Judge to oversee Topaz suicide probe. Jerusalem Post. Retrieved from http://www.jpost.com/servlet/Satellite?pagename=JPost/JPArticle/ ShowFull&cid=1249418652815.

McGrath, C. (October 22, 1995). The Triumph of the Prime~Time Novel. New York Times Magazine. 52.

Miller, D. E. (2000). Fantasies of Reality: Surviving Reality Based Programming. Social Polity, 31. 6-15.

Mitchell, J. (1992). Femininity, narrative and psychoanalysis, in David Lodge, (Ed.). Modern criticism and theory (pp.426-430). Singapore: Longman.

Rahamim, O. (2008, November 23). The link between Gilead Shalit and Yossi Bublil. Retrieved 20.7.2009 from http://www.103.fm/blogs/chapter.aspx?pun26r4Vq =GKF&Zrzor4Vq=LJJK

Tait, R. (2009, July 3). Find God, win a trip to Mecca (or Jerusalem or Tibet). The Guardian. 1.

Thompson, E. (2007). Comedy Verite? The observational documentary meets the tlevisual sitcom. The Velvet Light Trap, 60, 63-72.

In: Reality Television-Merging the Global and the Local
Editor: Amir Hetsroni, pp. 123-136

ISBN 978-1-62100-068-6
© 2010 Nova Science Publishers, Inc.

Chapter 8

REAL LOVE HAS NO BOUNDARIES? DATING REALITY TV SHOWS BETWEEN GLOBAL FORMAT AND LOCAL-CULTURAL CONFLICTS[1]

Motti Neiger
Netanya Academic College, Israel

INTRODUCTION

Globalization has become one of the most debated topics in social science research since the 1960s (Appadurai, 1990; Ritzer, 1995). During the late 1980s, mainly due to the collapse of the Soviet Union, the concept began to gain momentum in the scientific community (Srebreny-Mohammadi et al., 1997), and for over a decade a significant growth can be observed both in the research literature that refers to the phenomenon and in the number of researchers whose work deals with it (Guillén, 2001). One of the key questions regarding Globalization – the process that integrates different societies/cultures into a single cross-national unit – is whether local/national cultures are able to challenge the process of Globalization/Americanization ("The McDonaldization of Society", Ritzer, 1993) in a combined dynamics of "Glocalization" and Hybridity (Robertson, 1994; Ritzer and Ryan, 2003; Kraidy, 2002) or they must surrender their local identity in order to become integrated and compete in the capitalist market.

Reality television shows are an appropriate arena to address this question, as many of them make use of a successful global format while adapting it to the local audience and culture ("Mc TV", Waisbroad, 2004). Thus, the aim of this paper is to shed light on some of the tensions and conflicts that arise in the power field between the local and the global by a close reading of an Israeli reality television show, "*Of All The Girls In The World*"; this is a dating competition show, based on the formats of the successful reality shows, "The Bachelor" and "Joe Millionaire". These shows present a young single man (The Bachelor), who has to choose the "love of his life" from among a group of young women, who are

---

1 The author wishes to thank Avishai Josman,for his contribution to an early version of the paper, and Michal Hamo, Amir Hetsroni and Yonah Kranz for the thoughts and comments.

gradually eliminated through a series of contests involving various challenges and set meetings ("dates"), until one of the girls wins the heart and the money of the bachelor.

This study can serve as an insightful case regarding TV production processes, because it focuses on the conjunction between local audiences and the cultural adaptation of a global TV format that uses the universal theme and contents of romantic love ("the prince on a white horse"). Thus, the conflicts between local and global establish the show narrative, on the one hand, and illuminate the formation of identities in a globalized era, on the other hand.

REALITY SHOWS AND THE GLOBAL CONTEXT

In the last decade, reality shows have become a staple of primetime programming and, as a result, one of the most contested topics in television studies (Holmes and Jermyn, 2004), drawing on the many faces of the phenomenon, from different theoretical aspects and varied perspectives of production, text and audiences (Hill, 2005; Deery, 2004; Murray and Ouellette, 2004; Kraidy, 2005, 2006; Andrejevic, 2004; Tincknell and Raghuram, 2002).

The reality TV genre, in its various formats, suits the multi-channel television market, which is characterized by a decline in the market share of public channels (Webster, 2005) and fierce competition between commercial stations. Although some reality shows involve high production costs that commensurate with their equally high production values (e.g., filming *Survivor* in remote islands), most of the shows in this genre are low-cost productions. The fact that the producers do not need to pay the high fees asked by professional star-actors helps, on the one hand, to reduce production costs and, on the other hand, enhances the appeal of reality shows to audiences that can more easily identify with "people like themselves". Quite often, this turns reality shows into very profitable ratings-hits and serve as the "the talk of the day" for the viewers.

Reality shows take advantage of the viewers' desire for voyeurism (Calvert, 2000) and their search for "authenticity" (Holmes, 2004A), while providing some basic gratifications such as amusement, escapism and the opportunity to develop one's identity through parasocial interactions (Nabi, Biely, Morgan and Stitt, 2003).

Reality shows have spread all over the world, mainly through the influence of American cultural industries (Hill, 2000; Kilborn, 1994; Morley and Robins, 1995). Although many reality shows originate outside the USA, the genre is considered to be a paradigmatic example of Globalization/Americanization, in general, and the impact of this process on the media, in particular.

The main cause for this effect is that American production companies purchased successful formats from around the globe and then sold them to other countries. In the symbolic level, these programs represent the soul of the capitalistic ethos of the "American dream": individual competition and the ability of ordinary people to lift themselves up from anonymity and become rich and famous.

Despite the fact that many reality shows all over the world are virtually literal adaptations of the original format and basic contents, there are a few interesting cases where the producers change the original script and add a unique local socio-cultural angle. In the Israeli context, a likely example could be the TV show "The Ambassador". This show was based on the format of "The Apprentice", seeking for, in the localized case, a young articulated person who is the most qualified to act as an envoy of Israel in New York and - like an appointed diplomat - would need to make Israel's case before media outlets and college students (Neiger, 2005). Another case in point is an Israeli version of *The Amazing Race,* entitled *The End of the Road*, where the couples competing were "composed" of a religious and a non religious contestant, thus underscoring one of Israel's most sensitive social conflicts (Hamo, 2009). The case on which this paper focuses, the show *Of All The Girls In The World*, is similar in the sense that it brings socio-cultural conflicts and tensions to the fore, confronting ethnic, national and religious identities with the global theme of the show, which is "romantic love".

OF ALL THE GIRLS IN THE WORLD: THE RULES OF MEDIATED LOVE

Romance has become so central in Western culture that some define it as an ideology that is composed of common perceptions held widely by people, such as "there is such a thing as love at first sight"; "everyone has one true love"; "Love conquers all obstacles"; and "true love is timeless" (Knowx and Sporakowski, 1968; Sprecher and Metts, 1989). Such phrases together with "universal" cultural symbols such as "a rose", "a glass of wine", and "a diamond ring" constitute the conceptual-ideological imagery reservoir of the "romantic utopia". Illouz (1997) pointed to the tight connection existing between the post-modern era, romantic love, consumerism, leisure, and nature. Through her study, "Consuming the romantic utopia: Love and the cultural contradictions of capitalism", it is claimed that American mass media have played a central role in disseminating consumer romance throughout the world. The case of the universal theme sharpens the cultural debate regarding the power of global media to mold a uniform cultural identity among audiences in different countries. In this regard, one can ponder whether romantic reality shows dealing with contents that are clearly representative of Western culture necessarily lead to an erosion of local identities and to the creation of homogeneous Western identities.

Of All The Girls In The World is a dating-based reality game show, aired on Israel TV Channel 3, the most popular cable channel in the country, between August and November 2005. The show combines the formats of the American reality television shows "The Bachelor" – where a man would choose his "true love" potential wife from among 20-25 women – and *Joe Millionaire*, which builds on the illusion that the bachelor is very wealthy, with a unique national/religious component: Ari Goldman, a wealthy Jewish-American bachelor from New York, who holds a MBA degree from NYU and owns businesses all over the US, will "...meet 17 Jewish girls from Israel and abroad in order to find his true love. During the season, Ari will date the girls to learn as much as possible about them, while the girls will try to win his heart... All his life Ari declared that he would only marry a Jewish woman because of his and his family's strong linkage with Israel and Judaism".[2]

Applying semiotic and interpretative methods, this paper analyzes eleven episodes of the show, while examining the construction of the characters, the setting in which the series was

---

[2] The show website (in Hebrew) is available at http://habanot.nana.co.il/habanotsite/article.php?id=53, last consulted in August 2009.

filmed, the challenges the participants had to tackle, and the way conflicts were presented. In addition, the paper uses the information published in the press, the program website and blogs concurrently written by the contestants, as material for analysis. All of these served to answer the central research questions:

How are local and national cultural identities constructed and represented when encountered with a global television format, Western values and "universal" cultural symbols (in this case, those of the "romantic utopia")? Alternatively, from a different angle, how are the different tensions between local and global used to advance and promote the TV show's narrative?

The analysis will try to emphasize the complexity of the *glocal* cultural encounter by dividing it into four parts. Each part centers on one of the conflicts represented in the series, as expressed in the first round of elimination of contestants and in the cases of each of the three finalists:

1) *Cultural conflict:* The elimination of the first contestant, Leah, who comes from a traditional/conservative Bukharian family.
2) *Religious conflict:* The elimination of Maria, a non-Jewish contestant in the final episodes.
3) *National conflict:* The elimination of the last contestant (first runner-up), Netah, who wants to live in Israel rather than in the United States.
4) *The (apparent) age conflict:* The election of Mary (who sometimes uses her Hebrew name, Miriam), through which the romantic utopia is realized.

THE CULTURAL CONFLICT – THE ELIMINATION OF LEAH AND SETTING THE BOUNDARIES OF GLOBAL ROMANCE

Among the 17 young women who took part in the show, one – Leah Itzhakov, a 24-year old English teacher – was differentiated by her attitude towards the rules of Western romantic love as embraced by the producers and the other contestants. This is how she was presented in the show's website:

> "Leah lives in Tel Aviv with her older brother, Avi [...] Leah's family comes from Bukhara and she was raised according to the standards and moral values of their community. For example, Leah does not believe in pre-marital sex and she goes out only with young men introduced by her family and female friends [...]; Leah expects her husband to be a real gentleman that will respect her and open doors for her. Although she dresses as a secular girl, she maintains tradition and keeps the Sabbath. On the other hand, she loves Oriental music and practices belly dancing, Greek dance and reggae, and of course, Bukharian dance too".

Such a description of a woman poses a threat to Western romance, because it consecrates values that are completely different and rocks the very foundations of the television show: Seventeen girls competing for the heart of a man, the use of sex and flirting as a means of attraction, the capitalist world where money is the central value, the exaltation of entertainment outside home (she is an expert cook of traditional dishes), the informal atmosphere (she requires the man to dress smartly), and frequent dates as opposed to selective matchmaking. In spite of the fact that Western and traditional/conservative utopia apparently share the same ideal – the prince on a white horse – they differ on the tactics to fulfill it. The existence of the traditional/conservative option challenges the possibility of celebrating the Western romantic utopia. All too frequently, it reminds us that the very existence of such a TV-game broadcast nationwide and the acceptance of its rules might be a disgrace for the young women participating, that the girls may be trading on their bodies and the cost can be that the knight will finally cast them away with no "honor".

When presenting the contestants, the show host informs, "surprisingly, one of the girls arrived to the meeting chaperoned by her elder brother", and Leah continues: "My elder brother, Avi, used to change my nappies when I was a baby and took me to the day-care. The honor of the family is very important. I may not answer him back or be rude with him. What my brother says is sacred and therefore it is important for me that he gets acquainted with the young man".

Following their first meeting with Ari, the women go to his room by pairs, while the other girls stand close by, drinking champagne. The host says (V/O): "Unlike the other girls, Leah chose not to kiss Ari, but to politely shake hands with him".

Immediately after the meeting, Leah says to the camera, "I will not do anything I do not believe in. I don't know him and feel nothing towards him". It is understood that Ari does not accept her attitude and says, "she decided on this beforehand, she did not say 'let's see how it goes' but made up her mind in advance. This did not allow the meeting to develop at a natural pace". On the meeting itself, the contestants were asked to bring along photographs of themselves. Leah chose to bring a photo of her wearing a traditional Bukharian wedding dress. She said to Ari: "We dress like queens for the wedding. In our community, we say that if you want to be a queen you must make your husband a king. I hope to be able to make a king out of you".

On this first evening, six of the contestants are candidates for elimination and two of them would eventually be eliminated. Leah is the first to be chosen as a candidate. Like all the other candidates, she is given a chance to persuade Ari to change his mind. The production constructs a setting of romantic utopia that evokes French Impressionist paintings but contains local elements: The women are sitting on an esplanade, surrounded by flowers. The contestants are all wearing white clothes (dresses or slacks and shirts), and each of them is holding a blue parasol, that emphasizes the combination of Israel's national colors. The candidates for elimination try to entice Ari in various ways: one teaches him some hip-hop steps, another blows some balloons and fashions them into dog shapes, and yet another gives him a present. Unlike them, Leah tells Ari: "I never went out with someone just like that. Dates were always oriented to marriage and children". And she adds: "I've written a poem for you, but I won't give it to you. I'm keeping it". As a metaphor for her virginity, she saves herself. If he does not send her away, he will also win the poem. Moreover, in an interview immediately after the meeting, she stresses, "This is my honor". Ari dealt with the similitude between them when the evening came and she was the first of the candidates to be eliminated: "The most important things, like background and family, are things I can relate to. We come from worlds that are similar, yet also very different. I am not sure the gap between our worlds can be bridged".

Family and traditional values are important to Ari too, but in addition to the set of Western values and not as an alternative to them. Ari is seeking a Jewish bride. However, "the most Jewish" of all contestants is precisely the one he eliminates first, while the most "non-Jewish" contestant will be allowed to stay almost until the last stage, when only three finalists are left. All of this is possible because Jewishness is being considered in superficial-cosmetic terms. Maria, who is not Jewish by Jewish codex, does not undermine the foundations of Western romantic utopia, whereas Leah's religiousness does jeopardize them.

Leah's removal soon allows the show to proceed under the capitalist set of values: commoditized people, partying, playing, flirting – manifestly or furtively– glorifying consumer products and the romantic utopia based on capitalism become the focus, while conservative values are cast aside.

THE ETHNIC CONFLICT – THE ELIMINATION OF MARIA, THE NON-JEWISH CONTESTANT

The contestant who provoked the most perceptible tension was Maria Ivanova, a young woman aged 29 who immigrated to Israel from Bulgaria when she was 13. Until then, she had lived with her family in a small town close to a ski resort. The story of Maria and her family was told both in the show and in the show's website: "Both her parents are doctors and she too always dreamed to become a doctor, but in the end she renounced her dream to study home design, which she will finish next year. At the same time, she has a very developed spiritual side and is keenly interested in the subject. Now she lives in Tel Aviv and works as a bartender at a luxury hotel". We are further told that "when she came to Israel she learned Hebrew on her own, and only later she took formal Hebrew language lessons. English, on the other hand, she already spoke."

Yet the most important detail from the ethnic conflict aspect added to the show is that Maria's father is Jewish but her mother is not. This makes Maria a gentile (i.e., non-Jewish person) according to Jewish *Halacha* (Jewish orthodox law), and so she becomes the focus of conflict between the global-liberal values that reject restrictions on romantic love ("love has no rules"), on the one hand, and Jewish religion, generally, and Ari Goldman's family tradition, particularly, that set a strict rule, on the other hand. The fact that Maria's family used to celebrate the festivals of both religions (the "Hanukkah bush") will arise several times during the series, challenging the legitimacy of her participation in the show and despite Maria's repeatedly avowed allegiance to Judaism. In *Behind the Fence* (1909/1999), one of the most famous stories of the great Hebrew writer, Haim Nahman Bialik[3], the story tells about a forbidden love relationship between Noah, a Jew, and Marinka, a beautiful gentile woman, which mostly takes place through an opening in the fence separating their houses. Eventually, Noah returns to the Jewish fold and marries a Jewish woman, while Marinka and their baby boy watch them behind the same fence.

This story is a paradigm of the moral injunction against mixed marriages in Jewish culture – a taboo that combines both the desire for non-Jewish women and the social prohibition of institutionalizing relationships with them. "We date gentile women but we marry Jewish girls", says Ari. Powerfully drawn by the non-Jewish woman, Ari is physically attracted to her and sees her as a liberating option as well. However, he cannot reconcile the tension between Id and Super-Ego (both familial and tribal) – represented by his elder sister,

---

[3] As a poet, Haim Nahman Bialik (1873-1934) came to be recognized as Israel's "National Poet". He is considered one of the founders of modern Hebrew poetry as well as a distinguished literary editor and publisher.

Charlene, whom he calls to ask for her advice on whether he should let Maria stay on to the next stage. Already in the first episode of the series, Maria was one of the leading candidates for elimination and, actually, she was slated as such, but Ari revises the rules and leaves her in the show, to test the seriousness of her intentions. At the end of the first episode, Ari says, "Maria is just not a Jewish name", thus sharpening the conflict. The doubt might still have arisen if the contestant's name had been Danielle, Dorothy, or Sarah, but Maria, besides her "problematic" cultural and religious roots, also has a clearly marking name, a "label" so powerfully identified with Christianity.

Two decisive junctions shape the ethnic conflict: the first occurs on Maria's way to the final episode and the other on this episode proper. On the way to the final episode, a decisive point occurs during the main romantic meeting between Maria and Ari Goldman. Maria wrote about this meeting in her diary, which was published as a blog: "[...] all the questions began to engulf me: What if the food I cooked is not to his taste? And what if I feel that I do not want to be with him? If he feels pressured by what I prepared for him? And thousands more such questions – and the most difficult thing, disclosing to him and the whole nation something so personal as the issue of my conversion [...] I took him to the beach [...] and there, a bridal canopy *(chuppah)* was waiting for us [...] I waited for him beside the canopy, which in fact is where I am now in my life. Beneath the canopy was rabbi Amrany, my rabbi [the rabbi that accompanies her along the conversion process], who was waiting for us, and this symbolizes how I want my future to look, to be married according to Jewish Law. I was terribly afraid that Ari would approach us, see the *chuppah*, and say, "What's the matter with you"?

The meeting with the rabbi, who confirms the seriousness of Maria's intent to convert to Judaism, affords Ari the legitimacy of bringing her to be one of the three finalists. The last few episodes of the series were filmed in New York, where the contestants meet the Goldmans, and particularly Charlene, who is marked as "the family witch". Charlene goes out with each of the finalists. Right at the beginning of the "date", Maria tries to mollify her by giving her a present, a Star of David she brought from Israel. Later, Charlene puts Maria to a test: she takes her along to a designer shop and asks for her advice about the most appropriate outfit to wear to her son's *Bar Mitzvah* ceremony, which is to be held at a synagogue. When Maria finds a revealing dress to be the most appropriate, Charlene announces to her brother that it does not matter whether Maria would be converted or not: she does not really understand what it means to be Jewish. As Charlene sees it, more than a certificate of *kashrut*, Judaism is a set of defining cultural, traditional, and social values – and Maria definitely does not carry those values. Based on this assessment, Ari disqualifies Maria and is left with the two last contestants, Netah and Mary (Miriam).

THE NATIONAL-IDEOLOGICAL CONFLICT – THE ELIMINATION OF NETAH, THE CONTESTANT WHO WANTS TO LIVE IN ISRAEL

One of the conflicts repeatedly expressed during the series is that which reflects the meeting of the cultural global-American world with the cultural national-Israeli one. This conflict, attributed to Netah, stems from her strong attachment to her homeland, Israel, and hence the difficulty in leaving it to live abroad for the sake of her heart's chosen.

Netah Rilov, 24 was able to remain in the contest until the final episode, where lost to her Canadian opponent, Mary (Miriam) Inbar. Netah's persona represents the culture and lifestyle generally identified with the middle-upper middle class in Israel. Netah is not the only contestant whose character denotes the patriotic sentiment and the Israeli way of life in the 21st century. Together with her in the show, the viewers can meet figures like Natalie Chayut, 23, an officer in the Israeli Defense Forces and Keren Cohen, 24, who used to be a tour guide at a field school in the Negev desert, and knows every waterhole and every rock in the land.

In their words and behavior, both Natalie and Keren represent the traditional national ethos, grounded on the Zionist narrative that speaks of a dream that is common to all Diaspora Jews.

Netah – more than Natalie and Keren – represents the transition of Israeli culture from collectivist to neo-liberal society, in which social status is very significant in determining the social hierarchy. Despite the fact that it is her social status that gives Netah her strength, the show portrays her as the representative of the Israeli society in the series, because of the place she was born in: a remote farm in the Carmel hills, near the northern city of Haifa, and hence her primordial bond to the homeland – one that will go along with her throughout the series. The producers of the show emphasize the fact that Netah grew up without any of the technological amenities available in modern society. Therefore, Netah epitomizes the figure of the "pioneer"; her lifestyle and that of her family match the narrative built around the image of the Ashkenazi pioneers, who drained swamps, made the wilderness bloom and lead a frugal way of life – all for the sake of realizing the Zionist dream. In this context, it is important to stress the contradictory nature of the show that, on the one hand, emphasizes the unique background of the "Israeli girl" (army service, exploring the land on foot, attached to history) and underscores the Zionist connection between the people, the nation-state and the land, but, on the other hand, presents her ultimate goal as marrying an American and emigrating from Israel to the new world.[4]

However, whereas the Zionist myth establishes a link between the pioneers and the homeland, romantic utopia is what links Netah, the "pioneer", with Western consumerist culture. In Episode 7, Netah is given the opportunity to plan a unique romantic meeting for her and Ari. The production team asked her to write Ari a short letter describing what she has prepared for their coming meeting. Netah phrased her letter thus: "This is what happens when Israel and America meet. You will get a whiff of the smell of the soil. You will sit under a tree 2000 years old. You will eat healthy food and you better get ready for good Israeli rock'n'roll". From this letter, one can understand that Netah represents the rise of Israel's

---

[4] One might draw connections to the second, less successful, season of the American reality show "Joe Millionaire" ("The Next Joe Millionaire", 2003), filmed with European girls in Italy that, in the same sense, declared that for the chosen man who would fulfilling their romantic dream, they would happily trade their European passport for an American one.

upwardly mobile middle class, which enthusiastically embraces the culture of New Age, of commercial festivals and the various courses designed to attain a "spiritual balance".

This is not the "Israeliana" represented by Natalie and Keren, but one reflecting the upper middle class way of life. This is a way of life combining spirituality and consumerism. Netah epitomizes the Israeli neo-liberal spirit, which amalgamates well in the meeting between the global and the local in the 21st century. She plays according to the rules of the romantic consumerist game: she brings Ari to the farm where she grew up and begins the rendezvous in an olive grove in the midst of nature. As held by Illouz (1997), romance is embodied in the most perceptible form in the world of leisure and nature, away from the world of work, where the couple can be alone and communicate on a level that the world of work does not allow for. The meeting begins with Ari and Netah lying on an ornate canopy bed set in the middle of the olive grove. Instead of gathering food from the earth, as Keren did in one of the episodes (with the purpose of teaching Ari and the other girls about the plants of the Land of Israel), Netah chooses to bring products from her parents' shop to the meeting. She teaches Ari how to eat *houmous* with olive oil produced by her father. While eating the *houmous*, Netah and Ari drink red wine, and discuss whether she would be prepared to leave Israel to live with Ari in the United States

One possible conclusion is that Netah succeeded in the show because she combines the global and the local in a way that turns her into the ideal contestant, so much so that she is not eliminated until the final episode. For the viewers, her elimination only strengthens the fact that Netah is the embodiment of the national Israeli spirit. Although Netah does not really represent the traditional-modern national makeup, she succeeds in amalgamating all of the stereotypes of the "new Israeli", one who can enjoy a global culture blending spiritual experiences (e.g., traveling in the East, a bond with nature and the earth) with consumerist practices normally reserved to the upper middle class (e.g., taking yoga lessons, buying organic and naturist foods, outings in the midst of nature), while preserving the bond with their homeland.

THE (APPARENT) AGE CONFLICT – MARY'S TRIUMPH

In the final episode of the show, Ari must choose between Netah Rilov and Mary (Miriam) Inbar. After spending time with both of them in New York, he returns to Israel and at the end chooses Mary (Miriam).

Up to this point, three unsolvable conflicts were presented: the cultural conflict, the ethnic conflict, and the national-ideological conflict. The sole conflict that the romantic rhetoric succeeds in overcoming is the one created between Ari and Mary (Miriam). This conflict turns around the fact that Ari is 34-years old, while Mary is only 21.

Mary is the youngest contestant in the show. She came to Israel from Toronto, Canada, under a student exchange program, and is studying for a B.A. degree in Political Science at a local university. Mary (Miriam) attracted Ari's attention already in the first episode of the series, and he gave her a bracelet that ensures she would not be eliminated from the show at the first episode.

Like the other contestants who arrived to the decisive stages of the series, Miriam also plans a date for her and Ari; she decides to take him to a large amusement park. Like the other contestants, Miriam is asked to write Ari a short letter in which she hints to what he can expect to happen. She chose to insert in her letter some quotations from the American musical film *Grease* (1978), and signs with a red heart.

The choice of *Grease* evokes memories common to Ari and Miriam. Thus, for example, Ari is quoted saying: "These are my childhood memories". In Miriam's opinion, "This is something so American, it's Sandy and Danny, it's the most exciting love story". The romantic utopia as the embodiment of purity is in the center of their date from the very beginning. By means of the popular movie *Grease*, i.e., with the mediation of an American cultural text, Miriam establishes the cultural common denominator between her and Ari. If until now Miriam's young age has been a problem, mainly for Ari, their date succeeds in blurring the issue. This is something different from what can be observed on Ari's dates with Maria and Netah, where the focus was on the *differences* between both sides: the "date" between Ari and Maria began when she waited for him beside a bridal canopy, accompanied by the rabbi who assisted her with the process of conversion to Judaism. Therefore, the main issue clouding the relationship between them (Maria's religion) is in the center of the date. Similarly, the date with Netah begins with a discussion over whether she is prepared to leave Israel to live with Ari in the United States. Most of the meeting between Ari and Miriam deals with their preparations for the video clip they are going to make together. When the video clip is completed, Ari and Miriam can indulge in an intimate picnic on the lawns of the amusement park. The symbolic setting in which the picnic takes place is adorned with global romantic metaphors: a red-and-white checkered mat, a bottle of white wine, a basket of rustic country bread, goat cheese, and a vase holding a single red rose.

In this part of the meeting, Miriam gives Ari a letter her parents wrote. In this letter, they say they understand the place Jewish tradition has in Ari's family, and that the same tradition is an inseparable part of their own family. Further, they write that Miriam is a proper Jewish girl and Ari has nothing to worry in this respect. Here again, one can see the place Jewish tradition has in the global romantic meeting. It should be pointed out that romance, as portrayed in American media, seldom links between the lovers and their family. Illouz (1997) holds that in romantic commercials, the loving couple usually appears unaccompanied. Even if they are married, mostly they are shown without their children or family. In this show, however, family plays a pivotal role in constructing the romantic meeting, especially in the case of contestants for whom Jewish tradition is more important than the components of national Israeli identity.

Miriam also takes the opportunity to identify herself with Ari's sister, Charlene, who adamantly opposes Ari choosing Maria, the non-Jewish contestant. Miriam says she understands that upholding Jewish identity in the Diaspora is not easy and hence the concern of Charlene and her parents to prevent Ari from becoming assimilated.

At the end of their date, Ari asks Miriam whether she can see herself involved in a serious, long-term relationship with him. Here again, the issue of Miriam's age arises. Because she is so young, Ari doubts her ability to carry on a mature relationship. Miriam answers that she is ready for a long-term relationship, but refuses to commit to married life after such a short acquaintance.

Thus, the issue of Mary's age only arises at the end of the meeting. Before that, Miriam makes a point of discussing with Ari her advantages over the other contestants: contrary to Maria, she is a proper Jewish girl, and produces her parents' letter as proof. Contrary to Netah, she does not have any roots in Israel that may keep her from raising her children in another country. Nevertheless, the producers of the show try to suppress this dimension, depicting Mary's love for the Land of Israel. In one of the segments of the series they inform the viewers: "Miriam fell in love with Israel from the first day she came to visit, and since then she has made a point of spending two or three months every year in this country".

Despite her love for Israel, Miriam chooses to approach Ari by means of a popular text, known to both of them for the simple reason that they had been born and raised up in North America. In planning their date, Miriam does not try to challenge Ari's cultural toolkit (Swidler, 1986), but rather to take him back to the toolkit he knows from childhood. Actually, the romantic meeting between them was based on American setting and symbols, with no Israeli or Jewish content.

It is not surprising, then, that at the end of the show Ari should choose Miriam. Out of the triad of identities constructing the personality of the contestants in the show, Ari and Miriam have the most points in common: they both are partners in the consumerist romantic meeting and take an active part in it. Miriam's choice to make a video clip picked up from the movie *Grease* signals her identification with the Hollywoodian youth culture whose movies portray youngsters falling in love during the summer holidays, while vacationing with their parents away from home. Similarly, the meeting between Miriam and Ari occurs while they are staying in Israel. Moreover, they both identify themselves with Jewish tradition and the central role family and married life play in it and, contrary to most of the contestants, who were born or grew up in Israel, they both were born in a similar cultural environment.

Therefore, the apparent age conflict is easily resolved. Moreover, the issue of age appears in many classical legends (e.g., Cinderella), in which the rich prince chooses for his bride a poor, beautiful and innocent young girl.

DISCUSSION AND CONCLUSION

Like *The Bachelor* and *Joe Millionaire*, the Israeli reality television show *Of All The Girls In The World* ends with the triumph of romantic utopia. Therefore, one might posit the primacy of the cultural imperialistic thesis in both contents and format (which was adopted almost to the letter). The contestants in the show faithfully play according to the global-American courting ritual. The only contestant, who refused to abide by the rules of this ritual (Leah), was promptly removed from the show. However, despite these characteristics, the way viewers are led to the end of the series precisely illustrates the significance of the notion of hybridization (or *glocalization*).

The romantic ritual, even when encoded with neo-liberal Western values, cannot conceal the social relations of power represented by the contestants (most of whom stand for the local), and, therefore, Ari's choice at the end of the series reflects the hybridization between dominant local and global values and identity elements.

The analysis demonstrates the hybrid nature – based on a cluster of global and local identities – of a localized reality television show. The contestants who reached the final stages of the show represented, each from a different perspective, the meeting between the romantic global-American culture and its local-Israeli counterpart – a meeting that consists of an overlapping triad of identities: global-consumerist identity, Jewish identity, and national-Israeli identity.

These conflicts may play different roles for each factor involved in the process of creating the media text and its meanings: the producers, participants and audience:

1) Producers – Identity conflicts as narrative-promoting elements: Reality shows – especially those belonging to the sub-genre of competition – are based on narratives (Holmes, 2004b), which need conflicts to drive the story onwards and to keep the suspense alive from one episode to another. The producers of the show use this diversity and the conflicts arising from it to augment the tensions among participants as one of the main engines for building drama; indeed, diversity among the contestants ("pluralism" in representation guided by the producers) is a common production practice to achieve such a goal (Pullen, 2004, Hamo 2009).

2) Contestants – Identity conflicts as a tool for dialoguing with their "label": The representation in television in general and in reality shows in particular is superficial and stereotypical (Cavender, 2004; Pullen, 2004, Orbe, 2008). Although as the show progresses the participants reveal details about their life that go beyond their initial introduction ("a student, 24 years old, Tel Aviv"), these details usually do not constitute well rounded, complex characters, but rather reinforce the "labels" – in our case, "the pioneer" (or "the nature girl") for Netah, and "the non-Jewish contestant" (or "not quite Jewish …") for Maria. The contestants, in the story told by the production, relate to these "labels" and bolster the stereotypes rather than challenge them: Maria has chosen to bring Ari and her Rabbi together in order to "prove" her "Jewishness", and Netah met Ari under an old olive tree as a symbol of her bond with the land of Israel.

3) Audience – The identity conflict as a mirror for self-reflection: Reality shows serve various ends, provide diverse types of gratification, and are decoded and interpreted differently in different parts of the world (Hill, 2005; Reiss and Wiltz, 2004). Reiss and Wiltz (2004) have shown that "the more status-oriented people are, the more likely they are to view reality television and report pleasure and enjoyment. People who are motivated by status have an above-average need to feel self-important. Reality television may gratify this psychological need in two ways: One possibility is that viewers feel they are more important (have higher status) than the ordinary people portrayed on reality television shows... Further, the message of reality television— that millions of people are interested in watching real life experiences of ordinary people—implies that ordinary people are important. Ordinary people can watch the shows, see people like themselves, and fantasize that they could gain celebrity status by being on television. " (373-374). The identity conflicts assist in fulfilling the status-oriented pleasures as the audience can feel superior by mocking some of contestants conflicts (e.g., the conservatism of Leah) on the one hand, and better construct its parasocial relationship with characters and strength its identification with "people like themselves", on the other hand (e.g., the patriotism/nationalism of Netah).

As for the "heroes" of the finale, despite the successful romantic meetings Ari had had with Maria and Netah, when complex issues regarding common married life came to fore, it became apparent that the national and religious identity components occupy a central place in the identities of these two contestants. Therefore, the romantic utopia has failed in abridging this gap. The fourth conflict that of age differences is the only one that can be resolved, because the romantic utopia is able to contain it and even encourages it: The experienced rich prince conquers the heart of the young damsel and sails away with her, into the sunset, to his own country.

In Greek, the word "utopia" means "nowhere" (*ou* = not, no + *topos* = place) and is known as the title of a book by Thomas More (1516), which pictures an imaginary ideal community existing somewhere named "No Place". To exist, the romantic illusion must detach those who share in it from the sensation of a concrete place and take them from "the reality" of here and now to a realm of dreams. Thus, there is an oxymoronic nature to the adaptation that localizes the utopia, putting in place the illusion of "no place". This complexity and the conflicts it stirs up certainly are one of its most important narratological engines and make reality shows attractive for local audiences.

REFERENCES

Andrejevic, M. (2004). *Reality TV: The work of being watched.* Maryland: Rowan and Littlefield.

Appadurai, A. (1990). Disjuncture and difference in the global cultural economy. *Public Culture*, 2(2), 1-24.

Bialik, H.N. (1909/1999). Random Harvest and Other Novellas. Boulder, Colorado: Westview Press

Calvert, C. (2000). *Voyeur nation*: Media, privacy, and peering in modern culture. Boulder, Colorado: Westview Press.

Cavender, G. (2004). In search of community on reality TV: America's Most Wanted and Survivor. In S. Holmes and D. Jermyn (Eds.), *Understanding reality television* (pp.154-172). London: Routledge,.

Deery, J. (2004). Reality TV as advertainment. *Popular Communication*, 2(1), 1-20.

Guillen, F.M. (2001). Is globalization civilizing, destructive or feeble? A critique of five key debates in the social-science literature. *Annual Review of Sociology*, 27, 235-260.

Hamo M. (2009). Textual Mechanism for Complex Representation of Israeli identity in Reality Show. *Media Frames: Israeli Journal of Communication*. 3: 27-53. (in Hebrew).

Hill, A. (2005). *Reality TV: Audience and popular factual television.* London and New York: Routledge.

Hill, A. (2000). Fearful and safe: Audiences response to British reality programming. *Television and New Media*, 1(2), 193-213.

Holmes, S. and Jermyn, D. (2004). Introduction: Understanding reality television. In S. Holmes and D. Jermyn (Eds.), *Understanding reality television* (pp.1-32). London: Routledge.

Holmes, S. (2004A). But this time you choose!: Approaching the interactive audience of reality TV. International Journal of Cultural Studies 7(2), 213-231.

Holmes, S. (2004B). "All you've got to worry about is the task, having a cup of tea, and doing a bit of sunbathing": Approaching celebrity in Big Brother. In S. Holmes and D. Jermyn (Eds.), *Understanding reality television* (pp. 111-135). London: Routledge,.

Illouz, E. (1997). *Consuming the romantic utopia: Love and the cultural contradictions of capitalism.* Berkeley, CA: University of California Press.

Kilborn, R. (1994). How real can you get? Recent developments in 'Reality' television. *European Journal of Communication*, *9*, 421-439.

Knox, D. H., and Sporakowski, J. J. (1968). Attitudes of college students toward love. *Journal of Marriage and the Family*, 30, 638-642.

Kraidy, M. (2006) Reality Television and Politics in the Arab World (Preliminary Observations), *Transnational Broadcasting Studies*, *2*(1), 7-28.

Kraidy, M. (2005). *Hybridity or the cultural logic of globalization*. Philadelphia: Temple University Press.

Kraidy, M. (2002). Hybridity in cultural globalization. *Communication Theory*, *12*(3), 316-339.

Murray, S., and Ouellette, L. (2004). Introduction. In S. Murray and L. Ouellette (Eds.), *Reality TV: Remaking television culture* (pp. 1-19). New York: New York University Press,.

Nabi, R. L., Biely, E. N., Morgan, S. J., and Stitt, C. R. (2003). Reality-based television programming and the psychology of its appeal. *Media Psychology, 5*, 303-330.

Neiger, M. (2005). The Battlefield of Images: The TV Reality Show "the Ambassador" (HaShagrir) and Israeli myths. *Panim*, 31, 75-80.

Orbe, M. P. (2008). Representations of race in reality TV: Watch and discuss. *Critical Studies in Media Communication, 25*, 345-352.

Pullen, C. (2004). The household, the basement and The Real World: Gay identity in the constructed reality environment. In S. Holmes and D. Jermyn (Eds.), *Understanding reality television* (pp. 211-232). London: Routledge,.

Reiss, S. and Wiltz, J. (2004). Why people watch reality TV?. *Media Psychology, 6*, 363-378.

Ritzer, G. (1993*). The Mcdonaldization of society: An investigation into the. changing character of contemporary social life*. London: Prime Forge Press.

Ritzer, G., and Ryan, M. (2003). Towards a richer understanding of global commodification: Glocalization and grobalization. *The Hedgehog Review*, 66(11), 66-76.

Robertson, R. (1994). Globalization or glocalization. *Journal of International Communication*, 1(1), 33-52.

Morley, D., and Robins, K. (1995). *Spaces of identity: Global media, electronic landscapes and cultural boundaries*. London: Routledge.

Schiller, I.H. (1976). *Communication and cultural domination*. New York: International Arts and Sciences Press.

Sprecher, S., and Metts, S. (1989). Development of the 'Romantic Beliefs Scale' and examination of the effects of gender and gender-role orientation. *Journal of Social and Personal Relationships*, 6, 387-411.

Srebreny-Mohammadi, A., Winseck, D., McKenna, J., and Boyd-Barrett, O. (1997). Editors' introduction: Media in global context. In A. Srebreny-Mohammadi, D. Winseck, J. McKenna and O. Boyd-Barrett (Eds.), *Media in Global Context: A Reader* (pp. ix-xxviii). London: Hodder Arnold,

Tincknell, E., and Raghuram, P. (2002). Big brother: Reconfiguring the 'active' audience of cultural studies? *European Journal of Cultural Studies*, 5(2), 199-215.

Waisbord, S. (2004). McTV: Understanding the global popularity of television formats. *Television and New Media, 5*, 359-383.

Webster, G.J. (2005). Beneath the veneer of fragmentation: Television audience polarization in a multichannel world. *Journal of Communication*, 55(2), 366-382.

In: Reality Television-Merging the Global and the Local
Editor: Amir Hetsroni, pp. 137-149

ISBN 978-1-62100-068-6
© 2010 Nova Science Publishers, Inc.

Chapter 9

THE SEAL OF CULTURE IN FORMAT ADAPTATIONS: *SINGING FOR A DREAM* ON TURKISH TELEVISION

Sevilay Celenk
Ankara University, Turkey

My father never wore a suit. He watches you on television and says "Ibrahim Tatlises is wearing such a nice suit." May I ask you, Mr. Ibrahim, to give the suit that you are wearing to my father? (Peker and Peker, 2007, episode 8).

These words are from a viewer letter sent to the popular Turkish television show *Hayalin Icin Soyle (Singing for a Dream)*. No one, neither the participants, nor the viewers watching the program on the television, were expecting these words when the presenter announced that she would be reading a letter from a viewer. Nevertheless, this unusual request by a 10-year-old viewer was not so astonishing as it was only one of the many unexpected demands by the viewers, who wanted their own share of the glamorous world of Turkish television. In other words, this request was being expressed through such a locality and culturality, that it rendered ordinary its own strangeness. Ibrahim Tatlises, who is one of the most popular singers of Turkey, received many requests during this reality show, where he served as a jury member. In the letter mentioned above, he was being asked to take off what he is wearing and to give it to someone else. Still, this was not the only interesting expression of culture and identity, which rendered *Singing for a Dream* reality television talent competition overly local. The show, which had been adapted from a foreign format, had even established locality as an "excessiveness." At one point, Muazzez Abaci, who has been one of the most famous singers of Turkey for the last 50 years -though she was not as popular at all times- cheered a disabled contestant by saying "kurban olurum sana"—a traditional Turkish expression, which means "I would sacrifice myself for you." Neither was this one of the most interesting expressions of cultural difference. The tension between the content of *Singing for a Dream* and its foreign format forced it to become overly local and national—so as to eliminate any concerns of "foreignness." It even forced the program to drift into adopting a populist nationalist discourse.

Here, it should be said that the domination of foreign programs was actually never a serious threat to Turkish televisions. Even during the mid 1980s when nearly the half of programs on TRT (Turkish Radio and Television Corporation), the first and only public broadcasting corporation in Turkey, became of foreign origin, the situation was not considered as a threat to the national programing. Indeed, it soon became clear that the high percentage of foreign programs on national TV did not cause any serious shift in audience's taste towards these programs. They are then and always strongly drawn to indigenous programs. The stable interest of Turkish audience towards domestic contents stimulated a boom especially in production of domestic fictions. At the end of 1990s and the beginning of 2000s, the period when private television companies got thoroughly established, domestic television drama became the most favorite offerings of private television companies, as well as of state television channels of TRT. Even the hunger for a greater amount of television programs during the 1990s' spectacle multiplication of private national channels up to 16, besides four TRT channels, did not cause to any remarkable growth in program import (Celenk, 2002a).

Format development as one of the main creative aspect of television programming, has never been a remarkable ability of Turkish televisions though, whereas beginning with very early days of commercial television, adaptations from American or European program formats have always been quite popular. However the real triumph of format adaptations on Turkish television screens took place during the economic recession that hit all the major television companies alongside every other sector at the early 2000s. High production costs and fast-rising actors' fees brought television companies to consider program alternatives for prime time. Among those alternatives were reality talent shows, which seemed to be the most suitable replacement for domestic drama due to their remarkably lower production costs. In Spring 2001, a new hybrid genre of reality television, called *Someone Is Watching Us* (*Biri Bizi Gozetliyor* or BBG) appeared in Turkey. The program, aired by Show TV was the adaptation of *Taxi Orange*, a slightly different—Austrian—version of *Big Brother* which is a well-known reality-based competition program in Europe. Since that point the continued expansion of domestic drama in prime time started to be challenged by various imported samples of reality television. Copying BBG's achievement in capturing the interests of the audience, Turkish television institutions have attempted to broadcast several other adaptations of foreign competition programs (Celenk, 2002b). Among them, some the most popular ones were *Academy Turkey* (similar to *Star Academy*), *Turkish Pop Star* (similar to *American Idol*), *Dancing with the Stars*, and *Singing for a Dream*.

Singing for a Dream reality show, which will be analyzed in this article, is the Turkish version of *Cantando por un Sueno,* one of the shows of the Mexican television giant Televisa. The show is a hybrid of reality show and a talent competition. It includes a three-member jury, composed of some of the most popular persons in the music field, and one coach for each competitor. The coaches of the competitors are not as popular as the members of the jury, yet they are also famous singers. The contestants are young, ordinary people, who can sing, and who have various desires or dreams related to the tragedies that they have been through.

The show positions the contestants in the centre of narration. This is another reason why it is important in terms of this paper. This article basically focuses on nationalism, which is a result of the tendency to insistently emphasize cultural identity, national characteristics or superiorities. Nationalism is a tendency carried by many television programs, including those with adapted formats that bring ordinary people into the centre of narration. This drift towards nationalism in format adaptations becomes even more evident due to the tension felt by the contestants, who are seeking to express themselves in a format that is somewhat "foreign" to them. The contestants, who, all of a sudden, begin to share their private lives with the whole country, try to gain the sympathy of people under this tension. And in the meantime, they often cling onto national/cultural values or trends shared by the whole country. The jury members and the coaches of the contestants are constantly making populist comments on life and on the competitors, as they strive to deal with "being a jury member and a coach"—a position, which they are not accustomed to. When *Singing for a Dream* is the case in question, the discourse, through which, these comments could most easily be articulated, is the nationalist discourse that is becoming more and more popular. The season of the program that I looked into coincided with Turkey's cross-border operation in Northern Iraq. This coincidence facilitated a high-pitched nationalist discourse. It also made it more interesting to study the program closely.

Television format adaptations in general are foreign in terms of format, and the concern here is that this foreignness would lead to an estrangement and assimilation in content. Conversely to the concerns, this paper discusses that the dominance of local and national content in spheres of cultural production does not necessarily lead to the preservation of authentic culture and that it is necessary to pay attention to this issue for the purpose of understanding television. Specifically, such a discussion is necessary for the purpose of examining to what extent television provides different cultural identities with the opportunity to coexist.

TELEVISION, IDENTITY AND NATIONALISM

The issue of cultural identity has been dealt with by focusing on different factors in television studies. Researchers have studied certain television programs in their own countries (Aslama and Pantti, 2007; Tunc, 2002), or television channels in a country that broadcast programs targeting a certain ethnic community (Dhoest, 2007; Milikowski, 2000). Some of the studies took a comparative approach and studied the different versions of a format that in different countries (Skovmand, 1992). This research highlights the assimilation of cultural space, which is termed "Americanization" (Sparks and Tulloch, 2000).

There are studies that focus on the consequences of globalization on television broadcasts within the framework of a one-way program flow from countries with developed television technologies and production, to less developed nations. Yet, there are also studies that focus on a completely opposite flow. The opportunities of "identity," "sense of belonging," and "holding on to" presented to the diaspora population are analysed in these studies, which look into national televisions' broadcasts that target the diaspora population (Aksoy and Robins, 2000).

Another emphasis of the research is that, the phenomenon of globalisation, both by forming partnerships with national companies and actors, and by format adaptations. This creates a new phenomenon—glocal television (Richardson and Meinhof, 1999, p. 91). Taking it from here, for example, it is neither possible to see CNN Turk as the Turkish extension of the global television giant CNN, nor is it possible to see, for example, *Carki Felek* as the completely "estranged" translation of the international television format *Wheel of Fortune*.

Instead, it is necessary to apprehend the above mentioned experiences as glocal experiences. These experiences can neither be regarded as the conquering extensions of the global, nor as cultural facets that give away the resistance of the local.

The global television experience of our time points to a wide-scale interaction network. It speaks of global television industries and the television signal that travels from one continent to another through satellite broadcasts on the one hand, and program trading, on the other. Program trading, in turn, essentially takes place in three different ways in this complicated technological and industrial relations network: program importation, joint production, and format trade (Moran, 1998, p. 25). Imported programs often become the subjects of discussions over the colonization of cultural sphere and the identity problem. Joint productions represent sailing in relatively safer waters, because these productions are shaped through, more or less, equal participation by the two sides. Format adaptations are unable to abstain from criticism that they may lead to cultural estrangement—despite the fact that format adaptations gain such local facets that they make us forget that their original format is foreign. However, this criticism is generally not a criticism that touches on the tension that arises from the encounter of global format and local content in format adaptations. This encounter of programs that originate from one country with viewers from another country can only be successful if the tension it carries is minimized through "loose" formats that embrace different localities. The "loosest" format is probably the one that can easily be filled with local values and trends, and one that can be articulated with them.

As stated by Bonner (2003, p. 24), the important development in television programs in the 1990s is the rise of reality TV category together with the conceptual complexity relating to programs defined by this term. Bonner suggested the concept of *ordinary television* to define reality-based programs that place ordinary people as well as the quotidian into the centre of narration. They are composed of hybrid programs that fuse reality with show, and in which, different talents such as dancing or singing compete against each other; game shows; and programs such *Big Brother* and *Survivor,* that are based on continuously observing participants, who have been brought together for various reasons. Bonner (2003) defines people who participate in reality TV shows as *ordinary people*, but adds that this expression is not used in a belittling manner.

As Bonner highlighted (p. 53), ordinary television examples are rarely imported. It is more common to buy the right to adapt these programs from a licensed format. In many countries, the main reason for preferring to buy the format licenses and adapting them is the fact that formatted programs present the most compatible dress for the "cultural body." The formatted programs take the shape of this body exactly like a flexible dress. As long as format trade is a productive and profitable sphere of activity, licensed formats will present the ability to easily embrace different cultural lives. In short, format adaptations have the tendency to produce a heavy locality. Moreover, provided that suitable conditions are met, these programs can easily go beyond having the cultural tone of locality, and acquire the tendency to become excessive and to produce a "nationalist" emphasis. As this case becomes widespread in television programs, nationalism inevitably becomes the ordinary, and nationalist discourse inevitably becomes one of the daily means of expression of national or cultural identity.

The Semiosphere that Conquers Television

This state of excessiveness is closely related to the semiosphere that surrounds and conquers the television world. In this article, I am using the word semiosphere, which points to the sphere or layer that conquers the process of semiosis functioning through all sign systems, and mainly language, in its simple metaphoric meaning (Hartley, 2007). Here, semiosphere defines the semiotic sphere from which *Singing for a Dream* is produced—in other words, it defines the space of culture. The semiosphere that surrounds *Singing for a Dream* is a semiosphere that has absorbed various forms of nationalism. Therefore, the show is filled with indicators that are very convenient for the articulation of nationalism. The show has been characterized by a struggle for meaning carried on to render the semiotic space monolingual and to make the discourse uniaccentual. In other words, this is an ideology and practice of exclusion, which imposes the idea that a single and same meaning should be derived from concepts such as motherland, nation, patriotism, or a citizen's responsibility towards his/her country and nation. Single language and single discourse is the language and discourse of an aggressive nationalism that resists the possibility of becoming banal.[1]

As I have stated in the introduction to this article, the season of the show, *Singing for a Dream*, that I studied, coincided with a very critical period in Turkey's politics. The show, which was first broadcast on 18 December 2007, was simultaneous with the cross-border operation by the Turkish Armed Forces in Northern Iraq. Consequently, the Government was given the open authorization for a year to launch operations against PKK, which attacked Turkey from Northern Iraq (Arsu, 2007). One day before *Singing for a Dream* went on air, on 17 December 2007, the first group of soldiers was sent to Northern Iraq. The nationalist semiosphere that conquered the show was greatly entrenched through the Northern Iraq operation of the Turkish Armed Forces.

The operation in Northern Iraq was an event that had a "constitutive" nature not only in terms of the political atmosphere that surrounded the show that I studied, but also in terms of the construction of identity the task of social inclusion and exclusion that distinguishes "us" from "them" at the political-national level (Moran, 1998).

The nationalism that comes to light in *Singing for a Dream* is not a kind of nationalism that falls under the category of "banal nationalism" as defined by Billig (1995). His concept of "banal nationalism" is an inspiring concept for studies that focus on nationalism in the media and television (Aslama and Pantti, 2007; Dhoest, 2007; Law, 2001; Richardson and Meinhof, 1999). Although aggressive nationalism, which is characterized by a "lynch culture" is called "banal nationalism" in most scholarly works or in newspaper articles in Turkey, Billig has a slightly different definition for this concept: "Billig mostly situates banal nationalism in established nations, as opposed to the 'hot' nationalism of peripheral groups seeking to establish a new nation" (Dhoest, 2007, p. 70).

As Aslama and Pantti (2007, p. 53) advocate that Billig's concept of "banal nationalism" refers to the unnoticed practices and representations that make the daily reproduction of nations possible, and it involves the ongoing circulation and the use of the symbols, themes, rituals, and stereotypes of the nation. Billig suggests (as quoted by Aslama and Pantti, p. 53)

---

[1] In this article, I am using "resisting the possibility of becoming banal" to mean resisting the retreat from a clamorous and hot nationalism that shows signs of an explosion, and changing over to a banal nationalism that functions through silent, daily routines and practices.

"the metonymic image of banal nationalism is not a flag which is being consciously waved with fervent passion; it is a flag hanging unnoticed on the public building". According to the writers, the sense of national belonging and attachment is thus incited (p. 53).

In the light of above discussions, when we examine the Turkish versions of reality TV, we can obviously see that flags or symbols are far from being subtle, moreover, all the practices and representations that lead to the reproduction of Turkishness are extremely salient. This nationalism is expressed more passionately, the flag of nationalism is flaunted feverishly, and an exclusive discourse is adopted. Of course, this situation is related to the fact that national symbols, such as flags, have become more commonplace in the daily lives of people. The use of flags is no longer limited to public buildings. They are now "flaunted" on the outsides of apartments, cars and even in the houses. The fact that nationalism is on the rise in Turkey is being expressed in various ways (Belge, 2006; Bora, 2006; Ozkirimli, 2002).

Nationalism has taken on many different facets throughout history. Different forms of nationalism co-exist in any single historical conjuncture. Nonetheless, here, I am talking about a nationalism that meets at a common denominator and becomes popular. This is how Ozkirimli (2002) explained the rise of nationalism in Turkey: "nationalism is the system itself. As such, the most important function of the reproduction process has probably been placing nationalism into the centre and enabling it to become popular" (p. 716).

In fact, due to various political developments in Turkey over the last two decades, national identity began to be comprehended in line with the nationalist parameters of Turkishness. Even the simple and ordinary issues of daily life are turned into the subjects of a discussion over how much they are in favor of Turkey and Turkishness. The mentioned nationalist mentality has become notably dominant in the television discourse in especially programs that have ordinary people as participants and that easily incite viewers to take sides.

It is possible to say that this aggressive nationalism is a nationalism that is "on the way of becoming banal" even though it is not yet banal. In other words, this nationalism is almost the transition-period nationalism of a nation that is striving to adopt a stable route towards becoming a western country and a member of the EU.

In this regard, EU membership, as other political transformations, requires the re-imagination of nation in Turkey—in other words, its re-building. It should be taken into consideration whether this process, which marks a different kind of nation that is part of the international Western community, is witnessing the final throes of hot nationalism. As a matter of fact, Belge (2006) assesses rising nationalism as the loosening of faith for the nation state, and the nation state's holding on to nationalism and resisting withdrawing from history's stage. If that is the case, nationalism's resistance to cooling and becoming banal (Westernization), could bring together different nationalisms around a similar exclusion and aggression. It would not be an exaggeration to say that "ordinary television" is one of the channels through which, this kind of a process of becoming ordinary can best be observed, and also one that enables this process of becoming ordinary.

Singing for a Dream: The Reproduction of Identity

The hybrid shows that blend the components *of* reality TV with talent competitions have an important percentage among programs with a format license. *Singing for a Dream* is a hybrid show that can completely be considered within this category. The fourth season of this show, which is analyzed in this article, consisted of ten episodes that were broadcast between 18 December 2007 and 22 February 2008 on Star TV, one of the major terrestrial networks in Turkey. A total of eleven competitors—five women and six men —competed not only with their songs but with their stories, which attempted to fill the viewers' eyes with tears. On every episode, the "giant jury" composed of three of Turkey's most popular singers, Muazzez Abaci, Ibrahim Tatlises and Seda Sayan, assessed the competitors' performances. The viewers could support the competitors by sending SMS text messages. Such messages could put the competitors under protection and enable them to be shielded from the risk of being eliminated from the competition.

During the whole season of the show, national identity was being expressed and reproduced through two types of interactions. The first type took place through "commonalities and conventions" expressed a) in dreams and b) in glorifications. The second type took place through open and subtle "contradictions and clashes."

Commonalities and Conventions

a) Dreams: A dream often voiced by seven of eleven contestants was the dream of "buying a house": A house for their mother, children, and siblings. The dream of buying a house for their beloved ones exceeded the expression of a simple, real and concrete need, and became a metaphoric surface, over which flew the poverty, hopelessness and the insecure lives of large segments of the population; a level, on which was written the strong family bonds, as well as heavy responsibilities pertaining to the family, and the heavy burden people begin carrying on their shoulders starting from a very young age.

Another recurring dream, shared by the other four competitors was to solve their own health problems or the health problems of a relative. One competitor was dreaming of getting hand prosthesis for his wife. Another wanted a hip prosthesis for himself. Another hoped to be able to get the kidney transplant his little sister needed if he won the competition. The last one said he would buy a place to set up a tailors' shop for his mother.

It should be stated that the dreams of the competitors were sealed with "deprivations and needs that are special to Turkey." Another cultural commonality was the heartbreaking moments, during which these dreams were expressed. Whenever the competitors began to tell their stories, tears rolled down the eyes. This tearful act of pouring out their hearts in the competition continued throughout the season. Not only the competitors, but also the members of the jury were crying. A competitor, who was unable to make the jury cry, was not acceptable. In the most episodes, the members of the jury told off the competitors, who could not make them cry, by saying: "Your story does not touch my heart," "your story is not tragic enough," "you are wasting our time; you are keeping us here for nothing!" The competitors, who could not make the jury believe that they have suffered a lot, usually heard the words: "we don't have time for you!"

For example, on the show's first episode on 18 December 2007, Ibrahim Tatlises told competitor Ebru that he did not find her story dramatic enough. Tatlises said: "Make me believe you. Tell me you are a girl, you are a woman and you cannot work because you are harassed. Make me believe you." However, that woman was not talking about such a problem; she had quite different problems. Nevertheless, the issue brought up by Tatlises was an issue that would stir the attention of the viewers. Tatlises chose to explain the competitor's quietness by saying, in a belittling manner: "she doesn't have a sole, she isn't alive."

The things that the decision-makers most craved were tears. In Anatolia, if someone, who has lost a relative, does not wail and cry, people think that, that person is not upset enough (Gorkem, 2001). In the same way, if a competitor did not cry and thus, make others cry, it was decided that he/she did not have a soul and that he/she was not even alive. These young people, who were after meeting vital needs, had heavy responsibilities on their shoulders. They wanted to "rescue" their mothers, sisters, siblings or spouses.

This excessiveness took the competition out of its regular route and dragged it into a family or neighborhood atmosphere. Actually, here exists a tension caused by having to fit into various forms of dialogue, which we are not familiar with in the formal and informal relationships of our daily lives. For the competitors, this was the tension of sharing their private lives with the whole country. For the members of the jury, it was the tension of speaking and evaluating as "the jury," and basing this evaluation on a "system of values." Becoming "a jury member" or speaking before a jury is not an experience we have culturally "collectivized." In these forms of dialogue, which come with format adaptations, the same goes for the coaches of the competitors. Their position before the jury is to serve as the representatives of the competitors. However, even our court system does not allow for this kind of a representation. In Turkey, even lawyers do not advocate their clients using the strong and impressive language we see in foreign films. Moreover, they are not in front of a public jury, which they have to convince and gain their support. The rules of format adaptations, which don't appeal to a sense of "commonality" and the order of speaking and dialogue, forced the participants to overly grab on to—or even lean on to—the other "commonalities" of the culture. With the influence of the existing semiosphere, the mentioned commonality could mostly be established around a heroic patriotism.

b) Glorifications: Due to the coincidence with the above-mentioned cross-border operation, Singing for a Dream became a kind of arena, where all participants competed with each other to prove their patriotism. The program provided the participants with the opportunity of glorifying patriotism by three different ways;

1) by reproducing the existing myths on martyrdom,
2) by blessing the act of dying for one's country,
3) by continuously re-dictating how an acceptable/ideal citizen should be like.

The message that was being underlined through glorifications, creations of myths, and identifications was the idea that solidarity and unity—in other words, Nation—should "nowadays" urgently be reestablished.

On 1 February 2008, which was the first day of the cross-border operation, the program started with a widely known and popular military song (*Yaylalar Yaylalar*). The presenters,

The Seal of Culture in Format Adaptations 145

who were wearing black, began the program by expressing gratitude for the Mehmetcik[2] (Turkish soldiers) for their courage and devotion as they protect the country and the nation. All the contestants sang military songs throughout the night. The jury comments and presenter announcements were all sealed with an emphasis that the army and the nation are a single heart. Through a discourse dominated by a militarist vocabulary, it was being announced that all destructive actions against Turkey were doomed to perish in the face of this unshakeable relationship. The Turkish army and the Turkish soldiers were being blessed and the army-nation myth was being reproduced in almost every episode. However, in the first and final episodes, this nationalist language and discourse had complete domination due to the cross-border operation. In fact, as Altinay and Bora (2002) suggested, throughout the Republic's history the "army-nation" myth is a myth that has been reproduced in various forms by nationalist ideologues, leading military and civilian figures, academicians, educational institutions, the army, newspapers and televisions:

> One of the most significant characteristics of the ideologies of Turkish nationalism is the vastness of their military vocabulary and their centralism [...] "Turkish history" is a history of conquests and wars. "Martyrdom" is among the most glorified notions. (p. 143)

It was interesting to see that one of the competitors was told off by Ibrahim Tatlises on the same episode. Tatlises was angry with the song by a competitor named Fahrettin, because he had changed the words of an ordinary folk song and replaced the word "my hero" in the chorus section of the song with the word "my soldier." The competitors had been asked to find a military folk song that day and compete with that. The competitor had found the easy way out of this. Tatlises accused the competitor of "betraying" everyone and putting our soldiers in a helpless position. In fact, the competitors were continuously "being adjusted" throughout the program. The innocent changes a young competitor made to a song was seen as "betrayal" and "all hell broke loose." This was notable in terms of seeing the graveness of glorifications and identifications.

At the opening of the program's season finale (22 February 2008), a large flag was waving on the screen in the background. As the presenters spoke, their images were being mixed with the flag and reflected on the screen. After everyone gathered in front of the flag, a popular song, *Memleketim* (My Country) started playing. Though, immediately following the opening announcement, the presenters received the news that five soldiers had become "martyrs" in Northern Iraq. The presenters Gul Golge (female) and Cenk Eren (male) turned to the cameras and voiced their feelings about this news and conveyed the message of the program:

> Male presenter: It is very difficult to do this program. Five of our soldiers have become martyrs (the image of the flag in the background and the images of Cenk and Gul are mixed. The two look as if they have been wrapped up in the flag). It should be known that not a tiny bit of this land is going to be given away.
> Female presenter: This country has lived through many difficulties. Nothing happened to us. And nothing will. Nothing will happen as long as we stand erect.

---

[2] Mehmetcik is the general name used to define the Turkish soldiers, and which expresses the society's love and compassion for the soldiers.

Male presenter: They have reached the palm of martyrdom. They have become martyrs. Their place in heaven is ready.

A song followed this conversation. After the song, jury member Seda Sayan realized that the program flow is different than planned and asked for the reason. It was then understood that the presenters had been told to "slow down" the program. But there was a clear confusion on how this was going to be done. This confusion continued for some time. It was a strange moment during which everything seemed to be suspended. The strangeness of this moment could only be overcome by giving a break for commercials. This urgent break showed that, in fact, the show could not even establish a strategy for its own nationalist discourse and that the "words" that were being uttered were baseless. Therefore, the jingoism that dominated the show was thoroughly being displayed. It is possible to say that such a nationalist discourse, caricaturizes its own militarist and chauvinistic ideals and undermines these ideals only to leave them in complete emptiness.

CONTRADICTIONS AND CLASHES

Not only commonalities, but also contradictions and clashes between the participants contributed to the establishment of national identity and locality as the dominant tone of the program. In this program, there were things per se that contradicted with the structure and content of the program: What the viewers were expecting from this show was not to get the latest news on the Northern Iraq operation and to listen to such a univocal patriotism discourse. Neither were they expecting the contestants to be put in their places because of "inappropriate" demeanor or behavior. This situation, which was not in line with likely expectations, was the producers' response to the semiosphere that surrounded the production. In many countries, such shows feature excessive amounts of sensationalism that attract the viewers. However, the fact that in Turkey this exaggeration was more focused on "national" issues, rather than "individual" issues points at something that is unique about Turkey's national identity. .

It should be reiterated that the contradictory situations in the program were being exaggerated due to the haste to heal holes or cracks, and that not much time was spent on any state of sentiment. For example, in the midst of the eighth episode, broadcast on 1 February 2008 and dedicated to the Turkish soldiers, a show that was full of sad army folk songs, the mood suddenly switched and the program continued as a wedding atmosphere to celebrate the jury member Seda Sayan's wedding. A schizophrenic state of exaggeration was inevitable.

In addition to the contradictions there were also clashes, mostly between the members of the jury. Throughout the show, the most careful attention has been paid to the usage of a "proper" Turkish. Abaci, despite her modest personality, immediately and rather harshly reacted against any incorrect usage of Turkish by the other jury members and the contestants, but mainly by Tatlises—who was born to a family of Kurdish and Arabic origin in a cave-like shelter in Urfa, and yet had a splendid rise. Tatlises, today, is extraordinarily strong in the world of popular music and he is a magnificent performer. But, he has no education whatsoever. His Turkish is still broken after many years. This is not something that can be easily dealt with by, be it the other celebrities, or be it the "white Turks" of upper-middle and upper classes.

And Tatlises was almost stuck between his own strength and his vulnerable position brought about by his past and thus, continuously was in a state of tension that he had to keep under control. The mentioned tension between the two singers presented the viewers a highly concentrated picture of national contradictions and clashes. Among the three singers in this show conquered by a nationalist semiosphere, it was Tatlises, who was most intensely articulated in this semiosphere. The nationalist attitude one of Turkey's most popular singers had to adopt to render his ethnic origins invisible was noteworthy.

To sum up, the clash between the jury members, in part, was based on a class struggle and a struggle based on ethnic origin. This fact contributed to emerging tension between the light hearted entertainment nature of the universal format and the nationalist discourse of the Turkish version.

CONCLUDING COMMENTS

As noted at the beginning of the article, a number of studies have focused on format adaptations as the tools of foreign (mainly American) domination or hegemony over domestic television programming. Yet, quite a few studies have contested this idea and argued that format adaptations were not the examples of the local or the global, the East or the West, and that have pointed to the need to recognize them as something new (Coutas, 2006, p. 389). An often missing facet in these studies was, I suggest, the tendency of being excessive in identity statements as the result of the inherent tension between the global and the local, so that my aim has been to generate an awareness towards this tendency and to explore the problematic encounter between popular global form(at)s and cultural-local contents through the analysis of *Singing for a Dream*.

The nationalist discourse incited me to study the show, which was not in line with the program in terms of format and style. This popular reality talent-competition show juxtaposed young people living tragic lives with celebrities who had rarely been involved in any political activity or displayed an "open" political sensitivity. The inability of the young contestants to form any bonds between their tragic lives and the system, and the way they embraced a nationalistic ethos was very illuminating. The attitude of the celebrities was equally interesting. Despite their complete lack of reaction and criticism heretofore, they gave jingo speeches about the future of the nation and assured viewers that young soldiers would become martyrs without a moment's hesitation. This made the show more than just an ordinary entertainment program. *Singing for a Dream* became a conduit of popular, aggressive and exclusive nationalist speech.

There are a number of challenging issues which are beyond the scope of this article. However, it is important to understand that such programs are convenient for spreading their exclusive discourses since they incite the viewers to take sides and identify with "ordinary people" (just like them) who become instant heroes. Reality-talent shows are the most popular television programs alongside domestic drama in Turkey (Kilicbay, 2005) where television is the main source of popular entertainment (Celenk, 2005; Mutlu, 1999). Accordingly, it is vital to look closely to foresee the "excesses" caused by us nestling down in our own locality and in our own cultural identity.

REFERENCES

Aksoy, A., and Robins, K. (2000). Thinking across spaces: Transnational television from Turkey. *European Journal of Cultural Studies*, *3*(3), 343-365.

Altinay, A. G., and Bora, T. (2002). Ordu, militarizm ve milliyetcilik [Army, militarism and nationalism]. In T. Bora (Ed.), *Modern Turkiye'de siyasi dusunce: Milliyetcilik* [Political thought in modern Turkey: Nationalism - Turkish] (pp. 140-154). Istanbul: Iletisim.

Arsu, S. (2007, October 17). Parliament in Turkey votes to allow Iraq incursion. *The New York Times*. Retrieved March 18, 2008, from http://www.nytimes.com/2007/10/17/world /europe/17iht-18turkey.4.7929431.html

Aslama, M., and Pantti, M. (2007). Flagging Finnishness: Reproducing national identity in reality television. *Television and New Media*, *8*(1), 49-67.

Belge, M. (2006). *Linc kulturunun tarihsel kokeni: Milliyetcilik* [Historical origin of lynch culture: Nationalism - Turkish]. Istanbul: Agora Kitapligi.

Billig, M. (1995). *Banal nationalism*. London: Sage.

Bonner, F. (2003). Or*dinary television*. London: Sage.

Bora, T. (2006). *Medeniyet kaybi* [Lost of civilization]. Istanbul: Birikim.

Celenk, S. (2002a). Industrial developments: Turkish TV fiction in 1999. In M. Buonanno (Ed.), *Convergences: Eurofiction fourth report* (pp. 153-163). Napoli: Liguori Editore.

Celenk, S. (2002b). Indirect ways of foreign penetration: Turkish TV fiction in 2001. In M. Buonanno (Ed.), *Eurofiction: Television fiction in Europe* (pp. 149-150). Strasbourg: European Audiovisual Observatory.

Celenk, S. (2005). *Televizyon, temsil, kultur* [Television, representation, culture]. Ankara: Utopya.

Coutas, P. (2006). Fame, fortune, fantasi: Indonesian idol and the new celebrity. *Asian Journal of Communication*, *16*(4), 371-392.

Dhoest, A. (2007). Identifying with the nation: Viewer memories of Flemish TV fiction. *European Journal of Cultural Studies*, *10*(1), 55-73.

Gorkem, I. (2001). *Turk edebiyatinda agitlar: Cukurova agitlari* [Elegies in Turkish literature: Elegies from Cukurova - Turkish]. Ankara: Akcag.

Hartley, J. (2007). *Television truths: Forms of knowledge in popular culture*. Wiley: Blackwell.

Kilicbay, B. (2005). *Turkiye'de gerceklik televizyonu ve yeni televizyon kulturu* [Reality television and the new television culture in Turkey]. Unpublished doctoral dissertation, Ankara University, Turkey.

Law, A. (2001). Near and far: Banal national identity and the press in Scotland. *Media, Culture and Society*, *23*(3), 299-317.

Milikowski, M. (2000). Exploring a model of de-ethnicization: The case of Turkish television in the Netherlands. *European Journal of Communication*, *15*(4), 443-468.

Moran, A. (1998). *Copy cat tv: Globalisation, program formats and cultural identity*. Lutton, UK: University of Lutton Press.

Mutlu. E. (1999). *Televizyon ve toplum* [Television and society]. Ankara: TRT publications.

Ozkirimli, U. (2002). Turkiye'de gayriresmi ve populer milliyetcilik [Unofficial and popular nationalism in Turkey] In T. Bora (Ed.), *Modern Turkiye'de siyasi dusunce: Milliyetcilik* [Political thought in modern Turkey: Nationalism] (pp. 706-717). Istanbul: Iletisim.

Peker, Z., and Peker, O. (Producers). (2007). *Hayalin icin soyle* (*Singing for a dream*) [Reality television series]. Istanbul: Star TV.

Richardson, K., and Meinhof, U. (1999). *Worlds in common? Television discourse in a changing Europe*. London: Routledge.

Skovmand, M. (1992). Barbarous tv international: Syndicated wheels of fortune. In M. Scovmand and K. C. Schroeder (Eds.), *Media cultures: Reappraising transnational media* (pp. 84-103). London: Routledge.

Sparks, C., and Tulloch, J. (2000). *Tabloid tales: Global debates over media standards*. Lanham, MD: Rowman and Littlefield.

Tunc, A. (2002). A genre a la Turque: Redefining game shows and the Turkish version of wheel of fortune. *Journal of American and Comparative Cultures*, *25*(3/4), 246-248.

In: Reality Television-Merging the Global and the Local
Editor: Amir Hetsroni, pp. 151-162

ISBN 978-1-62100-068-6
© 2010 Nova Science Publishers, Inc.

Chapter 10

THE PRAISE AND THE CRITIQUE OF A NASTY FORMAT: AN ANALYSIS OF THE PUBLIC DEBATE OVER REALITY TV IN ISRAEL

Amir Hetsroni
Ariel University Center, Israel

Like in many other countries the success of reality TV formats such as "Big Brother" and "Survivor," which are based on closed-circuit television (CCTV) and feature constant surveillance of the contestants' daily life, performance of humiliating and sometimes explicit tasks, and unscripted battles between participants brought with it a heated public debate concerning the moral legality of the genre and a high pitched discourse revolving around the impact of the broadcasts. This paper reviews the controversy, which occupied a significant portion of the mass-media related public agenda in Israel over the years 2005-2008. Through a thematic content analysis of the arguments articulated in praise and critique of reality TV, I map the public discourse and contextualize the claims in more general longstanding debates. Thus, even though the data on which the analysis is based was collected in a single country, the mapping portrays the public discourse about reality TV in a way that can be relevant to various cultures.

A BRIEF HISTORY OF ISRAELI TELEVISION AND REALITY TV IN ISRAEL

Television arrived late in Israel. Construed as a "Zionist response" to Arab propaganda, the broadcasting started in 1968 with one public channel supported by a license fee and without any commercial advertising (Caspi & Limor, 1999). The programming mixed political discussions with shows devoted to national ethos. Only minimal attention was given to pure entertainment, which was by and large restricted to import shows (Oren, 2004). The situation remained so until 1993, when the first commercial station ("Channel 2") began to broadcast, joined nine years later by "Channel 10". Both of these channels rely exclusively on advertising for financial maintenance, but the content of their programming is regulated by strict franchise licenses, which dictate lineups that – compared to the public channel – contain more local programming and a heavier dose of entertainment, but in comparison with other western countries still offer more high brow shows like documentary films and serious drama (Caspi, 2007).

In 2005, the franchise license of Channel 2 has been extended by twelve years, which made the channel's management feel certain enough to dare with the presentation of more populist (and less welcomed by the regulator) formats, and paved the road to the proliferation of reality TV on the Israeli screen in the second half of the noughties. Of course, one cannot ignore other factors that contributed to this proliferation: the sharp decline in income that both Channel 2 and Channel 10 have experienced due to the growing competition from cable stations; the rise in production costs of documentary films and drama stemming partly from salary demands by the creatives of these shows and partly from regulation ordinances that dictated very high production values; and dwindling ratings because of the growing number of viewing options on the tube and off the tube (Noam, Groebel & Gerbarg, 2003).

The arrival of reality TV in the Israel, like the arrival of TV broadcasting, was belated. The genre took off slowly. The first reality show to be broadcast in Israel, "The Mole", sustained for only one season (2001) on Channel 2 and was considered a commercial flop, which deferred the production of further reality shows by three years. The next program, "Choose me, Sharon", a local non-franchised version of "The Bachelorette" was aired on Channel 2 in late 2004 and gained high ratings. While the show was not renewed, partly due to the fact that the American franchise owners threatened to sue the production company for copyright violation, its success indicated that international reality formats can work out in Israel, as proven by the success of "Survivor" (aired since 2007 on Channel 10), "The Models" (a local version of "America's next top model" which was aired for three seasons between 2005 and 2008 on Channel 10) and finally "Big Brother" (aired since 2008 on Channel 2). With a rating that approaches 40%, the finale episodes of Survivor and Big Brother are the non-news programs with the highest viewership in the last five years.

A REVIEW OF THE PUBLIC DISCOURSE REGARDING REALITY TV IN ISRAEL

The Israeli public had remained mostly indifferent to reality TV until the broadcasting of a Models episode which required the contestants to take part in semi-nude shooting. One of the girls, who refused to have her photograph taken due to religious convictions, was eliminated from the show. This fact, together with the emphasis put on the program on lean figure as a key factor to success in the catwalk runways, triggered hundreds of complaints submitted to the regulator's ombudsman, a number of Op-Eds and even a parliamentary hearing. On one side of the fence were voices demanding to censor and even cancel the show because of its supposedly pro-anorectic message. On the opposite side, producers and a few experts insisted that that the potential damage is too limited to justify any censorship

While until 2006 the debate over the potential damage of reality TV was overshadowed by a similar discussion revolving around the negative outcome of sexually charged advertising, the establishment of a moral code to advertisements (a sort of self censorship on

The part of advertisers) and the extended proliferation of reality shows on the screen in 2007 shifted the discussion from advertising to reality programming. The following sections review the critiques and praises that were raised in this discussion over reality TV. For a review of the debate about sex in advertising – see Hetsroni, 2009.

CRITICISMS AND PRAISES

To map the critical and praising voices concerning reality TV, I have assembled all articles that have been published on this topic in the three Israeli newspapers that enjoy the widest circulation: Yedioth Achronot (Ynet), Maariv (NRG) and Haaretz between June 2007 and June 2009 (the period in which reality TV was a "hot topic" on the agenda). To detect relevant articles, the newspapers' online editions were searched using "reality", "reality TV", "reality format" and names of specific shows as keywords. Only articles that appraise the genre (or appraise specific programs) were used in the analysis, which included – altogether – 47 articles that were content analyzed to detect the critical and praising themes.

The critical themes were:

a) Lack of authenticity
b) Lack of quality and dearth of educational values
c) Demolition of the local TV drama industry
d) Invasion of privacy
e) Stereotypical representation
f) Moral panic

The praising themes were:

a) Minimization of damages
b) Reality is a solution to the crisis in the TV production industry
c) Reality TV is art

I have assessed the share each theme occupies in the sample. However since single articles could (and in practice many of them did) contain more than one theme per-particle the overall number of themes (N=113) is larger than the total number of articles (N=47). The coding was performed by two students, who successfully passed a course in research methods and content analysis. After being trained by the author, they coded all the articles independently. To measure coding agreement, Cohen's Kappa coefficient was computed. Its value (κ=.808) points at adequate reliability. The next sections review and demonstrate the themes.

CRITICISM NO.1: LACK OF AUTHENTICITY

Ironically, one of the most recurrent allegations of reality TV charges that the format is, in fact, unreal. Instead of providing the viewer with an unobtrusive glimpse into other people's lives, the spectator is faced with a staged play, where roles and scripts are strictly followed. This critique posits reality TV as the less authentic relative of documentary films (Hill, 2005), a fact that may have historical roots as reality TV actually embarked in 1973 as a variant of documentary, when the American public channel (PBS) broadcast "American Family"– an eight hour so-called documentary series following the daily life of a real family across a period of one year (Andrejevic, 2004).

While this critique, which often appears in media scholarship, could be of significant interest to filmmakers, it is probably not so highly relevant to the ordinary viewer. In our database – lack of authenticity appeared only once (0.8% of the total sample).Unsurprisingly, responsibility for this appearance rests with a drama professor, who wrote an article in protest of the screen proliferation of reality shows (particularly "Big Brother") and in reference to the actors' union campaign against this show:

> Why has the actors' community reacted in such extensive anger to 'Big Brother'? Why has it not reacted similarly to thousands of programs and other non-sense shows? ...There is a reason to the special anger, and it is neither financial nor rating driven. It lays in one simple word - truth. Reality shows, more than any other TV garbage, say that they are true, when in fact they are an insult to the truth. Actors cannot remain bystanders, when the truth is insulted, since it affects the very heart of their creation. (Calderon, 2008).

CRITICISM NO.2: LACK OF QUALITY AND DEARTH OF EDUCATIONAL VALUES

This view criticizes reality TV for not being "good TV" in the eyes of high brow taste owners and for lacking an educational value. This line of criticism, which takes for granted an unequivocal hierarchy of taste, wherein reality programs occupy the lower ranks while drama and documentaries top the scale ladder, is common among academics and intellectuals, who make a significant bulk of the authors who express it publicly.

When compared with documentary films (see, for example, Bignell, 2005), reality programs are usually judged as not as thought provoking and not as mindful as the older genre. A similar line of argumentation typifies comparisons made with conventional TV drama. In both cases, the critical bottom line is that reality TV lacks the exquisite quality of high brow genres. An ample demonstration can be found in an article written by a TV critic that juxtaposed "Big Brother" and "Survivor" with a low-key drama about the life of twenty-something religious Israelis. Both series were aired simultaneously in 2009:

> There are some people in this country who want to see something other than 'Big Brother'. To make things clear, one day over a month in each year could and should be devoted to a reality show like 'Big Brother,' but in the rest of time we demand to see local drama, local sitcoms, movies, a game show here and there, and some imported dramatic serials. We have to explain that to broadcasters like I explain about food to my five-year-old child: I know that you like to eat Schnitzel with French Fries. I also find it tasty, but you cannot eat it every day all the day. (Tessler, 2009).

In the second part of this critique the author attacks the lessons that reality programs teach the viewers and contrasts these lessons with what should be considered an educational message (in his point of view):

> This is probably the last battle between light and darkness, between those who want to have their minds fucked up with 'Survivor,' where girls in a bikini and guys full of muscles play stupid social games and a wonderfully scripted well played drama.

Attacking reality TV for not carrying the "right educational message" is familiar from other countries. For example, Ellzey and Miller (2008) show how the American version of "The Apprentice" prioritizes street smart over book smart. In our sample, this critique appeared 17 times (15% of the cases).

CRITICISM NO.3: INVASION OF PRIVACY

This criticism, which was the focus of the attack on reality TV in Europe (Mathijs, 2002), is considerably less prevalent in Israel. I detected such arguments in only 9% of the themes that comprised the sample. This may reflect an actual cultural difference between Israel and Europe, as Israeli culture is less sensitive to privacy than European cultures are (Almog, 2004). However, when disruption of privacy did surface as a theme in Israel – the content and tone of the critique were not different from what we are used to in other countries:

> You want to see exposed bodies? Fine. Just don't say that you do it in the name of science or art. These bodies have had full life that you will never see. They had love and disappointments. They played and laughed. They were gloomy and happy. But what you see from them is not the reality. It can never be the reality. It is just plain voyeurism. (Levi, 2009).

CRITICISM NO.4: DEMOLITION OF THE LOCAL TV DRAMA INDUSTRY

This criticism postulates that the growing popularity of reality programming has been destroying the local TV drama industry, since broadcasters became reluctant to sign new contracts with creators of fictional drama. Scriptwriters, directors, films editors and professional actors, were unable to find work that is in line with their expertise in the new reality format. Untrustingly, the criticism which relates to the crisis in the local TV drama industry was voiced mainly by people from within this industry, as demonstrated in the words of an award winning actress:

> They cancel programs of drama and culture and flood us with garbage. 'Big Brother; is garbage and TV fattens us with garbage. (Almagor, 2008).

Even though the degrading view of reality shows as opposed to the sacred worshipping of dramatic programs, which sometimes feature similar core values, is not unique to Israel, this specific criticism, which appeared in 20.3% of the articles, has not been heretofore frequently voiced outside of Israel (Biltereyst, 2005). No less unique was the formal justification of the arguments, which was purely legal and charged the franchise owners of Channel 2 and Channel 10 with failing to meet the franchise demands when it comes to investing money in local drama and documentary films. The broadcasters' reply was that the definition of "drama" is outdated and that reality TV programming is a form of docudrama.

As happened in some European countries e.g. France and Greece (although there it happened under the flag of invasion of privacy – see Biltereyst, 2005), an anti-reality TV rally was organized by the Israeli stage and motion pictures actors union. The event took place in the night when the final episode of Big Brother's first season was broadcast. Less than 3,000 people attended the rally that included performances of well known actors and singers, while the final episode of 'Big Brother' drew more than 2.5 million viewers.

CRITICISM NO.5: STEREOTYPICAL REPRESENTATION

Although one can find criticism of reality TV for the way it misrepresents gender roles, stereotypically portrays minorities and confirms to traditional racial stereotypes in many countries (Andrejevic, 2004), in Israel the public discourse over this issue has been more excessive (present at 28.3% of the cases in the sample). One reason for the excessiveness could be that - compared to several western countries – Israeli society is still "in the making" and therefore the ethnic balance is more prone to be disrupted (Almog, 2004). However, I believe that another factor may have a greater responsibility for the exceptional visibility of this theme. As mentioned before, the explosion of reality TV in Israel occurred relatively late compared to the western world (2007-2009 vs. 2001-2002), but this was not merely a late arrival as it was an arrival following the public debate over the renewal of broadcasting licenses to operate Channel 2 (2004-2005). In that debate, the misrepresentation and lack of representation of minorities (women, Arabs, religious Jews, Jews of eastern origin) occupied a central role (Hetsroni, 2010). Thus, when reality TV finally took off, the airwaves were ready to absorb and discuss issues pertaining to the representation of racial, gender and religious stereotypes.

The most popular theme in this discourse related to the representation of Jews of eastern origin as "others", as reads the following commentary concerning one of the contestants in the first season of 'Big Brother':

> The editors of 'Big Brother' have chosen to preserve the power structures in Israel by stereotypically choosing participants. To portray the eastern Jew, they took an ostensible non-educational person. The mocked presentation of this person tells all about the role he was summoned to play (Lir, 2009).

Finally, we should add that the producers of reality programs, probably realizing that this kind of discourse may raise interest and expand the viewership, have been inciting the discussion by including people who hold extreme – sometimes practically racist – views in the cast along with people at whom their hatred is targetted. Examples include an out-of-the closet homosexual and a gay bashing person in the cast of 'Big Brother's second season and a supporter of exiling Arab minorities and an Arab in the cast of 'The Models' second season.

CRITICISM NO.6: MORAL PANIC

This criticism contends that reality TV is threatening the core values and norms of society and causes moral panic. Moral guardians (e.g. religious authorities, educators) and "socially accredited experts" (e.g. psychologists, sociologists) define the "deviants" from whom society should be afraid. These are groups of people who are closely related to alternative culture e.g. rock artists, drug users, motor cyclists. The media play a crucial role in whipping up public debate on the phenomenon, scapegoating particular "folk evils," and creating public anxiety (Thompson, 1998). In the case of reality TV, the role of the media is more self-reflective, since the panic's source is the media themselves.

In a way, moral panic is an umbrella category for many other criticisms, since the reason why reality TV threatens core social values and norms is often specified in the aforementioned categories. In practice, while moral panic was found in 14.2% of the articles, it rarely appeared independently. When it did, it tended to keep a non-directive tone, as can be seen in the following example:

> Clearly it is possible to create a different kind of television, but even if recurring success cannot always be gentle and subtle, it is still not an excuse to let go with the most extreme horrors of reality. These horrors need to be evacuated for our moral safety. (Aloni, 2008).

In different European countries, moral allegations dominated the popular protest against reality TV (Biltereyst, 2005). The explicitness of some of the shows, such as "Big Brother," was a source of special concern. In Portugal, for instance, a former health minister asked: "if a person can be jailed for exhibitionism in a town square or a municipal park, why are people permitted to do it on national television?" (Tremlett, 2001). In Israel, these issues have not been as highly prominent, possibly because none of the reality shows contained material as gratuitous as in Europe. Nonetheless, there were a number of expressions of distress, coming chiefly from feminist critics, who were not pleased with the depiction of women in reality shows as an "easy prey" or from the rare accidental exposure of intimate organs, which they interpreted as a signal of female subordination:

> Apparently, there is nothing wrong here other than low rate yellow journalism. This beauty model, who was shot without panties, is a celebrity, who willfully takes part in a show about celebrities whose essence is to peep into celebrities' life – so what is there to complain about? There is. The photographer, who took the shot, crossed the line, and the media that have shown the pictures crossed the line. A body line. A line of consent. Ms. Neumann [the celebrity model – A.H.] never initiated or requested to have a camera sneaking under her skirt. This cameraman enforced himself on her. This is sexual harassment. (Kazin, 2009).

We now review the praising themes. These voices were undoubtedly a minority, making up only 12.4% of the themes in the sample (14 appearances overall).

PRAISE NO.1: MINIMIZATION OF DAMAGES

The most common praise (present at 6% of the themes in the sample), which is actually not such an esteemed praise, was the claim that the protest against reality TV is motivated by business interests and political ideology and not supported by scientific evidence, and that the actual damage of the genre is limited. The following words from a communication researcher demonstrate this claim:[1]

> Without getting into the pleasure that is granted by watching 'The Models' (and it is not a public offense to admit that this show is a pleasure), and without delving into the philosophical question 'what is female beauty all about' (one is allowed to think that 180 cm spread across 50 kg are more attractive than 150 cm sitting heavily on 80 kg), the criticism, which claims that 'The Models' and similar shows encourage dangerous eating habits because they present skinny women in an attractive light – ignores the facts. A statistical correlation exists between obesity and TV viewing and not between anorexia and TV viewing. Couch potatoes are fatter and not thinner than the average. (Hetsroni, 2005).

PRAISE NO.2: REALITY IS A SOLUTION TO THE CRISIS IN THE TV PRODUCTION INDUSTRY

This praise contends that reality TV is, in fact, a viable option for TV producers to keep on working at a time of economic crisis, when dramatic serials and documentary films do not garner enough viewers that would justify their production. A demonstrating example can be found in the following words of an economic commentator:

> Every bubble is doomed to explode, and now it is time for TV creatives to realize that. Since the start of the economic crisis, broadcasters' incomes have been shrinking at in increasing pace... Rating wise, no drama can compete with a reality show. Targeting the anger and protest against the regulator is not able to change this simple economic fact. (Taig, 2009).

Unsurprisingly, such praise was often dismissed as "neo-liberal politics" (ignoring the fact that the actors' union critique of reality TV for demolishing their career can be regarded "Marxist rhetoric". The low prevalence of this praise (found in no more than 2.6% of the sampled arguments) is an indication as to how unpopular neo-liberal writing is in Israeli media critique.

PRAISE NO.3: REALITY TV IS ART

This praise is the most ambitious in scope. It does not merely attempt to justify the existence of reality programming in the context of producers' and broadcasters' needs to make a revenue amid the commercial failure of local drama or to postulate that any social or

---

[1] I am the author of the cited commentary.

personal damage brought about by the format is minimal. This line of praising boldly states that reality shows are not artistically inferior to any high brow genre:

'Big Brother' is not significantly different from 'No Exit' – Sartre's modernist existentialist play about three souls that are stuck in an elegant room, which is, in fact, hell, and try to understand where their life is going and who needs this life anyway. Sartre's passing thoughts about the meaning of life are not less dramatic or less insightful than those of 'Big Brother's contestants, who are certainly more authentic and have a closer connection to Israeli reality. The only advantage of Sartre is that he wears the "quality badge" of a Nobel Prize laureate, while 'Big Brother' contestants are construction workers, hairdressers and shop assistants. Talent wise, their dialogues are on par with Sartre. (Hetsroni , 2008)

CONCLUSIONS

Our analysis indicates that the critical tone was dominant in the public debate over reality TV in Israel, as it was represented in the press. Table 1 summarizes the figures pertaining to the prevalence of different themes.

Table 1. Critical and Praising Themes in the Public Debate about Reality TV in Israel (N= 113 Newspaper Articles)

Critical Themes	
Lack of authenticity	0.8%
Lack of quality and dearth of educational values	15.0%
Invasion of privacy	9.0%
Demolition of the local TV drama industry	20.3%
Stereotypical representation	28.3%
Moral panic	14.2%
Total share of critical themes	87.6%
Praising Themes	
Minimization of damages	6.0%
Reality is a solution to the crisis in the TV production industry	2.6%
Reality TV is art	3.8%
Total share of praising themes	12.4%

Critical arguments outnumbered praising contentions in a seven to one ratio. Despite this fact, the protest against reality TV rarely slipped into the streets and was unsuccessful in swaying large publics. The reason for that could be the low involvement of large publics, which is the norm among television viewers, who do not consider programming preferences a crucial part of their identity. However, it is also possible that the critical tone did not represent the opinion of the largest bulk of viewers, who willfully attend to reality shows despite the critical bashing of the format. In fact, a tacit assumption in most of the critical essays was that millions of viewers are being taken by the deceptions and tricks of reality TV (which the critic himself is able to see through) and that large audiences are being led astray by the shows' unsavory messages - to which the critic is immune.

The critiques often came in a mixed package, as 43 of the 47 articles that were included in our analysis featured more than one theme per article. The arguments tended to amplify rather than contradict one another. This means that praising contentions were rarely found when critical sayings were heard, and that criticism was not prevalent in articles where praises set the tone.

Complaints about stereotypical representation of minorities were the most frequent theme. This not a surprise in light of the salient presence of this theme in other public debates on media matters in Israel e.g. the content obligations of commercial TV stations (see Hetsroni, 2010). Another frequent theme was care about possible demolition of the local TV drama industry brought about by the success of reality programs. The high prevalence of this theme can probably be attributed to the fact that it had legions of loyal soldiers" (actors, directors, producers etc.) to carry the flag. In comparison with other western countries (see Biltereyst, 2005), Israeli critiques overrepresented arguments that relate to racial stereotypes and underrepresented arguments concerning invasion to the contestants' privacy. This pattern is in line with the Israeli society character that is less kin on privacy and more sensitive to racial conflicts (Almog, 2004).

If we put aside the exact percentages and the specific examples, we come out of this study with a comprehensive list of positive and negative arguments that can be used to analyze the reception of reality TV in different cultures. Furthermore, it is not unlikely that similar themes might be detected, when analyzing the critical reaction to the introduction of other populist TV formats. In the words of a sarcastic US TV critic, who rephrases Churchill: "Never have so many watched so much TV with so little good to say about it" (Poniewozik, 2003, p.65).

A Sweet End and an Unexpected Twist involving Academe and Showbiz

Since the mid 2000s I have been writing and commenting about reality TV in different arenas. It started with request from a newspaper to write down an "expert report" concerning the loud scandal that escorted the broadcasting of the local version of "America's next Top Model" ("The models") and continued with a long series of articles and reviews. My presented opinion was typically off-center from many of my academic colleagues, who used to criticize the genre for being too vulgar and not highly artistic, while I applauded many of the shows for their capability to provide harmless fun and offer sometimes brilliant dialogues in an easily digestible package that still contains a high degree of social relevance. I thought, and still think, that reality TV is the one of the very few forms of popular programming in which non-superpower media industries can excel, and that reality TV serves as an excellent conduit to connote local culture and bring to fore noteworthy social trends.

Some time after the current study was written I received one of the most unexpected phone calls. On the other side of the line was the chief executive producer of Big Brother, who told me that he and his team were loyal readers of my articles, and that they consider the option of having me as an expert in residence and on-camera commentator in the upcoming season. I said yes. My following adventures as an academic in la-la land can easily fill a number of articles. For the purpose of this paper, it is suffice to say that broadcasters and producers of reality shows do read and consider the positive and negative articles that are published about their programs in the press. Thus, the empirical and discursive analysis presented in this paper has an applied value that goes well beyond its theoretical contribution.

REFERENCES

Almagor, G. (2008, December 16). Ha-televizia malhita et yeladeinu be-zevel - Rea'ayon im Gila Almagor [Hebrew: Television feeds us with garbage- an interview with Gila Almagor]. Retrieved May 31, 2009, from http://www.inn.co.il/News/News.aspx/182863

Almog, O. (2004). *Preida misrulik: Shinui arachim bachevra hayisraelit* [Hebrew: Farewell Srulik: Value changes in Israeli society]. Haifa: University of Haifa Press.

Aloni, N. (2008, March, 16). La'atsor et zva'at ha-reality [Hebrew: Stop the reality horror]. *Ynet.* Retrieved May 31, 2009, from http://www.ynet.co.il/articles/0,7340,L-3518549,00.html

Andrejevic, M. (2004) *Reality TV: The work of being watched.* Lanham, MD: Rowman and Littlefield.

Bignell, J. (2005). *Big Brother: Reality TV in the twenty-first century.* London: Palgrave Macmillan

Biltereyst, D. (2005). Reality TV, troublesome pictures and panics: Reappraising the public controversy around reality TV in Europe. In S. Holmes and D. Jermyn (Eds.), *Understanding reality television* (pp. 91-110). London: Routledge.

Calderon, N. (2008, December 22). Ha-emet shel ha'ach hagadol [Hebrew: The truth of Big Brother]. *Ynet.* Retrieved May 31, 2009, from http://www.ynet.co.il/articles/0,7340,L-3641846,00.html

Caspi, D. (2007). *Slicha, takala: Deichata shel rashut hashidur hayisarelit* [Hebrew: Sorry, it's a mistake: The ebbing of public television in Israel]. Jerusalem: Tzivonim Publishing.

Caspi, D., & Limor, Y. (1999) *The in/outsiders: The media in Israel.* Hampton Press, Cresskill, NJ.

Ellzey, M., & Miller, A. (2008). Portrayals of Intelligence in Reality Television. In L. Holderman (Ed.), *Common sense: Intelligence as presented on popular television* (pp. 269-282). Lanham, MD: Lexington Books

Hetsroni, A. (2005, May 30). Duganiyot the mashmin? [Hebrew: Do 'The Models' make you fat?]. *Ynet.* Retrieved May 31, 2009, from http://www.ynet.co.il/articles/0,7340,L-3092217,00.html

Hetsroni, A. (2008, December 11). Ein lo a'ach [Hebrew: He has no brother]. *Ynet.* Retrieved May 31, 2009, from http://www.ynet.co.il/articles/0,7340,L-3636033,00.html

Hetsroni, A. (2009). So how much sex is there really on Israeli TV advertising: A longitudinal analysis. *Social Issues in Israel, 8,* 147-174.

Hetsroni, A. (2010). The representation of minorities and periphery on Israeli commercial television. *Cultural Trends, 19*, 81-91.

Hill, A. (2005). *Reality TV: Audiences ad popular factual television*. London: Routledge.

Kazin, O. (2009, March 23). Reality ze lo terutz le-hatrada minit [Hebrew: Reality is not an excuse for sexual harassment]. *Ynet*. Retrieved May 31, 2009, from http://www.ynet.co.il/articles/0,7340,L-3692119,00.html

Levi, M. (2009, March 25). Anashim metim, ze hakol [Hebrew: Dead people, that is all]. *Maariv-Nrg*. Retrieved May 31, 2009, from http://www.nrg.co.il/online /1/ART1/870/684.html

Lir, S. (2009, March 1). Girsat ha-celeb o girsat ha-flop? [Hebrew: The celebrity version of the flop version?]. *Maariv-Nrg*. Retrieved May 31, 2009, from http://www.nrg.co.il/online/47/ART1/860/155.html

Mathijs, E. (2002). Big Brother and critical discourse. *Television and New Media, 3*(3), 311-322.

Oren, T. G. (2004). *Demon in the box: Jews, Arabs, politics, and culture in the making of Israeli television*. New-Brunswick, NJ: Rutgers University Press.

Poniewozik, J. (2003, February 17). Reality TV is good for us. *Time*, 65-67.

Noam, E. A., Groebel, J. & Gerbarg, D. (2003). *Internet television*. London: Routledge.

Taig, A. (2009, May 10). Ha-yotsrim le-televizia nilchamim al parnastam aval mamshichim leikashel bemivchan ha-rating [Hebrew: TV creatives fight for their jobs but keep on failing in the rating test]. . *Haaretz*. Retrieved May 31, 2009, from http://www.calcalist.co.il/marketing/articles/0,7340,L-3281487,00.html

Tessler, Y. (2009, April 9). Hakrav ha'acharon bein or le-choshech [Hebrew: Last battle between light and darkness]. *Maariv-Nrg*. Retrieved May 31, 2009, from http://www.nrg.co.il/online/1/ART1/877/340.html

Thompson, K. (1998). *Moral panics*. London: Routledge.

Tremlett, G. (2001, May 28). TV watchdog bites after Portuguese live sex and nudity. *Guardian*, 23.

SECTION IV: CROSS-CULTURAL STUDIES

In: Reality Television-Merging the Global and the Local
Editor: Amir Hetsroni, pp. 165-187

ISBN 978-1-62100-068-6
© 2010 Nova Science Publishers, Inc.

Chapter 11

PERFORMING THE NATION:
A CROSS-CULTURAL COMPARISON OF *IDOL* SHOWS
IN FOUR COUNTRIES

Oren Livio
University of Pennsylvania, USA

Reality television provides wonderfully fertile grounds for investigating contemporary processes of globalization and localization, as well as their implications for the modern-day nation-state, due to the genre's simultaneous reliance on both universal, cross-national formats and the particular, localized customization of these formats for specific national contexts. As noted by Darling-Wolf (in press), "the genre's attraction is predicated on its successful adaptation of global formats to local environments," rendering it "the perfect exemplar of twenty-first century capitalism at its best – in all its glocalized, deterritorialized, indigenized and disjunctive messiness." This study examines the tensions and ambivalences of globalization and localization in their practical manifestations within the context of one of the most popular televisual formats in recent years, the *Idol* singing competition – focusing on the complex ways in which four different versions of the show (from the United Kingdom, the United States, Canada, and Israel) adapt the universal formula to accommodate different local cultural nuances and project divergent imaginings of a shared national identity. Rather than attempting to resolve the longstanding debate regarding the relative dominance of the universalizing versus particularizing impulses associated with modernity, I more modestly aim to identify some of the ways in which the tensions and struggles that accompany processes of globalization and localization are materialized in concrete cultural practice, and, more importantly, the ways in which these materializations are closely related to specific power relations and to the market-based motivations of show producers, who strategically appeal to perceived local, national identities in order to achieve popular and commercial success.

THE CULTURAL IMPERIALISM/PLURALISM DEBATE

Globalization has been broadly defined by Giddens (1990, p. 64) as "the intensification of world-wide social relations which link distant localities in such a way that local happenings are shaped by events occurring many miles away and vice versa." This "compression of the world into a 'single place'" (Robertson, 1992, p. 6) involves a range of interrelated economic, geographical, and cultural phenomena, all of which bear significant implications for the nation-state (Giddens, 2000). While the effects of globalization have been investigated in a wide diversity of disciplines (see Nederveen Pieterse, 2004, pp. 14-21 for a summary), within communication and cultural studies the focus has traditionally been on the degree to which globalization as carried via a set of cultural practices and media products has superseded the nation-state and eroded local identities and traditions through the imposition of an allegedly unified, Westernized discourse of capitalism, consumerism, and individualism on various local cultures (Barker, 1997; Durham and Kellner, 2006; Sreberny-Mohammady, 1996).

In general, theoretical perspectives regarding globalization can be positioned along a continuum ranging from the complete determinacy of globalization, and thus of cultural imperialism, at one end, to a belief in the power of local cultures to resist globalization, and thus to exhibit cultural pluralism, on the other. The cultural imperialism position builds on more general, economically-focused and critical models of imperialism to argue that the unequal allocation of resources and Western domination over the global means of technology transfer have led to an increasing global cultural homogeneity due to the unidirectional flow of cultural contents from the West to the Third World (Schiller, 1976, 1985; Tomlinson, 1991). According to this perspective, Western, capitalist values are explicitly and implicitly conveyed by cultural products originating in the West and exported to defenseless indigenous cultures, thus forever altering their local identities and traditions (Hamelink, 1983). In its scholarly heyday, a variety of studies informed by this perspective convincingly documented the actual trajectory of cultural imperialism and its influence on phenomena such as news (Galtung and Ruge, 1965), popular culture (Dorfman and Mattelart, 1975), professional values (Golding, 1979), and organizational structures (Katz and Wedell, 1977). While more recent studies have demonstrated the limitations of this approach, there is still much evidence for the growing cultural standardization that is taking place around the world in many forms, "from clothes to food to music to film and television to architecture" (Tomlinson, 1999, p. 83). The global ubiquity of corporate brands and icons of mass culture has in fact rendered these synonymous with the phenomenon of (most often American) cultural imperialism, as indicated by the use of neologisms such as "McDonaldization" (Ritzer, 1993), "Coca-colonization" (Howes, 1996), "Disneyfication" (Chan, 2002), and the hybrid "McDisneyization" (Ritzer and Liska, 1997) to describe it.

While acknowledging the potential standardizing force of global cultural practices and products, critics of the homogenization perspective have argued that it largely overstates their actual impact (e.g., Arnason, 1990; Hannerz, 1987; Nederveen Pieterse, 2004; Robertson, 1992; Smith, 1990), for several reasons. First, the mere fact that cultural products have been disseminated on a global scale does not necessitate their presumed ideological effects (Tomlinson, 1999, pp. 83-84). As many studies have shown, the reception of cultural forms is a multifaceted process in which audiences actively interpret, translate, and adapt the texts they encounter employing their own, localized cultural resources (e.g., Ang, 1985; Gripsrud,

1995; Liebes and Katz, 1990). Moreover, the complexities of modern culture are such that the "original" texts themselves – even within mainstream programming – are often laden with ambiguity, with a variety of competing and often contradictory embedded discourses and values (Barker, 1997).

Second, some scholars have called into question the seemingly taken for granted and inevitable success of American cultural products abroad. While American television, cinema, and music have often met with success on the global scene, audiences in many countries have demonstrated a continued preference for local cultural products, as in the well-documented cases of Latin American *telenovelas* (e.g., Allen, 1995), Bollywood cinema (Sreberny-Mohammadi, 1996), Japanese music (Barnet and Cavanagh, 1994), or Israeli television quiz shows (Hetsroni, 2004). Some of these products, most notably Latin American *telenovelas*, have in fact been successfully exported internationally, complicating the issue of the assumed unidirectionality of global cultural flow (Barker, 1997; Croteau and Hoynes, 2000). Complicating this issue even further is the fact that in many countries, the most commercially successful programs are local productions of global (most commonly American) television formats, with the final texts thus being characterized by both global and local elements.

Finally, critics of the cultural imperialism position have provided persuasive evidence demonstrating that Western cultural content "does not annul or overpower particular local traditions but rather transforms them, and is transformed through them, via a process of mutual accommodation" (Illouz and John, 2003, p. 203). This suggests that rather than imposing cultural homogenization, global culture in fact breeds new and elaborate forms of cultural heterogeneity. As global cultural products enter new societies, they are often indigenized and adapted to fit local values, worldviews, and traditions – a process that has been referred to as "glocalization" (Escobar, 2001; Robertson, 1992) or "hybridization" (Nederveen Pieterse, 1994, 2004). Over the past two decades, a seemingly endless number of studies have documented the intricate means through which forms of global culture are enabled by local forms of social stratification and identity (see, for example, Azaryahu, 1999; Darling-Wolf, in press; Feld, 1988; Garcia Canclini, 1995; Hannerz, 1987, 1989; Hetsroni, 2004; Hetsroni and Bloch, 1999; Ivy, 1988; Maynard, 2003; Smart, 1999; Thussu, 1998; Volcic and Andrejevic, in press; Watson, 1997; Yoshimoto, 1989). This suggests, as Appadurai (2006) has noted, that "if a global cultural system is emerging, it is filled with ironies and resistances" (p. 586).

Responding to these arguments, proponents of the cultural imperialism perspective have successfully updated their position so as to incorporate the criticisms leveled at them. While recognizing that the actual flow of Western values to subordinate nations is more complicated than originally conceived, they have suggested that the terms "hybridization" and "glocalization" fail to adequately take into account the inequality of power relations inherent to hybrid cultural forms (Dasgupta, 2005; Friedman, 1999; Kraidy, 2002). Whereas hybridity has traditionally been theorized as a complex amalgamation that incorporates varying doses of dominant global and subversive local discourses, it in fact often reflects traditional hegemonic relations, in that hybrid cultural forms' local components are fabricated and manipulated by global media institutions (Dasgupta, 2005). Within this updated view, cultural hybridization is thus conceived not as a random or ideologically neutral pastiche of contradictory elements, but rather as a new, increasingly sophisticated form of the "elite gaze" (Friedman, 1999, p. 236; see also Croteau and Hoynes, 2000, pp. 352-354; Tomlinson, 1999).

The approach taken in this study builds on this view of cultural hybridization as neither homogenization nor heterogenization, exclusively, but rather as a mixture of both of these (Rantanen, 2005, p. 116), yet in ways that cannot avoid being influenced by the asymmetrical power relations that are inherent to global communication dynamics. This theoretical perspective on power was recently advanced by Kraidy and Murphy (2008, p. 351) in order to anchor "manifestations of power in concrete contexts, cultural codes, and social relations." It is such contexts, codes, and social relations that this study sets out to explore.

GLOBALIZATION AND REALITY TELEVISION

While the implications of globalization have been investigated with regard to a wide variety of cultural products and practices, there is little doubt among scholars as to the centrality of the mass media, and in particular television, in enabling the globalizing nature of modernity and in facilitating the construction of complex cultural identities encompassing both universalizing and particularizing components (Barker, 1997; Rantanen, 2005). Television at once "constitutes, and is a consequence of, the inherently globalizing nature of the institutions of modernity" (Barker, 1997, p. 13). The increased transnational flow of television programs from the West to the Third World since the 1970s has served as the starting point for many early examinations of the homogenizing effects of global media (e.g., Silj, 1988, pp. 22-58; Schiller, 1976; Wells, 1972), as well as for later critiques of the cultural imperialism thesis and theorizations of the resistant local audience (e.g., Ang, 1985; Liebes and Katz, 1990; Miller, 1992).

In recent years, a growing number of these analyses have focused on reality television, broadly defined as "programs where the unscripted behavior of 'ordinary people' is the focus of interest" (Bignell, 2005, p. 1), but in fact a rather elusive concept that encompasses a wide range of programs and one that is used by different people and different discourses to refer to diverse programming formats (Bignell, 2005; Biressi and Nunn, 2005; Brenton and Cohen, 2003). Reality TV has been particularly useful for studies of globalization (and its discontents) due to the fact that from the outset, it has been a global phenomenon, with broadcasters and format producers quick to capitalize on successful formulas by marketing them on the international scene, sometimes in their original form but most often in localized, culturally adapted inflections (other than the *Idol* franchise, similar examples include *Big Brother*, *Wife Swap*, *The Apprentice*, and *The Biggest Loser*, all of which have had several international incarnations; see Bignell, 2005).

While these transnational flows of homogeneous formats seem to promote a global agenda, however, internationally franchised reality TV is also infused with a strong sense of the local. Its contestants are always reflective of each country's native population; producers customarily tailor program formats to fit local tastes; and it constantly projects an image of democratic involvement that is inescapably associated with the political characteristics of the nation-state, due to the reliance on audience involvement and voting (Bignell, 2005; Darling-Wolf, in press; Holmes, 2004; Punathambekar, in press). Thus reality TV, with its complex negotiations of both universalizing and particularizing elements, serves as a convenient site for investigating how these negotiations are concretized, and for examining the frameworks and assumptions associated with globalization.

THE *IDOL* FRANCHISE IN THE UK, US, CANADA, AND ISRAEL

One of the most successful television formats in recent years, the *Idol* franchise – jointly owned by production company FremantleMedia and multimedia company 19 Entertainment – has at the time of writing been sold to over 40 countries and territories, and has been a resounding popular success in most (Bignell, 2005; Holmes, 2004). Essentially a repackaging of an old television formula, the audition/variety show, and incorporating elements from contemporary reality TV formats and flashy production techniques (Brenton and Cohen, 2003), *Idol* shows all over the world stage a singing competition between a number of previously anonymous contestants vying for the audience's votes and culminating in the hoped for, and often achieved, "development of stars, singles, and albums that have more of an autonomous existence outside the televisual text itself – as products of the pop music industry and as a highly visible element of popular music culture" (Holmes, 2004, p. 150).

Despite some differences between *Idol* shows around the world, the basic format of all franchised *Idol* programs is similar. In the first stage, auditions are held around the country in which a large number of potential contestants are screened and selected by a panel of judges, generally for their singing talent but occasionally for their humorous potential as performers to be ridiculed on prime time. In the second stage, the remaining contestants undergo additional auditions, at the end of which a group of finalists is selected. These contestants then compete every week, performing their versions of generally well-known songs and critiqued by a panel of celebrity judges, with advancement in the competition dependent upon receiving the viewing audience's votes. The competition inevitably culminates in an extravagant final show, in which the national *Idol* is crowned by viewers and receives various prizes – the most significant (and consistent across countries) of which is a recording contract with the sponsoring company. In this paper I focus on four *Idol*-type shows from four different countries:

- *Pop Idol* (UK) – Originally derived from reality pop programs in New Zealand (in 1999) and Australia (in 2000), *Pop Idol* is nevertheless widely recognized as the first program in the *Idol* franchise (Holmes, 2004). Receiving its launch on British network ITV1 in 2001, *Pop Idol*'s first season was an astounding success, with over 13 million viewers (a 57% audience share) watching the season finale and over nine million votes cast by viewers over two hours (Dann, 2003; Holmes, 2004). A significant decline in ratings over the subsequent season, however, coupled with the departure of the show's star judge, Simon Cowell, to the American spin-off, led to the show's being suspended indefinitely following the second season.[1] Nevertheless, several of the contestants on the show, including both winners, have achieved great success on the charts and become household names in the UK.

- *American Idol* (US) – Broadcast on the Fox network since its 2002 debut and one of the top-rated shows in the history of American television, in terms of its continued success, *American Idol* has been the longest running and, arguably, the most successful of the *Idol* shows (Dann, 2003; Stern, 2004). Unlike its British predecessor, *American Idol* succeeded in increasing its viewership (as well as the

---

[1] In late 2006, after ITV's license to produce *Pop Idol* had expired, talks concerning a relaunching of the show on a different channel were announced (West, 2007). At the time of writing, however, a revival has not materialized.

number of votes cast by the audience) every season up to its fifth season, during which the show averaged 31.1 million viewers per week (Shaw, 2006). Since season 6, the ratings have shown a steady if minor decline in viewership. Most winners of the show, as well as several other contestants, have gone on to enjoy successful careers, mostly in singing and acting.

- *Canadian Idol* (Canada) – Like its American counterpart, CTV's *Canadian Idol*, launched in 2003, was highly successful in its first few seasons, averaging over two million viewers per show and over three million votes per week, thus making it, at the time, the most consistently watched English Canadian television program since electronic measurement was introduced (Canadian Idol Audience Surge, 2004). In recent years it has seen something of a decline, and in December 2008, CTV announced that the show would be temporarily suspended for the 2009 season. Several winners of the show have released well accepted hit singles and highly successful albums.

- *A Star Is Born* (*Kokhav Nolad*) (Israel) – While not officially licensed as part of the *Idol* franchise, even Israeli producers of the show acknowledge that it is essentially a local version of the popular format (How Is a Star Born, 2006). The program, broadcast on Israel's commercial Channel 2, has consistently been the most highly rated show on Israeli television since its launching in 2003, and has won a number of popular awards, including 2003's "TV program of the decade" award voted by the readers of Israel's most popular entertainment weekly, Pnai Plus. Nearly all winners of the show, and several additional contestants, have gone on to have successful musical and acting careers, with some – most notably the winner of the first season, Ninet Tayeb – achieving the status of cultural icons and superstars (Libsky, 2005).

METHOD

The analytical focus of this study is a discursive analysis of the four aforementioned *Idol* shows. The sample included: (a) the two complete seasons of the British *Pop Idol*, originally broadcast from 2001 to 2003; (b) seasons 1 through 5 of *American Idol*, originally broadcast from 2002 to 2006; (c) seasons 1 through 3 of *Canadian Idol*, originally broadcast from 2003 to 2005; (d) seasons 1 through 6 of the Israeli *A Star Is Born*, originally broadcast from 2003 to 2008.[2] In addition, and in order to provide an additional common baseline for comparing the four shows, official compilations summarizing specific *Idol* seasons released on DVD or VHS tape in three countries were also viewed.[3]

Obviously, in any analysis many different representations may be brought to the forefront, and the *Idol* case is no exception. In this case, my interest in the *Idol* franchise as a vehicle through which to examine questions of globalization and localization led to a focus on

---

[2] While complete seasons of the shows were obtained, on a few rare occasions certain episodes were missing or technically unwatchable. These rare exclusions are insignificant with regard to the ability to make useful analytical generalizations.

[3] These included – from the UK: *Pop Idol: A Star Is Born* (season 1 compilation VHS tape) and *Pop Idol: Living the Dream* (season 2 DVD); from the US: *The Best and Worst of American Idol, Seasons 1-4* (three-DVD box set) and *American Idol: The Search for a Superstar* (season 1 DVD); from Israel: *A Star Is Born: We Will Not Stop Singing* (season 1 DVD) and *A Star Is Born: The Greatest Moments* (season 2 DVD).

the construction of national identity and the imagined community of the nation, at the expense of other, and certainly no less important, features. This said, the notion of national identity in itself is quite nebulous (see Billig, 1995, p. 7), and what falls under its rubric may obviously be contested. To facilitate the demarcation of the scope of the analysis, I thus employed the general discursive framework suggested by Billig (1995), which focuses on the routine and banal ways in which nationhood is constantly flagged in everyday features of contemporary life, including popular culture and the mass media. Significantly, Billig's approach does not assume that the discourse of nationhood can ever be unified or coherent, but rather that it always embraces divergent discourses within its public manifestations. Thus national identity – while consistently flagged and occasionally flaunted – is constantly engaged in a struggle with various other discourses, including those of a more global, universalizing nature (see also Darling-Wolf, in press). The four *Idol*s discussed should thus not be seen as essentialist, mutually exclusive expressions of unique and cohesive national identities, but rather as complex sign systems in which traces of competing and sometimes contradictory elements can be located. The following analysis looks at several of the discursive strategies through which these elements are articulated.

FLAGGING THE NATION: REIFICATION AND INTEGRITY

Hetsroni (2004, 2005) has noted how changes in the titles of TV shows may reflect the attempts made by producers to cater them to local tastes. If its title is anything to go by, the original, British, *Pop Idol* did not consciously set out to situate itself within the discourse of national identity (Cowell, 2003). It was only following its transatlantic journey that *Pop* became *American* and *Canadian*, thus explicitly projecting the notion of a unified, coherent national identity. This does not mean, of course, that the British show was necessarily any less invested in the projection of such an identity, but rather that with no existing other to distinguish itself from, this process was rendered less explicit.[4] In the US and Canadian versions, however, the nation is constantly imagined – and articulated – as one integrated whole, and nowhere is this more evident than in the constant address to the nation as one reified entity. In *American Idol*, host Ryan Seacrest, the judges, and the contestants themselves repeatedly invoke the unified, mythic nation in statements such as "Well done, America," "Thank you, America," and "I'd like to congratulate America for getting it absolutely right." The same occurs in Canada, as illustrated by judge Zack Werner's statement, "Here's where Canada and I disagree" or the standard plea preceding commercial breaks, "Don't go anywhere, Canada." Interestingly, this type of reification is completely absent from the British version, where the viewing public is simply addressed as "You."[5]

In both the American and Canadian *Idol*s, the reification of the nation serves a similar purpose, that of projecting an image of solidarity. When considered in conjunction with other discursive markers of national identity, however, it becomes clear that the two cases exhibit

---

[4] Interestingly, even the shows' introductory title sequences make this difference apparent. While the same computer-generated, androgynous idol figure appears in the British, American, and Canadian versions, the US intro replaces the airplanes in the British version with waving American flags, and in the Canadian intro, the idol appears against the background of iconic structures of the national landscape.

[5] Since *Pop Idol* was discontinued after its second season, one can only imagine how it might have repackaged itself in light of the revisions made by its transatlantic successors.

slightly different inflections of what this solidarity means. In the US, the need for solidarity specifically addresses a post 9/11, War in Iraq context. As such, it is aligned with a nationalistic, militaristic, but essentially self-gazing brand of patriotism: "America" as defined against itself and its internal unrest, and one whose main concern is projecting *political* unity or at least mutual political interests among its citizenry. *American Idol* took special pride in its explicit flag-waving, with members of the Armed Forces occasionally featured among the show's contestants and their sacrifice for the nation underscored. Thus, for example, during the show's second season (in which one of the contestants was a member of the US Marines), host Ryan Seacrest proclaimed "a special announcement for America," declaring that a charity single would be released featuring a group performance by *Idol* contestants of the songs "I'm Proud to Be an American" and "God Bless the USA." In the background, an image of President George W. Bush appeared requesting that "God continue to bless America."

In contrast, the Canadian brand of solidarity as projected on its local show is fundamentally *cultural* and "other-gazing": "Canada" as defined against (and distinguished from) its neighbor to the south, and in light of its continuous search for a distinct cultural identity. While an attempt is made to portray this identity as extremely multicultural and inclusive of all Canadians, with the contestants' ethnic backgrounds often mentioned, as projected on *Idol* this identity is almost exclusively English Canadian. While the show's producers make sure to hold auditions in Montreal every year, very few contestants from Quebec have made it to the final round. *Canadian Idol* is not broadcast in French, nor do contestants perform French language songs. It is also far less successful commercially in Quebec than in any of the other provinces.[6] Yet the show's producers, hosts, and judges constantly attempt to sidestep this problematic by viewing the presence of some French Quebecers on the show as evidence of its all-encompassing inclusiveness. When Quebecer Eva Avila was crowned the winner of *Canadian Idol*'s fourth season, an executive producer of the show utilized her success to claim that "It's clear that Canadians of all ages from all backgrounds and all regions have embraced *Canadian Idol*. (…) This is truly a national program" (CTV Media Site, 2006).

While *American Idol* and *Pop Idol* contestants consider English language songs to be the default, and these are performed by the contestants with no reference to their national origin,[7] *Canadian Idol* (as well as the Israeli *A Star Is Born*) takes particular pride in advertising the unique local character of its music – even when the nature of that uniqueness remains extremely vague – and theme shows often focus on Canadian artists.[8] In one show, songs by Canadian artists were even referred to as "Canadian anthems." As judge Farley Flex explained to a contestant on a show featuring the music of Canadian rock group The Barenaked Ladies:

---

[6] As one CTV executive has noted, *Canadian Idol* has more viewers in Newfoundland than in Quebec, despite the fact that Quebec's population is approximately 14 times larger.

[7] Overall, however, the diversity of songs performed as well as the divergent stylistic musical backgrounds of the contestants on both *Pop Idol* and *American Idol* are certainly meant to reflect and demarcate the limits of each country's cultural identity.

[8] At the same time, other theme shows highlight the songs of internationally renowned singers with no Canadian connections, such as Elvis Presley or Stevie Wonder. This illustrates the constant tension between the global and the local, but also serves as a means for aggrandizing the local, since successful local artists are put on par with global superstars.

A lot of times when people talk about the identity of Canadian music, one of my references has always been the Barenaked Ladies as being a unique Canadian band, and I think that's the direction you're heading in as well. I think you're gonna be identifiably a Canadian rock 'n' roll artist.

Paradoxically, however, this nebulous local identity is forever in need of a global certificate of approval, for success is always defined in terms of international, rather than domestic, appeal. Thus, Canadian artists are introduced using references to the fact that they have made "a massive impact on the world stage," "turned homegrown appreciation into world domination," or "struck gold in the US." Similarly, in the Israeli *A Star Is Born*, first season runner-up Shiri Maimon's success in the pan-European Eurovision Song Contest was constantly evoked. In the UK, and even more so in the US, this is unnecessary; local and global success are essentially equated, with the default presumption being that any star is a global star.[9]

The Israeli version's articulation and performance of national unity deserves special attention. Interestingly, this is not normally accomplished through the iteration of the country's name. The Israeli title does not mention the nation by name,[10] and explicit addresses to the audience rarely make use of it. Instead, the program projects unity through its unique format and accompanying discourse. As aforementioned, *A Star Is Born* is not officially part of the *Idol* franchise. Rather, it was developed as a hybrid between *Idol* and a previous program, *Lo Nafsik Lashir* ("We Will Not Stop Singing"), which had been quite popular. In fact, throughout its first season the program even included its predecessor's title and was officially called *Lo Nafsik Lashir – Kokhav Nolad* ("We Will Not Stop Singing – A Star Is Born"), though the first phrase was dropped in subsequent seasons.

While generally similar to the *Idol* format, the composite program retained certain components of the original *Lo Nafsik Lashir* as well. *Lo Nafsik Lashir* was effectively not a competition at all, but rather a singing program that invoked the Jewish Israeli tradition of *shira betzibur* ("singing together") – a ritual of singing nationalistically-tinged folk songs in unison, which in the nation's formative years had been an integral part of the education system and a central cultural symbol of the nation-building enterprise (Regev and Seroussi, 2004). Its renewed public popularity in recent years (after years in which it had been largely ridiculed) was seen as a reaction to the feelings of national threat experienced by many Israelis following the outbreak of the second Palestinian Intifada in September 2000, and as a nostalgic desire for a national unity that was feared to have been lost. However, although allegedly a unifying enterprise, *shira betzibur* had in fact always been highly exclusive and identified solely with the music of the then dominant *Ashkenazi* ethnic group – Jews who had immigrated to Israel from European countries. The more oriental music created by *Mizrahi* Jews from Arab countries (*musica mizrahit*) was largely excluded from the canon. The

---

[9] Billig (1995, pp. 150-151) illustrates this pattern in the case of sports, where local American teams are crowned "World Champions" after winning competitions with titles such as the "World Series." Despite the fact that these are in fact local competitions, symbolically they transmit an ethos of global domination. As Billig (1995) notes, "It is a cultural pattern which well fits a nation seeking world hegemony" (p. 151).

[10] The Israeli title also does not include the term *Idol*. While this is partly due to its being an unlicensed version, it is also likely that this is a result of the highly loaded connotations of the term in Jewish culture, where it is associated with the unforgivable sin of idolatry. *A Star Is Born* explicitly attempts to project an image of Jewish unity, as discussed in this section, and the use of such a term might potentially alienate Israel's sizable religious population.

invocation of the *shira betzibur* ritual in *A Star Is Born* thus reflects an explicit political agenda of unification, albeit one that has traditionally been associated with oppressive Ashkenazi hegemony.

A Star Is Born deals with this loaded history by preserving the unifying mission of *shira betzibur* while extending the range of music it encompasses so as to include songs that are associated in style with that of Mizrahi Jews, and thus "proving," so to speak, that ethnic discrimination in Israel is a thing of the past. The result is an explicit performance of Jewish unity, with the producers taking particular pleasure in encouraging Ashkenazi performers to select Mizrahi songs, and Mizrahi performers to select Ashkenazi songs.[11] The fact that most winners of the show thus far have been Mizrahi is often mentioned, particularly when concerns over continued ethnic discrimination are voiced in public discourse.

A Star Is Born is the most explicit of the four shows examined in this study with regard to its aims at inclusiveness and proportional ethnic representation. While issues of class and (mostly) race have also surfaced in the British, American, and Canadian *Idol* shows, with a few exceptions such discussions have mostly taken place outside of the actual shows (e.g., McKinney, 2006; Showalter, 2003; Thrupkaew, 2004), and their manifestations on the show are often implicit or toned down. The issue of representativeness on *American Idol* is relatively unspoken – constantly hovering in the background, always present in the selection of contestants and musical styles, but rarely brought up directly.[12] On *Canadian Idol* it is more overt, particularly with regard to immigrant-related markers of identity, but less so with regard to race, and in any case dealt with caution.[13] Conversely, the Israeli version is quite blunt, with the judges often explicitly referring to the contestants' ethnic origin, usually with the goal of demonstrating how the show represents Israeli diversity and a joint commitment to the common cause.[14] As judge Ronny Braun stated in season 1, "We have here immigrants from Russia and someone from Georgia and someone from Iran and people from Yemen. We have here all the people of Israel." The Israeli show also parades its aspirations for representativeness by consistently including soldiers, immigrants from Ethiopia, and observant orthodox Jews among its cadre.

This constant flaunting of the nation's unity in diversity renders one omission exceptionally glaring: that of Israel's Arab citizens, which comprise approximately one-fifth of the population. Over the show's first seven seasons, only one Arab has progressed beyond the preliminary auditions and made it into the show's lineup, in which 145 contestants have participated thus far.[15] While constantly invoking the notion of a united *Israel*, therefore, what is articulated in practice is a united *Jewish*, Hebrew culture. This is hardly irregular in Israeli

---

[11] Significantly, however, Mizrahi songs are generally of the hybridized, toned-down variant that has become part of the Israeli mainstream, whereas purer forms of *musica mizrahit* are still largely absent from the show.

[12] Discussions of racism on *American Idol* certainly surface occasionally, but normally this occurs outside of the discourse of the show itself. Thus, for example, Elton John, who had served as a guest judge in a season 3 episode dedicated to his songs, labeled the *Idol* audience "incredibly racist" for its voting tendencies. However, this criticism was voiced more than a week after his appearance on the show.

[13] Issues of gender and sexuality, while extremely important, are regrettably left out of the discussion here, as they are extremely complex and more ambiguously linked to the construction of national identity as it occurs on the shows.

[14] This is also probably reflective of the Israeli tendency towards expressing thoughts with directness, frankness, and little regard for civility. See Katriel (1986) for elaboration on this topic.

[15] Interestingly, this is not reflective of all Israeli reality television. Arabs have in fact been winners on shows less explicitly engaged with questions of national cultural identity, such as an Israeli variant on *Big Brother* and the Israeli version of *The Next Top Model*.

culture (see Avraham, 2003), but in light of the explicit and uninhibited discourse of national identity on *A Star Is Born*, it is particularly blatant. The Israeli *Idol* show, straightforwardly put, excludes the Arab population and renders its voice virtually inaudible, as though it simply did not exist. Unlike the explicit discourse with regard to other ethnic groups, the absence of Arabs is normally not referred to even implicitly. The view of Israel as being exclusively Jewish thus becomes the common sense, that which is so obvious it does not even need to be spoken of. When an Arab contestant finally did appear on *A Star Is Born* in its fifth season,[16] the judges repeatedly emphasized the fact that she had been selected solely for her singing talents, thus attempting to project a form of ethnic-blindness that often verged on the ridiculous. This attempt to negate the contestant's national identity reached its culmination after she was eliminated, with one of the judges rising from his seat and screaming, repetitively, "We like her only because of how she sings!"

The reproduction of national identity always involves an imagining of a shared place, or homeland, along with the demarcation of that place's spatial boundaries (Billig, 1995, pp. 74-78). The virtual exclusion of Arabs from *A Star Is Born* is closely related to the tension that exists in Israeli society with regard to the extent of the collective's borders (see Rosen-Zvi, 2004): rather than advancing a conception of territorial, political space, that is governed by the country's borders (and thus necessarily inclusive of Arab citizens), *A Star Is Born* promotes a perception of tribal space that is guided by Jewish ethnicity. In the British, American, and Canadian versions of *Idol*, it is quite clear that eligibility to vote is contingent upon being present within the country's political borders. In Israel, on the other hand, not only is voting from abroad tolerated, but it was in fact encouraged by including audible comments from residents abroad detailing who they had voted for in the show's first season.[17] Moreover, in *A Star Is Born*'s fourth season, not only the right to vote, but even the right to *compete* was extended to Jews all over the world, rather than being limited to Israeli citizens. With support from the Jewish Agency for Israel, an organization whose goals are to forge closer ties between Jewish communities abroad and Israel and to encourage Jewish immigration to Israel, the show's producers conducted auditions for Jewish contestants in their home communities abroad, with the eventual winner being flown in to Israel to participate in the competition. In later seasons, this opportunity was extended beyond Israel's borders to both Jews without Israeli citizenship and to Israeli citizens residing abroad, thus blurring the porous borders of the collective community to an even greater extent. To sum up, while national identity in the British, American, and Canadian *Idol*s is closely tied to a sense of geographical integrity and linked with a particular physical space, in Israel it is much more amorphous, encompassing both political-territorial and tribal-diasporic components.

Idol shows are, by definition, individualistic. The competition is between individuals, and the ultimate goal is the selection of one star. At the same time, this sense of individualism is in tension with the contradictory impulse towards collective identity, exhibited both on a micro-level scale, with the contestants constantly referred to (and describing themselves) as

---

[16] This contestant, Miriam Tukan, was a member of Israel's minority Christian Arab population, which comprises only nine percent of the Arab population and is generally considered by the Jewish population to be less threatening than the majority Muslim population. Thus, it appeared that even when the Arab population was represented, an attempt was made to do so in ways that were least controversial to Jewish eyes.

[17] Intriguingly, this arrangement is the exact opposite of the four countries' conceptions of political democracy as illustrated by election laws. British, American, and Canadian citizens can vote in national elections from abroad; Israeli citizens cannot.

being "one happy family," and on the macro-level of the nation, with the persistent flagging of shared components and characteristics of an imagined national identity. This tension mirrors, to a large extent, that between the local and the global, and the four *Idol* shows unsurprisingly differ in their respective negotiations of it. Once again, while small differences can be observed between all four shows, the most obvious discrepancy is between the Israeli version and the others.

In the British, American, and Canadian *Idol*s, contestants sing alone. Group performances are limited to certain early audition stages and festive, non-competitive later stages. Moreover, individuality, originality, and uniqueness are constantly invoked, with contestants being urged to "make the song their own" and "take risks." When asked what the show set out to discover, *American Idol* judge Randy Jackson replied:

> Hopefully we'll find a unique voice that can have like a long long career that we can all stand and say, you know what? They don't sound like anybody else, they got their own style, their own vibe, their own vocal talent, they got their own "it" thing.

Thus, while celebrating the close-knit relations between the contestants, these three *Idol* shows nevertheless articulate the conviction that individuality is ultimately superior. This discourse is obviously not absent from the Israeli *A Star Is Born*, which is also a competition among individuals, but the show's producers do virtually everything in their power to disguise this characteristic, promoting instead – as already demonstrated in other aspects – a sense of collective identity. On *A Star Is Born*, contestants must work together all the time. In the show's first three seasons, each competing contestant was backed by two of the other contestants in every round, and in later seasons, contestants sang duets every week – thereby rendering the competition a group project (in the season finale, up to five eliminated contestants sang with the finalists).

The implications of losing on *A Star Is Born* are also very different: eliminated contestants always receive prizes as well, and the losing finalists (and sometimes even contestants eliminated before the final) normally receive their own recording contracts. Finally, the atmosphere of collectiveness is not limited to the contestants, but includes various other players, as exhibited in certain spatial features. Unlike the British and North American *Idol*s, in Israel the backing band is part of the show, with its members present on stage, introduced individually every week, and taking part in various scripted scenes; the contestants' families are often invited onto the stage during the reading of the results; and the audience at home is also seen as part of the show: during songs, karaoke-style lyrics are often flashed on the bottom of the screen, inviting the viewers to sing along.[18] Overall, unlike the spirit of competition that dominates its global counterparts, *A Star Is Born* touts itself as "the happiest show on television," a celebratory, saccharine carnival of imagined solidarity, unity, and kinship. In a way, the show's producers almost seem disappointed that they must ultimately declare a winner.

---

[18] This of course also stems from the tradition of *shira betzibur*, as discussed earlier.

UTILIZING THE JUDGES: RESISTANCE AND DEFERENCE

Judges play a central role in all *Idol* shows. As the first hurdle the contestants must navigate in order to reach the final stages of the competition, their impact is obvious. In their continuous explicit and implicit labeling of potential contestants, the judges are engaged in defining the perceived limits of legitimate national identity from the outset. It is once the show reaches its post-audition stages, however, that the judges – and the uses the producers make of them – become even more subtly invested in the construction of national identity.

Both spatially and discursively, all *Idol* shows are constructed as a battle between the contestants and the judges, with the host and the audience on the side of the contestants, and one judge in particular being singled out as the meanest of the bunch. Rhetorically, the judges are often referenced in a cheekingly insulting manner, as when host Ant (Anthony McPartlin) remarked on *Pop Idol*: "We were looking for the most challenging musical minds we could find. Unfortunately all we could afford was the most musically challenged minds in the country." The hosts are in charge of countering any criticism directed by the judges toward the contestants, and thus defusing any bad vibes or redirecting them toward the judges themselves (see Smith, 2004). This routine often involves a direct interpellation of the audience, explicitly positioning it on the side of the contestants and against the judges. In phrases such as "The judges don't decide, America does," or "it's not up to you, it's up to Canada," the audience is at once presented with the promise of both agency and resistance (see Holmes, 2004, p. 168, for similar statements from the British version), unlike other reality shows where agency is constructed as relatively neutral (see Tincknell and Raghuram, 2002). The judges themselves cheerfully partake in this construction. As judge Simon Cowell said on *American Idol* when a contestant he had ridiculed was voted through by the audience: "This is what makes this competition so fantastic: the public voted against bigmouth."

While this construction of social relations is similar in all *Idol* shows (naturally, to somewhat different degrees), where the shows take divergent paths has to do with the role of the contestants themselves in these relationships. This is of course extremely important, because it is with the contestants that the viewers are meant to identify. The participants' reactions to criticism by the judges thus signal what behavior is ultimately desired. Obviously, the argument here is not that all contestants within each country respond in the exact same way, but rather that general patterns can be recognized, and – more importantly – that these patterns are encouraged by the producers through a complex structure of foregrounding.

The British and American versions of *Idol* clearly endorse resistant behavior by contestants, which is constructed as a symbol of individuality. Defiant responses are played over and over again, as are the live audience cheers and the hosts' encouraging comments. Downright impertinence is also acceptable, as long as it is done sassily. As contestant Darius Danesh addressed Simon Cowell on *Pop Idol*: "Simon, I think with your waistband being so high you might want to undo it a notch, because it might be restricting the blood flow to your head."

This type of response is noticeably absent from *Canadian Idol*, where contestants repeatedly thank the judges even in the face of the harshest criticism. As research has shown, Canadians highly value politeness and consider it to be one of their defining characteristics (Baer, Curtis and Johnston, 1996). On the average, Canadians are also less individualistic and less achievement oriented than Americans, and display a higher regard for deference to authority (Friedenberg, 1980; Lipset, 1990). A defiant contestant might run the risk of being seen as un-Canadian, thus alienating the viewers and severing the constructed tie (and similarity) between the audience and the contestants. On *Canadian Idol*, it is thus normally left only to the host to respond to the judges, and this is often done through playful banter, similarly to the American and British cases.

In the Israeli *A Star Is Born*, the celebratory atmosphere of national unity dictates that the judges' role as critics be minimized. In the show's first season, the judges did not even pass judgment on performances in most shows, and while in subsequent seasons they did, this was often done after public voting had already taken place, thereby rendering any potential clash with the audience largely irrelevant.[19] During its first season, *A Star Is Born* did not even have a malicious, Simon Cowell-type judge, and the festive mood was rarely marred by any overly acerbic comments. This feature was added in the show's second season, but public objections to this judge's nastiness led to an obvious wane in the severity of criticism and to her eventual replacement. Thus Israeli contestants did not usually have the opportunity to respond to harsh criticism – such was simply not leveled at them. At the same time, it must be acknowledged that the Israeli tendency toward outspoken, forthright speech (see Katriel, 1986) sometimes did make itself manifest in comments by both judges and contestants, although this was not a common occurrence.

The presence of Simon Cowell on both *Pop Idol* and *American Idol* enables an intriguing comparison with regard to the role he plays on the two shows. In both versions, Cowell is constructed as the brutal, in-your-face judge who "says it like it is." In the UK, this is simply seen to be a personal character trait, bearing little relation to any constructed national identity. When fellow judge Pete Waterman expressed his reservations over Cowell's style, these were framed as personal: "I don't mind your opinion, but you say it in such a way sometimes." In the US, however, Cowell's meanness is constructed as the British "other" against which Americanness is reverse-mirrored. Cowell's British identity is constantly referenced in various ways, from the recurring mockery (and imitation) of his accent, through random statements invoking his foreignness,[20] to explicit discussion of how his allegedly "British" qualities are different from, and inferior to, the qualities associated with the American national identity. In an emblematic moment on *American Idol*'s first season, judge Randy Jackson actually got up from his seat, towered over Cowell and challenged him to a fight after the British judge had called a contestant a "loser." Constructing Americanness through its discursive other, Jackson exclaimed: "Simon, you can't call people losers. (...) This is America, we don't do this to people, we don't insult people. (...) Dude, you're in America now. Dude, you can't do this." In another characteristic scene that was shown repeatedly, a contestant who had been lambasted by Cowell responded: "You know what, at least I'm from a country where people brush their teeth twice a day." And Cowell himself partakes in his own othering, humouredly recognizing its local cultural import. In explaining his relatively placid remarks to one contestant, Cowell stated: "If I'd said what I wanted to say, I think I might be thrown out of America." Given Cowell's constructed identity, the contestants' cheeky rejoinders therefore become not only expressions of individuality, but also a symbolic

---

[19] This pattern eventually changed, with judges' comments preceding the audience vote, in line with the other *Idol* shows.

[20] A prime example of this was judge Paula Abdul's comment on Cowell, "He's British, he's crazy, he's got mad cow disease."

means of asserting a distinctly American identity, and one which the local audience can thus identify with.

SUMMONING DEMOCRACY: CIVIC PARTICIPATION AND SOCIAL MOBILITY

Positioning the audience and judges as adversaries and interpellating the viewers as voters cater to one of the features most frequently associated with reality TV, the invocation of democracy at work (e.g., Bignell, 2005; Biressi and Nunn, 2005; Punathambekar, in press). Public and academic discussion surrounding *Idol* shows has often focused on this aspect, either celebrating its participatory nature (e.g., Cornfield, 2004; Fritz, 2002; McKinney, 2006; Paskoff, 2003; van Zoonen, 2004) or cautioning against its illusory democratic pretensions (e.g., Price, 2003). The shows consistently summon this type of discourse, overtly equating voting for contestants with democratic participation ("you have the power to choose"), employing election metaphors, and constructing actual voting as a courageous and committed act of civic participation and responsibility. As the host of *A Star Is Born* made clear in addressing the audience in one episode:

> You are the ones who have to crown one of our incredible talents. The voting booths have been open since 8:00 AM, and the voting is going on. All of you are registered to vote, and the ballot boxes are located in the most convenient spot: in your own living room at home. So use your democratic right and make an influence.

Audiences are "rewarded" for voting not only by seeing their favorite win, but also by being implicitly addressed, and thanked, by the winner through the lyrics of the winning song, written especially for the competition. In all *Idol* shows but the Israeli version (where no such song is composed) this song cleverly functions at once as both a traditional love song, addressed to a romantic partner, and as an expression of gratitude to the viewers for their support. Thus, for example, *Pop Idol*'s original winning song, "Anything Is Possible," proclaimed:

> Your love's made me see that anything is possible.
> Possible cause you believe in me.
> I never believed it that a dream could come true,
> But if anyone has changed my mind, then baby, it's you.

As these lyrics (and the song title) demonstrate, the notion of democratic participation is closely related the Horatio Alger, rags-to-riches myth, or to what Dyer (1998) has termed the democratic success myth – the idea that "society is sufficiently open for anyone to get to the top regardless of rank" (p. 42). As Dyer notes, this perception of social mobility incorporates and reconciles several contradictory elements: the ordinariness of the star, the conviction that the system rewards talent, the belief that lucky breaks can happen to anyone, and the insistence that hard work and professionalism are necessary for stardom. These elements are all blended in the discourse of all *Idol* shows, albeit in doses that vary by culture. In addition, the definition of what social mobility in fact means also differs from country to country.

In *American Idol*, the discourse is one of extremes, with the emphasis placed on the tremendous discrepancy between the contestants' humble beginnings and their eventual lives as superstars. Introductory lead-ins state: "Right now they could be parking cars or even waiting on tables. (...) By the end of the summer that person's life will change forever," and contestants selected to "go to Hollywood" – the location of the final auditions, but more importantly a symbolic representation of the myth of stardom[21] – routinely admit that they've never even been on a plane before. The emphasis on the contestants' initial ordinariness deliberately blurs the boundaries between them and the viewers, thereby invoking the audience's own aspirations of success (Holmes, 2004). Thus each and every viewer may believe that he or she, too, might some day experience the transition "from kids next door to overnight celebrities, from modest homes to a multi-million dollar mansion."

As this account demonstrates, the American version of mobility is deeply associated with notions of both fame and financial success, as well as the lifestyles that these concepts are associated with. The show is constructed as a liminal, transitional stage where contestants get to experience their potential future: living in luxurious locations, driving fancy cars, shopping in Beverly Hills, going to events, and mingling with celebrities. When asked what the winner would get out of the competition, Simon Cowell responded: "What they always wanted: fame, stardom, a ton of money. It's what it's all about." Fame, fortune, and celebrity status are thus constructed as obligatory American ambitions; the possibility that one would not desire them is never raised, and the contestants themselves repeatedly mention them. At the same time, the contestants' professionalism and down to earth roots are never forgotten, as it is these qualities that guarantee their authenticity, downplay the negative connotations of fame, and render them accessible to the audience (Holmes, 2004).

Traces of the same types of discourses can be found in all other *Idol*s, but their degrees greatly differ. In the UK and in Israel, money is seldom mentioned; rather, the winner is merely promised a life of fame, with the show serving as the vehicle. As in the US version, the contestants get to meet and work with celebrity artists, but they are much less involved in consumer activities. Their ordinariness is likewise emphasized, and modesty is often preached. As judge Pete Waterman explained on *Pop Idol*, conveying the stardom dialectic: "The superstar will be the most reluctant part of this show. The people that show off will not be the superstar." In Israel, an oft-broadcast scene showed one of the contestants declaring that his dream was simply to go have lunch with a famous Israeli singer as a matter of course, demonstrating the emphasis on celebrity status rather than money. Similarly, the theme song that accompanied season 1 of *A Star Is Born* was a well-known Israeli oldie, "Illusions," which treats money and fame with a combination of hope and disillusionment – the verses referring to one's dreams of "living in a luxurious apartment, believing you're a millionaire, waking up in the afternoon next to a girl or two, never having to work, never serving anyone, holding your head up in the clouds, traveling to Paris or Rome, living like Casanova," only to be countered by the cynical chorus, which proclaims that these are all "illusions that may come true only in one's dreams."

In Canada, both money and fame are dealt with uncomfortably. This seems to confirm research findings that Canadians highly value hard work and thrift, rather than excessiveness (Baer, Curtis and Johnston, 1996). Discourses of social mobility are still evident, as when a

---

[21] The British, Canadian, and Israeli versions obviously lack such a resonant cultural myth. In these competitions, contestants merely "go through to the next round."

season 3 finalist's past as a part-time auto mechanic was repeatedly mentioned, or when contestant Casey LeBlanc was referred to as "a small town girl with big city dreams," but these dreams are normally expressed as rather low-key. When asked why they want to become Idols, contestants give responses such as "I hope to grow as a musician" or "To be a role model for kids." The discomfort of explicitly addressing money and fame is often punctured by the use of humor. Thus, when asked why it was important to win the contest, one of the hosts replied: "Because it sure beats losing."

CONCLUSION: COMMERCIAL NATIONALISM IN PRACTICE

Any comparison between texts from different countries runs the risk of bogging down to an expression of cultural stereotyping and generalizations, and the analysis presented in this study is no exception. This is certainly not its intention. The texts of the four *Idol* shows examined demonstrate that local national identity is a complicated and often ambiguous structure, which often involves the simultaneous articulation of contrary elements both global and local, universalizing and particularizing. More importantly, however, the complex structuring, foregrounding, and backgrounding strategies employed by the shows' producers demonstrate that the cultural differences between the four versions are not the inevitable result of some innate national characteristic representative of each culture. While some differences undoubtedly reflect real cultural variation, most are more reflective of the strategic and intentional customization by show producers to cater to what they perceive to be local tastes. Neither globalization nor localization is an agentless process. Rather, each is a "social and cultural practice, implemented by actors, with intentions, motivations, and goals" (Illouz and John, 2003, p. 201). In this case, the producers' main motivation is commercial.

Idol shows in all four countries examined in this study are commercial enterprises, with the ultimate goal of appealing to audiences and advertisers and selling related products (see Stern, 2004). The fact that they are so sensitively tailored – often changing style and format from season to season and even within seasons in light of perceived public opinion – demonstrates that producers in different countries feel that different projections of national identity may be more effective for achieving their similar commercial goals. In this respect, the discourses of the shows are representative of what Seo (2008) and Volcic and Andrejevic (in press) have referred to as commercial nationalism – the ways in which nationalist appeals migrate from the realm of political propaganda to commercial appeal. The projection of a unique national identity by each *Idol* show, and the marketing of this identity by show producers in ways that appeal to the respective national viewing audiences, lead to both identification and – more importantly from the producers' point of view – consumption, both of the televisual text itself and of its associated products, which include voting for those contestants whose national appeal is most successful. As Volcic and Andrejevic (in press) note, commercial nationalism thus goes hand in hand with the participatory promise of reality television, with nationalism mobilized "not as a top-down imposition but as the reflection of the aggregated desires of individual consumers." This does not mean, of course, that these desires are false or entirely manipulated. Rather, they simultaneously reflect both individual opinions and tastes and the commercial, nationally-tinged elicitations by producers. The performance of national identity on *Idol* shows may thus be viewed dialectically as representing both real cultural differences and imagined or perceived local traditions, which are strategically employed for the promotion of ideologies of consumption. These ideologies are themselves underpinned by a variety of global and local discourses on individuality, collectiveness, democracy, social mobility, agency, and resistance, among other elements.

STUDY LIMITATIONS

The limitations of the sample and analysis must be acknowledged on several levels. First, the focus on the four countries selected, which was of course influenced by availability and language comprehension, presents certain constraints on any attempt to generalize. Investigations of globalization and its resistances have customarily, and logically, focused on strategies for localization in relatively peripheral, often underprivileged, areas of the globe. In contrast, this study examines *Idol* shows primarily in Western society,[22] and across cultures that exhibit cultural proximity, defined as similarities between countries in value systems, historical background, language, ethnic makeup, standards of living, educational and political systems, religion, and climate (Hofstede, 2003; Straubhaar, 1991). Exploring the impact of globalization in arenas that indeed share many of these features is thus somewhat counter-productive. This said, however, the fact that some significant differences were found even in these contexts is illuminating, as well as being compatible with the relevant research literature. Indeed, while previous comparisons of TV shows between culturally proximate countries have found them to be more similar than those broadcast in culturally distant countries [for example, contrast Skovmand's (1992) comparison of quiz shows in Scandinavian countries with Cooper-Chen's (1994) comparison across 50 different countries], significant differences have been found even between cultures that are considered to be similar (see Hetsroni, 1999, 2001 for dating games and quiz shows, respectively). As Hetsroni (2005) notes with regard to different versions of *Who Wants to Be a Millionaire*, "what might seem, at first sight, like a blind adoption of global success is, in fact, a culture-sensitive mechanism that befits the transnational format to local preferences, and is fine-tuned to notice even relatively small cultural differences" (p. 107). This certainly appears to be the case for the contexts examined in this study. At the same time, additional explorations of *Idol* shows in other countries and languages would obviously be extremely worthwhile.

Second, it goes without saying that *Idol* shows do not stand in for all cultural production. Some similarities and differences between the shows have no implications whatsoever with regard to the globalization debate, and some simply reflect the nuances of their specific format and the differences between the people involved in their production and performance. The findings of this study are not reflective even of all the reality programming – to say nothing of broader cultural fields. While the popular success of the American, Canadian, and Israeli *Idol*s would seem to indicate that their respective producers have accurately identified the winning formula with regard to the performance of national identity, it must be acknowledged that the present study deals solely with the cultural texts themselves, rather than with not immediately noticeable aspects of production or audience reception. Audience

---

[22] While Israel is not geographically a part of Western society, Israelis have been shown to share a common worldview with, and to embrace similar values as, Americans (see Hofstede, 2003; Rebhun & Waxman, 2000).

responses are inevitably complex and multilayered, and assuming psychological or cultural effects simply from ratings tables is a risky endeavor indeed. Ratings, popular success, and audience participation notwithstanding, viewers may in fact be partaking in practices of consumption desired by the shows' producers for reasons altogether different from nationalistic sentiment and identification. For this reason, further explorations of the production and reception aspects of *Idol* shows – as have already been carried out in some contexts (e.g., Punathambekar, in press; Volcic and Andrejevic, in press) – would undoubtedly be of immense value.

REFERENCES

Allen, R. (Ed.) (1995). *To be continued... Soap opera around the world.* London: Routledge.

Ang, I. (1985). *Watching "Dallas": Soap opera and the melodramatic imagination.* London: Methuen.

Appadurai, A. (2006). Disjuncture and difference in the global cultural economy. In M. G. Durham and D. M. Kellner (Eds.), *Media and cultural studies: Keyworks (Revised Edition)* (pp. 584-603). Malden, MA: Blackwell.

Arnason, J. P. (1990). Nationalism, globalization and modernity. *Theory, Culture and Society, 7*, 207-236.

Avraham, E. (2003). *Behind media marginality: Coverage of social groups and places in the Israeli press.* Lanham, MD: Lexington Books.

Azaryahu, M. (1999). McDonald's or Golani junction? A case of a contested place in Israel. *Professional Geographer, 51*, 481-492.

Baer, D., Grabb, E., and Johnston, W. (1990). The values of Canadians and Americans: A rejoinder. *Social Forces, 69*, 273-277.

Barker, C. (1997). *Global television: An introduction.* Oxford: Blackwell.

Barnet, R. J., and Cavanagh, J. (1994). *Global dreams: Imperial corporations and the new world order.* New York: Simon and Schuster.

Bignell, J. (2005). *Big brother: Reality TV in the twenty-first century.* Houndmills, UK: Palgrave MacMillan.

Billig, M. (1995). *Banal nationalism.* London: Sage.

Biressi, A., and Nunn, H. (2005). *Reality TV: Realism and revelation.* London: Wallflower Press.

Brenton, S., and Cohen, R. (2003). *Shooting people: Adventures in reality TV.* London: Verso.

"Canadian Idol Audience Surge Anchors CTV's Mid-summer Nights' Ratings Dream" (2004). Retrieved April 15, 2006 from http://www.channelcanada.com/Article535.html

Chan, J. M. (2002). Disneyfying and globalizing the Chinese legend Mulan: A study of transculturation. In J. M. Chan and B. T. McIntyre (Eds.), *In search of boundaries: Communication, nation-states, and cultural identities* (pp. 225-248). Westport, CT: Ablex.

Cooper-Chen, A. (1994). *Games in the global village: A 50 nation study of entertainment television.* Bowling Green, OH: Bowling Green State University Popular Culture Press.

Cornfield, M. (2004, August 1). What the parties could learn from "American Idol." *Campaigns and Elections, 25(7)*, 30.

Cowell, S. (2003). All together now! Publics and participation in American Idol. *Invisible Culture, 6*. Retrieved February 20, 2008 from https://urresearch.rochester.edu/retrieve/2446/IVC_iss6_Cowell.pdf

Croteau, D., and Hoynes, W. (2000). *Media/society: Industries, images, and audiences.* Thousand Oaks, CA: Pine Forge Press.

CTV Media Site (2006, September 18). *3.4 million cheer on Eva Avila as tri-lingual Quebecer wins Canadian Idol.* Retrieved February 20, 2008 from http://www.ctvmedia.ca/idol06/

Dann, G. (2004). American Idol: From the selling of a dream to the selling of a nation. *Meditations, 1*, 15-21.

Darling-Wolf, F. (in press). World citizens "à la française": Star Academy and the negotiation of "French" identities. In M. M. Kraidy and K. Sender (Eds.), *Real Worlds: Global Perspectives on the politics of reality television.* New York: Routledge

Dasgupta, S. (2005). Visual culture and the place of modernity. In A. Abbas and J. N. Erni (Eds.), *Internationalizing cultural studies: An anthology* (pp. 427-438). Malden, MA: Blackwell.

Dorfman, A., and Mattelart, A. (1975). *How to read Donald Duck: Imperialist ideology in the Disney comic.* New York: International General.

Durham, M. G., and Kellner, D. M. (2006). Introduction to part VI. In M. G. Durham and D. M. Kellner (Eds.), *Media and cultural studies: Keyworks (Revised Edition)* (pp. 579-583). Malden, MA: Blackwell.

Dyer, G. (1998). *Stars.* London: BFI.

Escobar, A. (2001). Culture sits in places: Reflections on globalism and subaltern strategies of localization. *Political Geography, 20*, 139-174.

Feld, S. (1988). Notes on world beat. *Public Culture, 1(1)*, 31-37.

Friedenberg, E. Z. (1980). *Deference to authority: The case of Canada.* White Plains, NY: M. E. Sharpe.

Friedman, J. (1999). The hybridization of roots and the abhorrence of the bush. In M. Featherstone and S. Lash (Eds.), *Spaces of culture: City-nation-world* (pp. 230-255). London: Sage.

Fritz, B. (2002, October 4). Revolution televised. *The American Prospect.* Retrieved July 1, 2009 from http://www.prospect.org/cs/articles?article=revolution_televised

Galtung, J., and Ruge, M. (1965). The structure of foreign news. *Journal of Peace Research, 1*, 64-90.

Garcia Canclini, N. (1995). *Hybrid cultures: Strategies for entering and leaving modernity.* Minneapolis: University of Minnesota Press.

Giddens, A. (1990). *The consequences of modernity.* Stanford, CA: Stanford University Press.

Giddens, A. (2000). *Runaway world: How globalization is reshaping our lives.* New York: Routledge.

Golding, P. (1979). Media professionalism in the Third World: The transfer of an ideology. In J. Curran, M. Gurevitch and J. Woollacott (Eds.), *Mass communication and society* (pp. 291-308). Beverly Hills, CA: Sage.

Gripsrud, J. (1995). *The "Dynasty" years: Hollywood television and critical media studies.* London: Routledge.

Hamelink, C. (1983). *Cultural autonomy in global communications*. New York: Longman.

Hannerz, U. (1987). The world in creolization. *Africa, 57*, 546-559.

Hannerz, U. (1989). Notes on the global ecumene. *Public Culture, 1(2)*, 66-75.

Hetsroni, A. (2001). What do you really need to know to be a millionaire? Content analysis of quiz shows in America and in Israel. *Communication Research Reports, 18*, 418-428.

Hetsroni, A. (2004). The millionaire project: A cross-cultural analysis of quiz shows from the United States, Russia, Poland, Norway, Finland, Israel, and Saudi Arabia. *Mass Communication and Society, 7*, 133-156.

Hetsroni, A. (2005). The quiz show as a cultural mirror: Who wants to be a millionaire in the English-speaking world. *Atlantic Journal of Communication, 13*, 97-112.

Hetsroni, A. and Bloch, L. R. (1999). Choosing the right mate when everyone is watching: Cultural and sex differences in television dating games. *Communication Quarterly, 47*, 315-332.

Hofstede, G. H. (2003). *Culture's consequences: Comparing values, behaviors, institutions and organizations across nations*. Beverly Hills, CA: Sage.

Holmes, S. (2004). Reality goes pop! Reality TV, popular music, and narratives of stardom in Pop Idol. *Television and New Media, 5*, 147-172.

Hosterman, A. R. (2004). *The American Idol as mythic hero: The creation of a cultural icon*. Paper presented to the 90th annual convention of the National Communication Association, Chicago, IL, November 11, 2004.

"How Is a Star Born" (2006). *Mymusic Magazine*. Retrieved April 15, 2006 from http://www. mymusic.co.il/MyMusic/magazine/20060403055310.htm (Hebrew)

Howes, D. (Ed.) (1996). *Cross-cultural consumption: Global markets, local realities*. London: Routledge.

Illouz, E., and John, N. (2003). Global habitus, local stratification, and symbolic struggles over identity. *The American Behavioral Scientist, 47*, 201-229.

Ivy, M. (1988). Tradition and difference in the Japanese mass media. Public Culture, *1(1)*, 21-29.

Katriel, T. (1986). *Talking straight: Dugri speech in Israeli sabra culture*. Cambridge: Cambridge University Press.

Katz, E., and Wedell, G. (1977). *Broadcasting in the Third World*. Cambridge, MA: Harvard University Press.

Kraidy, M. M. (2002). Hybridity in cultural globalization. *Communication Theory, 12*, 316-339.

Kraidy, M. M., and Murphy, P. D. (2008). Shifting Geertz: Toward a theory of translocalism in global communication studies. *Communication Theory, 18*. 335-355.

Libsky, K. (2005, August 17). Watching the stars. *Rating, 322*, 26-38 (Hebrew).

Liebes, T., and Katz, E. (1990). *The export of meaning: Cross-cultural readings of "Dallas."* Oxford: Oxford University Press.

Lipset, S. M. (1990). *Continental divide: The values and institutions of the United States and Canada*. New York: Routledge.

Maynard, M. L. (2003). From global to glocal. How Gilette's SensorExcel accommodates to Japan. *Keio Communication Review, 25*, 57-75.

McKinney, D. (2006, February 3). Idols true and false. *The American Prospect*. Retrieved July 1, 2009 from http://www.prospect.org/cs/articles?article=idols_true_and_false

Miller, D. (1992). The *Young and the Restless* in Trinidad: A case of the global and the local in mass consumption. In R. Silverstone and E. Hirsch (Eds.), *Consuming technologies: Media and information in domestic spaces* (pp. 163-182). London: Routledge.

Nederveen Pieterse, J. (1994). Globalisation as hybridization. *International Sociology, 9*, 161-184.

Nederveen Pieterse, J. (2004). *Globalization and culture: Global mélange*. Lanham, MD: Rowman and Littlefield.

Paskoff, M. (2003, May 23). Idol worship: What American politics can learn from American Idol. *The American Prospect*. Retrieved July 1, 2009 from http://www.prospect.org/cs/articles?article=idol_worship

Price, A. (2003, January 24). Popular demand. *The American Prospect*. Retrieved July 1, 2009 from http://www.prospect.org/cs/articles?article=popular_demand

Punathambekar, A. (in press). Reality TV and the making of mobile publics: The case of Indian Idol. In M. M. Kraidy and K. Sender (Eds.), *Real Worlds: Global Perspectives on the politics of reality television*. New York: Routledge.

Rantanen, T. (2005). *The media and globalization*. London: Sage.

Rebhun, U., and Waxman, C. (2000). The Americanization of Israel: A demographic, cultural and political evaluation. *Israel Studies, 5*, 65-91.

Regev, M., and Seroussi, E. (2004). *Popular music and national cultures in Israel*. Berkeley: University of California Press.

Ritzer, G. (1993). *The McDonaldization of society*. Newbury Park, CA: Pine Forge Press.

Ritzer, G., and Liska, A. (1997). 'McDisneyization' and 'post-tourism.' In C. Rojek and J. Urry (Eds.), *Touring cultures: Transformations of travel and theory* (pp. 96-109). London: Routledge.

Robertson, R. (1992). *Globalization: Social theory and global culture*. London: Sage.

Rosen-Zvi, I. (2004). *Taking space seriously: Law, space and society in contemporary Israel*. Burlington, VT: Ashgate.

Schiller, H. (1976). *Communication and cultural domination*. New York: Sharpe.

Schiller, H. (1985). Electronic information flows: New basis for global domination? In P. Drummond and R. Paterson (Eds.), *Television in transition: Papers from the first International Television Studies conference* (pp. 11-20). London: BFI.

Seo, J. (2008). *Manufacturing nationalism in China: Political economy of "say no" businesses*. Paper presented at the 49th annual convention of the International Studies Association, San Francisco, CA, March 26-29, 2008.

Shaw, J. (2006, April 21). The biggest idol ever. *Entertainment Weekly, 873*, 34-39.

Showalter, E. (2003, June 30). Window on Reality: American Idol and the search for identity. *The American Prospect, 14(7)*. Retrieved July 1, 2009 from http://www.prospect.org/cs/articles?article=window_on_reality

Silj, A. (1988). *East of Dallas: The European challenge to American television*. London: BFI.

Skovmand, M. (1992). Barbarous TV international: Syndicated wheels of fortune. In M. Skovmand and K. L. Schoder (Eds.), *Media cultures: Reappraising national media* (pp. 84-103). London: Routledge.

Smart, B. (Ed.) (1999). *Revisiting McDonaldization*. London: Sage.

Smith, A. D. (1990). Towards a global culture? *Theory, Culture and Society, 7*, 171-191.

Smith, M. J. (2004). *And your American Idol is... Ryan Seacrest!? Inventing the postmodern multimedia icon.* Paper presented to the 90th annual convention of the National Communication Association, Chicago, IL, November 11, 2004.

Spencer, M. (1992). Continental divide: The values and institutions of the United States and Canada (Review). *Theory and Society, 21,* 610-618.

Sreberny-Mohammadi, A. (1996). The global and the local in international communications. In J. Curran and M. Gurevitch (Eds.), *Mass media and society (Second Edition)* (pp. 177-203). London: Arnold.

Stern, D. (2004). *Fans (Ford) focus on the product: American Idol and the future of television advertising.* Paper presented to the 90th annual convention of the National Communication Association, Chicago, IL, November 11, 2004.

Straubhaar, J. D. (1991). Beyond media imperialism: Asymmetrical interdependence and cultural proximity. *Critical Studies in Mass Communication, 8,* 39-59.

Thrupkaew, N. (2004, May 27). Fantasia island. *The American Prospect.* Retrieved July 1, 2009 from http://www.prospect.org/cs/articles?article=fantasia_island

Thussu, D. K. (1998). Localising the global: Zee TV in India. In D. K. Thussu (Ed.), *Electronic empires: Global media and local resistance* (pp. 273-294). London: Arnold.

Tincknell, E., and Ranghuram, P. (2002). Big Brother: Reconfiguring the "active" audience of cultural studies? *European Journal of Cultural Studies, 5,* 199-215.

Tomlinson, J. (1991). *Cultural imperialism: A critical introduction.* London: Pinter.

Tomlinson, J. (1999). *Globalization and culture.* Chicago: The University of Chicago Press.

van Zoonen, L. (2004). Imagining the fan democracy. *European Journal of Communication, 19,* 39-52.

Volcic, Z., and Andrejevic, M. (in press). Commercial nationalism on Balkan reality TV. In M. M. Kraidy and K. Sender (Eds.), *Real Worlds: Global Perspectives on the politics of reality television.* New York: Routledge.

Watson, J. L. (1997). *Golden arches east.* Stanford: Stanford University Press.

Wells, A. (1976). *Picture-tube imperialism? The impact of U.S. television on Latin America.* Maryknoll, NY: Orbis.

West, D. (2007, January 23). Foxy hopes for "Pop Idol" return. *Digital Spy.* Retrieved July 1, 2009, from http://www.digitalspy.com/realitytv/a41951/foxy-hopes-for-pop-idol-return.html

Yoshimoto, M. (1989). The postmodern and mass images in Japan. *Public Culture, 1(2),* 8-25.

In: Reality Television-Merging the Global and the Local
Editor: Amir Hetsroni, pp. 189-209

ISBN 978-1-62100-068-6
© 2010 Nova Science Publishers, Inc.

Chapter 12

MOBILE MAKEOVERS: GLOBAL AND LOCAL LIFESTYLES AND IDENTITIES IN REALITY FORMATS

Tania Lewis
La Trobe University, Australia

'These twelve people have one thing in common—they're about to change their lives forever!': so goes the dramatic voiceover for *The Biggest Loser* (TBL), one of the recent big success stories in reality TV formats. A competitive weight-loss show, the format has captured the imagination of audiences worldwide with local versions of the US program broadcast in the UK, Australia and the Middle East (where the show has been rebadged *The Biggest Winner*). While the format has varied somewhat over different seasons and between countries, the basic premise involves a group of overweight people being brought together in a *Big Brother*-style house to compete for a large sum of money by losing the highest percentage of their starting body weight, initially as part of competing teams and later as individuals.

So how might we understand the apparently global cultural currency of a reality TV show aimed at helping obese people lose weight with the (not always gentle) guidance of various 'lifestyle experts' and under the gaze of a watching public? To date a number of rationales have been offered to explain the transnational mobility of reality TV. One important context for the rise of this mode of programming has been global shifts within the television industry. In particular, an increasingly deregulated market, a fragmented audience and a plenitude of competing viewing choices has put pressure on the industry to produce cheap abundant programming. An economic analysis, however, is only part of the story. The global currency of such formats has also been read as marking broader cultural and social shifts around the world in particular the growing prevalence of US-inflected late capitalist models of consumption, lifestyle and selfhood.

In his book, *Big Brother: Reality TV in the Twenty-First Century* (2005), Jonathan Bignell asks whether the transnational mobility of reality TV indicates the universalization of a 'western' preoccupation with 'personal confession, modification, testing and the perfectibility of the self'. Certainly the widespread uptake of these formats around the world and across different cultures—from Chinese-Singaporean reality-based home renovation shows like *Home Décor Survivor* to the Panamanian version of *Extreme Makeover* (*Cambio Radical*)—can be seen to point to the global currency of certain types of consumerist and (neo)liberal models of selfhood and citizenship. While the embrace of reality TV and its associated modes of selfhood and lifestyle consumption might be a globalized one, the failure of some formats in certain countries and the popularity of indigenised versions of imported formats indicates that this mode of programming is still strongly shaped by local conditions.

Focusing in particular on reality-based makeover programmes, this essay discusses the global spread of the reality TV format, examining the way in which it has been shaped by both globalising forces and domestic concerns and contexts and drawing upon examples from a range of different sites from Australia to Asia. Arguing against a one-size-fits-all approach to reality TV, the essay brings together a variety of critical frameworks for understanding the global spread of the format. The chapter is divided into two parts: in part one, I locate the rise of reality TV within the context of broader shifts within the television industry before then going on to outline some broader socio-cultural contexts and theoretical frames for understanding the global flow of these formats. In part two I trace the emergence of the reality makeover show in the UK and US, concluding the chapter by examining the way in which the makeover format has 'travelled' to and been reworked in two different TV markets, namely Australia and Singapore.

INDUSTRY AND SOCIAL TRENDS UNDERPINNING THE RISE OF REALITY TV

Television in the late 20th and early 21st century has become increasingly enmeshed in everyday concerns and social processes. The recent rise of lifestyle makeover shows on primetime TV for instance can be read as one symptom of a broader set of shifts in televisual culture towards a growing focus on 'the real'. As Frances Bonner points out (2003, p.28), since the 1980s television has been increasingly concerned with the mundane and 'the ordinary', as reflected in its growing focus on domestic space and the lives of members of the public. Likewise, writing in the 1990s about the growing role of 'true-life-story' genres in the US, UK and Europe, Ib Bondeberg (1996) argues that this has occurred as part of a broader embrace of privatized modes of discourse, with the camera increasingly turning to focus on the intimacies of people's lives and relationships. The recent history of television can thus be summed up as being distinguished by a preoccupation with the 'everyday terms of living' (Corner, 2004, p.291).

From 'True Life' to Reality: Recent Shifts in the Televisual Landscape

The reasons for this turn towards the everyday are complex and multifaceted. As noted, one argument within TV scholarship is that the turn to the real reflects economic developments within the industry itself. In particular, the shift since the late 1980s towards modes of relatively cheap, 'unscripted' television focused on ordinary people can be seen as an attempt to deal with an increasingly deregulated market and a fragmented audience, with free to air networks now competing with pay TV for viewers' attention, offering audiences an

Mobile Makeovers: Global and Local Lifestyles and Identities in Reality Format   191

abundance of programming choices (Bonner, 2003; Ellis, 2000). Concomitant with these economic transitions there have been a range of shifts in the policy underpinnings and 'culture' of the industry around the world. While the nature of these developments have varied considerably in different national contexts, in general the TV industry globally has been marked by the adoption of more populist modes of address, a shift that for some reflects the 'democratization of an old public service discourse' (Bondebjerg, 1996, p.29) while for others suggests a growing and problematic convergence between public service and commercial concerns.

The transition over the past two to three decades to a more 'democratized' approach has been accompanied by some interesting innovations in the realm of genre. Just as news around the world has been marked by the growing prominence of 'tabloid journalism' (Hill, 2005, p.15) and 'info-tainment' values (Thussu, 2007), the realm of entertainment television has increasingly embraced forms of 'dramatized factual television' (Bondebjerg, 1996, p.27), including the emergence of 'docu-soaps' like the UK's *Airport*. In broad terms, then, there has been a growing hybridization of TV genres, incorporating the fictional and the melodramatic into documentary and factual forms, and culminating in the more recent rise of purportedly new formats such as 'reality television'.[1]

Of course reality television is not necessarily generically 'new' in the sense that it can be seen to borrow extensively from a range of pre-existing forms of programming, blending together for instance the competitive element of US game shows and the humiliation of their Japanese counterparts with the voyeurism, melodrama and confessional dimensions of talk shows and soaps. Today's reality shows, however, bring together these older generic elements in ways that speak to distinctly contemporary concerns. As Helen Wood and Beverley Skeggs note (2004, p.205), contemporary lifestyle and reality shows are distinguished by a shared focus on 'interrogations of selfhood under the pressures of particular conditions'. The reality formats that have come to dominate the last decade rely on increasingly contrived scenarios, from bringing together diverse people to live together in an artificial community under hothouse conditions to encouraging people to undertake major personal lifestyle transformations under the intrusive gaze of the TV camera.

Another reason cited for the contemporary success of reality TV has been its ability to travel as a 'format' into a range of different television markets. The deregulation of the television industry around the world in the 1980s and 1990s and the emergence of a multi channel environment has produced a situation where the pressure for product has encouraged local producers to create programmes that can potentially move across a range of markets (Moran, 1998; Waisbord, 2004). This situation has seen a relative challenge to US hegemony in global TV traffic and trade as TV formats increasingly emerge from the UK and Western Europe as well as from smaller players such as Australia and Mexico (Magder, 2004; Moran and Keane, 2006; Waisbord, 2004).

---

[1] Reality television has become a catch all phrase used to classify a range of popular factual forms. However, as Corner points out (2004), the term first emerged in the US context where it was initially associated with low budget, actuality-based television of the *COPS* variety. While the term had a certain strategic edge in the US context—where it came out of the pressures of a highly competitive deregulated industry—as he notes it would have been unlikely for the term to have emerged from the UK or Europe where there was already a long tradition of 'real' modes of TV in the form of social observational-style documentaries. British and European TV production companies have nevertheless been highly successful at producing, marketing and selling formats internationally under the reality TV banner with the Dutch company Endemol producing the urtext of reality shows *Big Brother*.

As Waisbord comments (2004, p.359), the rise of format television has resulted in a situation whereby 'around the world, television is filled with national variations of programs designed by companies from numerous countries'. For instance, Endemol, a company that originated and continues to be based in Holland, first created the global reality TV phenomenon *Big Brother* for the Dutch market and then sold the format to numerous countries. Reality programs offered up as format 'shells' such as the garden makeover show *Ground Force*, which first aired in the UK, have been shown to have considerable transnational mobility and selling power, as they are amenable to being readily 'indigenized', even in the case of programs emerging from non-English markets (such as the Dutch market), and are relatively risk free having been previous tried out on an audience (Waisbord, 2004).

Socio-Cultural Contexts for the Reality 'Turn'

The success of reality TV cannot be purely reduced to a question of industry economics. The format's cross-cultural appeal is also linked to its unique blend of domestic melodrama, personal confession and transformation, which in turn speak to a number of wider socio-cultural developments in late modernity, particularly in relation to emergent contemporary forms of cultural citizenship and public selfhood. Here I want to briefly map out some of these broad shifts before discussing how some of these trends have been played out via the reality format in different cultural settings.

Given its links to various forms of feminine-coded TV genres (melodrama, talk shows, lifestyle TV), one important context for situating certain kinds of reality TV, especially those shows centred on the home and on the labours of transformation (from home makeover shows to 'life transformation' shows like the British programme *Wife Swap*), is in relation to questions of gender and domestic space. In particular, the kind of public displays of intimacy central to many reality TV shows can be seen as reflecting the breaking down of distinctions between the public world of work and the privatized realm of the home in late modernity. As Skeggs and Wood (2009, p.120) argue, with the increasing contemporary focus on managing one's self and lifestyle, 'the paradigmatic form of new work is *domestic work.*'

Another way in which questions of value enter into the picture in relation to reality TV's depiction of domestic labour is through class and taste. McRobbie argues that the feminine skills increasingly valued and rewarded within contemporary neo-liberal societies—and here she is talking not only about domestic skills but also more broadly about feminine forms of 'cultural capital' related to self-care, the management of interpersonal relations and presentational aesthetics—are those tied to middle class competencies and values (McRobbie, 2004). Certainly, many reality shows, particularly makeover formats, can be seen to model bourgeois and at times haute bourgeois lifestyles and values for their audiences. Reality TV however has an often complex and contested relationship to such middle class values. With its focus on ordinary people and their lives and its genealogical links to feminized forms of 'trash TV', the format can be seen to be marked by a relative democratization of its mode of address and concerns, and a seeming embrace of diversity as witnessed by the number of working class accents and faces now represented on a variety of reality shows (Redden, 2007, pp.155-156; Taylor, 2002, p.486).

At the same time many reality TV shows are often implicitly underpinned by middle class frames of value. Discussing reality-based personal makeover programs like the British fashion show *What Not to Wear*, Gareth Palmer notes that class and social distinction is a particularly pervasive feature of this type of television and its brand of popular expertise (2004). On such shows, celebrity experts like Trinny and Susannah (the posh fashionistas on *What Not to Wear*) function as models or cultural intermediaries for 'good' taste and style.

Regulated Realities: Consumer-Citizens and Enterprise Selves

Alongside the promotion of middle class forms of selfhood and modes of living, many lifestyle-oriented reality formats also work to model and naturalize certain forms of consumption. Part of a broader shift towards the inclusion of advertising and commodities within TV content itself, or 'below the line' advertising as it is known in the industry, reality shows from *The Biggest Loser* to *Queer Eye for the Straight Guy* work 'to alert viewers to the existence of more products and services for their utility in the endless project of the self' (Bonner, 2003, p.104), from gym equipment and personal training to home décor and fashion.

What reality programming also often 'sells' to the audience are not just products but ways of living and being. As I've argued elsewhere in relation to reality-based makeover television (Lewis, 2008a), in recent years there has been a shift in these formats to a more overtly educational approach to personal and lifestyle change, one concerned with emphasizing responsible modes of consumption and citizenship and with transforming oneself into a 'good' consumer. We've seen a growing range of lifestyle makeover formats focused for instance on personal and family health, behaviour and relationships, from more overtly pedagogical formats such as *Honey We're Killing the Kids* (which aims to transform the unhealthy lifestyles of families and in particular their children) to competitive game show-style weight loss shows like *The Biggest Loser*. In attempting to transform the overweight and behaviourally 'aberrant' people featured on these shows into ideal citizens, the emphasis on health, fitness and behavioural change is closely tied here to regulating people's modes of consumption. In particular (and somewhat ironically, given that many of these shows are aired on commercial television), these issues are often framed in terms of curbing 'excessive' consumption from over eating to watching too much television.

In promoting certain models and rules for living—regulated consumption (*The Biggest Loser*), middle class taste (*Queer Eye for the Straight Guy*), entrepreurial modes of selfhood (*The Apprentice*), community-mindedness (*Extreme Makeover—Home Edition*)—reality shows can be seen to play an important role not only in teaching the public personal life skills but also promoting and validating certain models of the 'good citizen'. The rituals and new traditionalism of lifestyle-oriented reality programming in particular can be seen to mark a convergence between questions of lifestyle choice and a broader model of selfhood, an ethical or moral model emphasizing the role of personal and domestic lifestyle management as a site of pleasure *and* responsibility. This conflation of lifestyle and consumer choice on reality TV with responsibility sees a growing connection between the self, the home and the everyday, and more public, community-based concerns, the personal aspirations of the self-improving lifestyle consumer being refigured as those of the citizen.

An important critical approach that has sought to contextualize the rise of the lifestyle-oriented consumer–citizen—and that offers a particularly useful way of thinking through the performance and management of selfhood on reality TV—can be found in Nikolas Rose's work (1989; 1996). Influenced by Foucault's conception of modern power and governance as being played out through the 'freedoms' associated with liberal selfhood, Rose argues that the rise of neoliberal governments in many nations in the 1980s (in particular the UK and US), alongside the emergence of a wider 'enterprise culture,' has seen a shift in the dominant paradigms through which we conceptualize modern citizenship. In particular, the figure of the self-governing citizen, an individual who is constructed as enterprising and self-directed, has become a cultural dominant. This has occurred in the context of the state increasingly seeking to devolve questions of social and political responsibility to the level of the individual consumer–citizen, a situation shored up by a 'therapeutic culture' that pairs freedom and moral development with self-mastery and self-development. Thus, in neoliberal settings, the personal, health and relationship advice increasingly offered on lifestyle-oriented reality shows like *The Biggest Loser*, for example, can be seen to be attempting to fill the gap left by the state as it passes on responsibility for what would once have been 'public health' concerns like obesity onto the self regulating consumer-citizen (Ouellette and Hay, 2008).

A crucial aspect of the ethos of self-governance here is the increasing rationalization of subjectivity via the figure of the expert—something we can see played out via the various 'gurus' featured on reality programming, from child behaviour specialists and nannies (*Honey We're Killing the Kids*, *Supernanny*) to personal trainers (*The Biggest Loser*) and specialists in etiquette and grooming (*From Ladette to Lady*). On such shows and in our broader daily lives then 'the conduct of everyday existence is recast as a series of manageable problems to be understood and resolved by technical adjustment in relation to the norm of the autonomous self aspiring to self-possession and happiness' (Rose, 1996. p.158).

Such processes of rationalization and self-regulation go hand in hand with an increasing focus in society on self-surveillance and confession (Andrejevic, 2004), where the 'gaze' of the expert is turned inwards upon the self. Confessional modes of neo-liberal self-surveillance are particularly central to reality formats. Reality television is often strongly reliant on personal self-disclosure, a process that is often done straight to camera, inviting the audience to both identify with and judge the self-surveillant subject. These shows are also heavily reliant on the use of more literal 'technologies' of surveillance, such as hidden cameras aimed at capturing anything from fashion bungles to moments of 'anti-social' behaviour. Technologies that again can be seen as an extension and normalisation of therapeutic conceptions of selfhood as a site 'opened up for expert judgement, and normative evaluation, for classification and correction' (Rose, 1989, p.244).

Overall then, reality TV—from competitive formats like *The Biggest Loser* to personal makeover shows like *Queer Eye for the Straight Guy*—can be seen to emerge out of a complex conjuncture of social, cultural and economic factors. Linked to the growing push toward modes of reflexive, consumer-based individualism, reality TV can also be seen as emerging out of a neo-liberal culture of self-governance and self-surveillance. Despite claims that the self-governing consumer-citizen now exists in a post-traditional realm marked by the growing irrelevance of social categories like class, the ideals and norms held up on reality shows are often underpinned by class-based (particularly middle class and lower middle class or 'aspirational') models of taste and lifestyle.

MAKEOVER CULTURES

In mapping out these broad social and political contexts for the rise of reality TV the question becomes how and to what extent such concerns are articulated to reality formats in a range of cultural settings. Much of the discussion and debate concerning the ideological dimensions of reality TV to date has tended to be focused on the US and the UK. While in recent years reality formats have taken off around the world there has been relatively little cross-cultural analysis of this phenomenon. As noted, the assumption for some media scholars has been that the global mobility of these formats can be seen as marking the broader currency around the world of Western-inflected cosmopolitan and neo-liberal models of consumer-citizenship (Bignell, 2005; Miller, 2007). Certainly, a glance across the wide range of reality formats now found on primetime TV schedules around the world indicates that such concerns are certainly not only central to American formats like *The Biggest Loser*. They also form the ideological underpinnings of local formats in such culturally disparate sites as China where reality television has been a big winner with audiences. Hunan Satellite TV for instance has become a very successful producer of shows like *I am the Champion* in which (like *The Amazing Race*) two-person teams consisting of one parent and one child compete at a military training camp while also having to cooperate with each other to avoid elimination.

The mobility of these kinds of competitive reality formats can be read as part of a broader globalisation of neoliberal models of good citizenship. At the same time, television formats represent sites marked by complex negotiations between globalising forces and domestic concerns and contexts (Hetsroni, 2005; Moran, 1998). As noted, reality shows—whether sold as format shells to be localized for a domestic market or shown in their original form—do not necessarily succeed in all TV markets. For instance while many popular factual formats travel readily across cultural borders, as Bignell notes (2005), cases like the relative disinterest in *Big Brother* in the US and *Survivor*'s relatively poor ratings in Britain indicate that one should be cautious in over generalizing the universal appeal of such formats.

In the next section of this essay, I want to use the example of the reality makeover format as a way of examining some of the issues around the global mobility and local reception of reality shows. Makeover shows—from home renovation to personal weight loss shows—are a particularly generative subcategory for thinking through such issues, tied as they often are to both globalising models of consumer-citizenship *and* national/local lifestyle concerns, practices and models of modernity. In this next section then I want to discuss a range of 'television makeover cultures' as Mischa Kavka (2006) has termed them. Opening with a discussion of the rise of makeover TV in the UK and US (two countries associated with the 'origins' of reality TV), I will then move on to discuss two contrasting examples of the globalisation of the format, Australia and Singapore.

The 'Makeover Takeover' on British Television

While the reality makeover format is often popularly associated with US TV culture, industry commentators and numerous television scholars point to the UK in the 1990s as a defining moment in the emergence of the format as a major primetime player, with Rachel

Moseley (2000) describing British television in the 1990s as having undergone a 'makeover takeover'.

A number of reasons have been offered for the rise and popularity of the makeover format in Britain. Moseley (in Brunsdon et al., 2001, p.32) suggests that it marks an important shift in the gendering of the mode of address of television. Arguing that the format is 'the most visible marker' of a more general mainstreaming of feminine makeover culture, she contends that the broad-based popularity of makeover television reflects the fact that men are now engaging with a range of once feminine-coded activities.

While previously found primarily on daytime TV where it often featured as segments on magazine shows aimed at women, the 1990s saw the makeover expand into a full length format and move into primetime schedules on UK television. However, where daytime TV makeovers often focused on issues of personal style and fashion, the first successful makeover formats were shows oriented towards investing in and improving the home rather than the self. Thus one of the first breakthrough makeover shows on primetime UK TV was the home renovation-game show format, *Changing Rooms*[2] (broadcast on the BBC in 1996 and later sold into a number of international markets).

While acknowledging the importance of a gender-based analysis, Gareth Palmer (2004) has argued that the rise of makeover programming in the UK speaks centrally to shifts in British class culture and in particular the recent growth of an aspirational, petit bourgeoisie. As noted, Palmer reads the tips provided by the new echelon of experts that emerged on both home shows and fashion makeover formats like the BBC's *What Not to Wear*, which first aired in 2001, as offering strongly class-inflected modes of guidance around questions of style, taste and social distinction—a focus that has remained a particularly prominent feature of makeover programming in the UK but, as I have suggested above, can also be seen to underpin many reality formats around the world.

A range of shifts in industry policy arising from an increasingly competitive, deregulated British TV market also provided another important context for the 'primetiming' of the makeover format in the 1990s. The BBC in particular was under pressure to expand its public service broadcasting model to embrace a broader, more democratized approach to its audience (Hill, 2005). In the mid 1990s a BBC strategy report emphasized the role of factual entertainment formats in responding to audience needs, and in enabling the broadcaster to 'find inventive ways to cover the leisure pursuits of our audiences' (Moseley, 2003. p.104), extending upon the more traditional modes of factual programming that had always been a central to the BBC 'brand'. In response, the BBC introduced shows like *Can't Cook Won't Cook*, *Style Challenge* and *Real Rooms*, the latter two expanding upon segments that had featured in daytime magazine programming.[3]

While the primetime program-length makeover format drew directly from existing makeover segments on women's daytime magazine shows, it also emerged out of what was a long tradition on British television of modes of leisure and advice programming aimed at both

---

[2] *Changing Rooms* was the brain child of British lifestyle TV guru Peter Bazalgette, who also created groundbreaking lifestyle formats such as Ready Steady Cook and Ground Force. Bazalgette's TV production company eventually became part of Endemol UK, and as chairman of the company he introduced Big Brother to British television audiences.

[3] Again the central role of Peter Bazalgette's production company Bazal in producing many of these formats should be noted, with Bazalgette often described in the British press as the man behind the 1990s lifestyle TV 'revolution'.

men and women. Brunsdon (2003, p.9) notes, however, that the contemporary 'engagement with consumerism, makeovers and game shows' has seen a dramatic transformation in the form, aesthetics and mode of address of British lifestyle programming, with a shift away from teaching the audiences 'how to' skills to a focus on the spectacle of transformation itself and the emotional responses of the ordinary people on the show; thus, many of the pleasures offered to the audience by these new formats are those associated with melodrama (Brunsdon, 2003; Moseley, 2000). Such a shift again supports the argument that the mainstreaming of the makeover (and reality TV more broadly) has occurred, both in the UK and elsewhere, partly in concert with a broader cultural re-valuing of once feminine concerns around the private and domestic and with a growing focus on 'real life' emotions and relationships on television and in the public sphere more broadly.

Another important generic influence has been social observational television—a genre that has been particularly popular in the UK. From Paul Watson's *The Family* (BBC, 1974), which followed the daily lives and struggles of a working class British family to Roger Graef's *Police* (BBC, 1982), these early fly-on-the-wall, observational formats displayed the public educational concerns of documentary—those of representing social diversity and difference and documenting pressing social issues. More recently the reality makeover show in Britain has been marked by something of a return to the educational and social concerns of the social observational form. For instance, the past couple of years have seen a variety of shows emerge out of the UK around the intersection of health, lifestyle and parenting issues, with *Supernanny* being one of the more successful and well-known of these British formats. Shows like *Honey We're Killing the Kids*, which offers parents and in particular their children a complete lifestyle makeover, combines social observational elements and melodramatic spectacle with a strongly didactic approach to issues of diet, health and family relations. While such shows draw strongly on social documentary traditions, the educational approach here is far from sociological, tending instead to focus on the emotional dimensions of people's lives and to reduce social issues to questions of individual lifestyle choice. As discussed above in relation to reality TV more broadly, this is television aimed at training a neo-liberal citizenry to manage their own social welfare—with the DIY premise of early lifestyle shows here turned upon the self and the family and marked by a strong focus on self-surveillance and regulation.

US Television and the Personal Makeover Show

In contrast to the UK, US free-to-air networks were relatively slow to embrace the reality makeover show as a primetime format. While *Changing Rooms* saw makeover TV break into primetime schedules in 1996 in the UK, the breakthrough show on US network TV—the surgical makeover programme *Extreme Makeover* shown on the ABC—didn't make its appearance until 2002 (Kavka, 2006). Instead, US primetime television in the 1990s was preoccupied with a range of other early reality-style formats from low budget, actuality-based television of the *COPS* variety to shows like cable network MTV's *The Real World* (1992), a clear precursor to the *Big Brother* format.

Any discussion of the role of makeover TV and reality formats more broadly on American television however must also take the cable industry into account. Over the past couple of decades, an increasingly deregulated media environment has seen cable playing an increasingly significant role in the US market, as in many other markets around the world. As Miller (2007) has noted, alongside the relatively diminished power of the networks, there has been a shift in the US away from traditional public service modes of factual and informational formats to a growing focus on lifestyle and consumption with cable being central to this process. Since the early 1990s, cable channels like HGTV and The Learning Channel (TLC) have played an important role in popularizing lifestyle formats in the US (Everett, 2004). The big success story for US cable TV however has been the home makeover format. Two years before network TV showed *Extreme Makeover*, TLC achieved high ratings (rivalling the free-to-air networks) with the first US reality-based home makeover show *Trading Spaces*, a US adaptation of the BBC's *Changing Rooms*.[4]

Despite the popularity of *Trading Spaces* on cable TV, arguably however what has broadly distinguished US makeover culture and its mode of lifestyle advice from its British counterpart has been its preoccupation with personal, particularly body makeover shows. While there has been considerable trans-Atlantic traffic of both program formats, ideas and personnel over the past decade,[5] Kavka (2006, p.220) contends that the US focus on personal makeover programmes emerges out of and speaks to a distinct cultural context, marked by an 'unwavering belief in positive transformation'. Thus, where UK makeover shows tend to often be domestic in focus playing out class and taste-based concerns, US shows are underpinned by fantasies of individual perfectibility.

While today reality-based lifestyle makeover shows in the US and UK share much in common, often airing local versions of the same formats (*Supernanny*, *The Biggest Loser*), arguably such shows speak to rather different audiences and are framed by distinctive televisual and cultural traditions. The centrality of the self-help movement to US culture is one important and distinctive cultural frame here. Biressi and Nunn (2005) argue that the rise of makeover shows can be linked to the growing sway of a self-help oriented 'therapeutic culture' in which broader social issues such as obesity are reduced to questions of personal transformation and self-improvement. While this drive to self-improvement ethos has made its influence felt around the world, as Heller (2006) notes such beliefs are deeply rooted in the American ethos and its preoccupation with the possibilities of self-transformation and upward mobility (Bratich, 2007, p.10).

The genealogy of US makeover TV can in many ways be seen to reflect this broader cultural mythology. While British makeover formats on TV emerged out of the context of a long history of DIY, cookery, fashion and hobbyist programming, contemporary US lifestyle and makeover shows can be seen to draw upon genres of programming that feature personal transformations as a central trope, such as 1950s daytime shows like *Glamour Girl* and *Queen for a Day*. As noted, another influential TV genre here has been the talk show. In the 1990s, the makeover trope reappeared on women's daytime TV in the US in the form of makeover segments on talk shows directed by various guest experts. While often oriented towards fashion and 'transformative glamour', Bratich (2007, p.7) notes the early 1990s also saw a

---

[4] In the 1980s there was also some attempt on cable TV to introduce the advice culture and transformative ethos of women's service magazines more broadly onto television via 'video magazines'. The popular women's magazines *Cosmopolitan*, *Woman's Day* and *Good Housekeeping* all produced home and lifestyle-oriented advice shows. By and large though these 'video magazines' were not successful as cable audiences at that time were not large enough to cover the expense of producing these shows (McCracken, 1993: 293-6).

[5] Indeed, Bignell (2005: 39) contends that there has been somewhat of a 'reverse colonization of US television by British programmes and producers in the Reality TV arena'.

spate of 'makeover-themed daytime talk shows' in which for instance 'deviant teens' were transformed on 'boot camp' style episodes. The US talk show thus worked to pave the way for the development of reality-based makeover shows, bringing together the makeover transformation of women's daytime TV—often directed on talk shows at people marked out as 'deviant'—with the personal confession, and with a particular focus on self-surveillance in the name of community 'values'. The 1990s also saw the emergence of actuality footage-based crime, medical and accident shows on primetime TV in the US, modes of programming which evolved over the 1990s into more scripted reality formats often marked by competitive individualistic themes (Crawley, 2006), the exemplar here being *Survivor*.

Today, elements of all these modes of programming can be seen to come together to varying degrees in the US personal makeover format. Shows like *Queer Eye for the Straight Guy* for instance combine reality television and lifestyle advice (in this case delivered by five gay men) with the consumer-driven transformations of women's daytime TV shows in the 1950s and the talk show's morally charged focus on personal confession. While body makeover shows like *The Biggest Loser* bring the 'warts and all' reality-lifestyle format together with a more competitive, boot camp approach in order to both transform and reform overweight citizens into slimmer, more go-getting versions of their former selves—speaking again to the centrality of entrepreurialism and personal promotion to US makeover culture.

Australianising the Makeover

Like the UK, Australia has a strong tradition of lifestyle TV and in particular home and garden-oriented shows, the latter modes of programming reflecting the centrality of home ownership to Australian culture. Since the late 1990s, however, older forms of magazine-style lifestyle programming have been boosted by a growing number of lifestyle makeover formats, with Australia importing a range of reality makeover shows from the US and UK as well as producing its own successful shows (such as the highly successful competitive renovation format *The Block* (2003), which has sold into a range of territories around the world). Whether oriented towards renovating one's home or oneself, such shows are often highly consumerist in orientation, teaching audiences about new products and services. These modes of programming also share a concern with providing life lessons in middle class taste and distinction, often modelling forms of consumption that are linked to particular kinds of normative, morally-inflected lifestyle practices e.g. maintaining a healthy body. Australian primetime TV today thus features a mixture of lifestyle makeover shows oriented towards (often) consumerist modes of self- and life-improvement—including international imports (*How to Look Good Naked*), locally produced versions of international formats (*The Biggest Loser*) and home grown formats (*Domestic Blitz*).

Along with the influence of US and UK makeover formats, Australia's take on the lifestyle makeover show has also been shaped by a range of distinctive cultural concerns (Lewis, 2009). For instance, while Australian lifestyle shows are often concerned with teaching audiences about middle class forms of taste and style—with many Australian TV producers actually referring to these modes of programming as 'aspirational television'—in contrast to US or British television, there is a reluctance to portray overly bourgeois lifestyles on Australian lifestyle TV. Older-style lifestyle magazine shows like *Better Homes and Gardens* are thus directed towards a very ordinary and somewhat banal version of 'middle

Australia'. Rather than presenting audiences with the kind of highly bourgeois models of lifestyle offered up by US gurus such as Martha Stewart or UK lifestyle shows like *Grand Design*, the forms of consumerism, style and taste promoted on *Better Homes and Gardens* lie at the rather middle brow, mildly aspirational end of the lifestyle spectrum.

While lifestyle makeover shows would seem by definition to be more strongly aspirational, oriented as they are towards upward mobility through the complete and often radical transformation of bodies, homes and lifestyles, Australian examples are often relentlessly ordinary, making use of friendly familiar hosts with strong Aussie accents (such as the popular TV personality and carpenter Scott Cam). In contrast to many British makeover shows, where class conflict between the experts and participants is often a central feature of the narrative, Australian programs tend to de-emphasise class concerns through the use of more ordinary hosts (in contrast to 'posh' lifestyle gurus like the UK's Trinny and Susannah).

Domestic Blitz, an Australian reality-based makeover show currently airing in a primetime slot on the free-to-air commercial network Channel Nine is a good example of this kind of 'feel-good', non-conflictual variety of programming. The format of the show is that of the classic house and garden makeover complete with a good looking makeover team and two hosts, blonde attractive presenter Shelley Craft and carpenter Scott Cam, the epitome of the 'Aussie bloke'. In the 'before' part of the narrative, the audience is guided through the existing house and garden by the show's host-experts, with its 'problems' in terms of taste and utility diagnosed along the way, before being shown a dramatic montage of the makeover process. The owners are then taken on a (usually highly) emotional tour of the transformed house complete with 'before' and 'after' shots to remind the audience of the radical nature of the transformation. Rather than depicting the ordinary people on the show as lacking a sense of style or taste (a common feature on UK and some US makeover shows), on *Domestic Blitz* the emotional dimension of the narrative emerges from the fact that the makeover recipients are people who are framed by the show as 'deserving Aussies' who have been 'doing it tough'. For example, in one episode the house and garden of a mother of three with a passion for charity and community work, who suffered from depression after the death of her husband, is given a compassionate makeover. The makeover on this show thus teaches the audience about taste, style and consumption while giving them lifestyle tips at the same time as it emphasises the emotional needs of the makeover-ees, as this summary on the show's website suggests:

> We went for a classic 'Hamptons' look in the lounge room. We've used cool colours on the walls to create a comfy lounge area. White shutters have been added to the windows and we've created a photo feature wall and installed a massive floor-to-ceiling bookcase. We hope this will create the perfect retreat for Margaret.

The promotion of consumption (the show's website has a list of suppliers, and audience members regularly comment on the website and ask for details of paint colours, stockists of furniture, etc.) and teaching of taste here are seamlessly blended into the makeover recipient's broader personal lifestyle, while the show insists on portraying itself as offering a kind of community service for those in need.

In its focus on supporting families and community and on the potentially therapeutic aspects of the makeover process, *Domestic Blitz* draws strongly upon elements of US makeover formats such as *Extreme Makeover: Home Edition* in which the show's producers work with local builders to provide a home makeover for families that have faced some sort of recent or ongoing hardship. Such shows in turn draw on a long tradition in the US of beneficent television (e.g. the fifties show *Queen for a Day*) (Watts, 2006). However, *Extreme Makeover: Home Edition*, like many US makeover shows, has a stronger focus on spectacle, heightened emotion and sentimentality ('a dream come true' 'making a difference one family at a time'), while the hosts and experts are rather more polished and glamorous than the more ordinary hosts on Australian TV.

Another point of difference between Australian and US approaches to the makeover is the latter's relative emphasis on personal makeovers. While Australian audiences have recently embraced personal makeover shows like *The Biggest Loser*, they tend to be less comfortable with the highly aggressive competitive individualism that is often central to US game show-style makeover and reality formats. And, as noted, Australian reality TV tends to be more concerned with constructing a familiar, neighbourly mode of address rather than with emphasising social differences. Reflecting these concerns, the focus on the Australian version of *The Biggest Loser* (particularly in the first series) has tended to be less on US-style individualism than on losing weight for one's family and the community. This is not to say that there are no issues of competitive individualism at work in Australian makeover formats—aspirationalism is a key mantra in Australian lifestyle culture. But the discourse of 'getting ahead' tends to be framed in terms of aspiring to a kind of average-ness, a preoccupation that speaks to a broader cultural mythos of social egalitarianism.

'Oriental Vogue': Hybridising Culture on Singaporean Reality TV

In recent years lifestyle makeover formats have not only 'travelled' extensively within linguistically and culturally congruous TV territories, they have also started to make an appearance in less culturally proximate sites such as Asia. As noted China has embraced a range of reality formats as well as lifestyle shows (Xu, 2007). Likewise Singapore, the object of focus for this essay, has produced a number of locally made makeover shows.

One important context for the rise of reality and lifestyle television in Asia is the broader explosion of media consumption and rise of lifestyle consumer practices across the region in the past few decades. The media and entertainment sector in Asia, for instance, is one of the world's fastest growing industries, with television being by far the sector's dominant player—television now reaches 97% of the population in China, for instance (China Media Monitor Intelligence). The growing role of television in the region has occurred hand in hand with the liberalisation of economic and, to a varied degree, state structures. One of the corollaries of these processes has been a rapid rise in social mobility and the emergence of new forms of consuming 'middle classes' (Chua, 2000; King, 2008).

In this context, far from just being cheap, disposable television, reality shows in Asia are playing a significant role in promoting certain lifestyle behaviours and, concomitantly, social identities, offering not just consumer advice but lifestyle guidance in a period of shifting cultural and social mores. As noted, the rise of reality makeover TV speaks to a range of broader social shifts in neoliberal western states reflecting in particular the increasing dominance of an individualistic, consumer-driven approach to lifestyle issues in which late modern selfhood is seen as endlessly malleable—a project to be worked on and invested in. If the rise of lifestyle TV in the west can be linked to these broader economic, cultural and social shifts, however to what extent can these developments be applied to Asian contexts marked often by different cultural and political traditions and models of modernity? The Singaporean case study that follows explores these issues through examining the sorts of values and models of lifestyle, selfhood and citizenship being offered up on local makeover shows.

Unlike many of its Asian counterparts, Singapore—as an advanced capitalist, ex-British colony—already has a well developed consumer and lifestyle oriented media culture, with advertising, radio and print media in Singapore offering up a complex combination of 'western' and 'Asian' lifestyle imagery. In contrast to the TV industry in the Anglo-American context where deregulation and privatisation is the norm, broadcast television in Singapore is highly regulated, falling primarily under the jurisdiction of the state owned collection of companies known as Mediacorp. Mediacorp thus operates all three Singaporean terrestrials—Channel 5 (English-language), Channel 8 (Chinese) and Channel U (Chinese)—as well as the TV12 specialty services: Suria for Malay audiences, Kids Central, Arts Central and the Tamil-language Vasantham Central.

While TV broadcasting in Singapore is dominated by Chinese programming, a review of the evening schedule indicates that lifestyle-reality formats feature on most TV channels. In January 2008, for instance, the 8-10pm slot on Mediacorp's main Chinese channel Channel 8 featured Chinese-language shows like *Home Décor Survivor 3* (a home makeover show) and *Good Food Fun Cook* (a reality-style cooking show).[6] Arts Central (which features a variety of mostly foreign, English language programming often with Chinese subtitles) offered a regular 9-10pm lifestyle slot including a rather glossy, locally produced but strongly western-inflected game show-style cooking show, *The Food Bachelor*, in which a group of attractive, ethnically diverse young men (chosen to reflect Singapore's multicultural community) with minimal cooking skills compete for the opportunity to host their own cooking show. The Malay channel Suria meanwhile offers shows like *Cari Menantu*, described as 'a reality program that gives a spouse-to-be a crash marriage preparatory course' and a D.I.Y. home décor show *ID Kreatif*. While reality shows feature on all of the Mediacorp channels, the majority of Singaporean lifestyle programming are made for Chinese audiences and hence the focus in this next section is primarily on Chinese reality formats, and in particular makeover shows.

Like early forms of the makeover on UK and US television, makeover TV in Singapore has often focused on beauty makeovers and has been aimed at female audiences (e.g., Channel 8's *Beautiful People* aired in 2003).[7] More recently though the makeover format has started to evolve and diversify on Singaporean TV, targeting a broader audience through the emergence of Chinese-language shows like the eco-makeover programme *Energy Savers* and the highly popular *Home Décor Survivor* series, which first aired in 2005 with the spin off *Junior Home Décor Survivor* coming in at number five in the top twenty TV programs for the ratings period of March 2008.

---

6 Mediacorp's other Chinese channel, Channel U also has some lifestyle shows but these cater for more of a niche audience, addressing in particular university students and high school students with shows like *Campus Yummy Hunt* where the hosts head to different campuses to find the best and cheapest food outlets.

7 At the time of writing, a new full length beauty show was being aired on Mediacorp 8 at 8.30pm on Friday. *Follow Me to Glamour* is a reality style 'outdoor game show' based around the search for suitable candidates to undergo beauty make-over sessions in public.

One point to note here is that, aside from beauty shows, makeover shows focused on individual personal transformations have not featured so strongly in local Singaporean programming. Constrained by low budgets, an arguably more communitarian approach to social identity and a relative cultural reserve—amongst Chinese audience members at least—in relation to exposing oneself and one's lifestyle on television, the spate of plastic surgery, behavioural and body makeover shows of the ilk of *Extreme Makeover* and *The Biggest Loser* that have taken off more recently in the west have yet to emerge on Singaporean TV.

The types of makeover shows that have started to become popular on primetime television then are ones oriented towards renovating the home rather than the self. The *Home Décor Survivor* series, for instance, is a popular home show which borrows heavily from Anglo-American makeover formats, offering a kind of Chinese-Singaporean version of *Changing Rooms* with a touch of *Queer Eye for the Straight Guy* (albeit with the overtly gay elements and the personal makeover taken out). Featuring a competitive game show element, two teams each led by a young male host (comedian Mark Lee and Bryan Wong, known as Mediacorp's 'hosting king') vie to make over the interior space of two homes while staying within a budget of $6000. Like many western home makeover formats, the show combines a class-inflected education in modernist taste, style and aesthetics (in one show there is a particular emphasis—highlighted by English words popping up on screen—on 'modernism' and on creating spaces that are 'funky' and ' industrial looking'), with a focus both on DIY and thriftiness, and consumerism. Thus, the teams are seen creating one-off art works and wall stencils for the home interiors of the show's participants while home owners are also taken to various stores to buy furniture (with prices and the name and address of stores provided to the audience), with frequent adverts in the break for the show's sponsor, a home wears store.

While educating the audience in design and aesthetics, the overall tone of the program is one of youthful informality, with the young hosts and makeover team engaging throughout in cheeky banter and comic hi-jinks, presenting themselves and relating to each other in a manner that is distinctly 'student-ish' as opposed to 'respectable' or 'serious.' Thus the show offers a fairly soft and accessible form of lifestyle pedagogy, modelling forms of middle-class cosmopolitan taste in ways that are clearly targeted toward 'ordinary Singaporean youth': students and young families living in small, standardized government flats.

While the show borrows heavily from Anglo-American home makeover formats, it also draws upon the aesthetics and conventions of hybrid Japanese-Chinese variety TV. In particular, the show has a comic, zany feel with pop-up coloured images and words exploding onto the screen accompanied by comic sound effects. The group presentation, slapstick humour, incessant cheeky cross-talk, 'busy' screen aesthetic and dense sound scape contribute to an overall feel of *renao*, a positive term meaning 'lively; busy; noisy; fun' that encapsulates the feel aspired to by much Chinese-language variety-style television.[8] Likewise, the content of the show blends European design tips and global cosmopolitan style with local concerns. The focus is mainly on renovating the small Housing Development Board flats in which most Singaporeans live with an emphasis on hybridising modern design with traditional aesthetics; one episode is themed 'Ethnic Fusion' with the team's goals being to blend ethnic Peranakan style with modern design while another focuses on 'oriental vogue'.

---

8 Thanks to Fran Martin for this insight.

Energy Savers, which aired in 2008 on Channel 8 on Thursdays at 8.30 pm, is another home makeover show of sorts, although one concerned with transforming the energy consumption of Singaporean households. Like *Home Décor Survivor* it adopts a reality-based, competitive game show format with the show's central 'challenge' being for the twelve participating households to reduce their energy consumption by at least 10 per cent while thinking up 'creative ways' for saving electricity along the way, with the best household winning $5,000. The show's male and female hosts are young attractive Singaporean personalities and, like *Home Décor Survivor*, the show's tone is highly comedic and playful with rapid comic voice-overs and the liberal use of pop-up images and words (complete with 'zany' sound effects) to emphasise particular household tips or energy consumption issues.

However, while the show aims for a light variety-style feel its agenda is rather more educational than *Home Décor Survivor*; the hosts guide the audience through an audit of the households, noting the range of appliances they own and their current energy use and then offering suggestions for reducing energy consumption. The households on display here range from a young couple with a baby living in a Housing Development Board flat to larger, more affluent families living in freestanding houses suggesting the show is aimed at a rather larger cross-section of the Singaporean public than the more youthful audience of *Home Décor Survivor*.

While *Energy Savers* is similar in feel to other info-ed/variety shows on Singaporean television, the format's focus on reducing energy consumption aligns it with a range of recent lifestyle makeover shows coming out of the Anglo-American context from competitive weight-loss shows like *The Biggest Loser* to eco-makeover formats like Australia's *Eco-house Challenge*. While such shows are ostensibly entertainment-oriented makeover formats they can also be seen to promote neo-liberal models of good consumer-citizenship in which community concerns such as obesity and the global oil crisis are treated as issues that can be dealt with at the level of individual consumer behaviour and self-regulation (Lewis, 2008b).

Such a show however also needs to be understood in the context of Singapore's rather distinctive mode of authoritarian capitalism. Possessing a neoliberal market alongside a strongly regulatory state, Singapore is marked by a form of neoliberalism rather different from its western counterparts, one that as Harvey (2005, p.86) notes blends capitalism with Confucianism, nationalism and a 'cosmopolitan ethic suited to its current position in the world of international trade'. Singaporean entertainment-based television then, while addressing consumers as self-governing citizens and consumers, is also strongly shaped by state dictates around cultural values (such as ensuring that hosts speak standardised Mandarin). While there has been a distinct pedagogical 'turn' on Anglo-American reality television, the bottom line for programmers in these settings (the BBC being somewhat of an exception) tends to be commercial and ratings driven. While such concerns are also important for Singaporean TV producers, the public educational elements of Singaporean shows are more overt; shows are often packaged in terms of their benefit to the community while reality TV producers often take into account government concerns and campaigns around lifestyle issues when they are creating shows aimed at promoting good citizenship.

Another very popular form of programming in Singapore which again speaks to the question of television's articulation to hybridising formations of cultural modernity in Singapore is food TV. Food programming has been a long standing genre in a range of Asian TV markets with the Japanese game show format *Iron Chef* even being exported to the US and remade as *Iron Chef America*. Alongside Japan, Anglo-American trends in lifestyle TV

Mobile Makeovers: Global and Local Lifestyles and Identities in Reality Format   205

have also arguably had an influence on the more recent rise in the region of the celebrity chef and reality-style, entertainment-oriented cooking shows more generally, shows which, while not strictly speaking makeover shows, are marked by a 'transformational aesthetic' and by a concern with teaching ordinary people about middle class forms of taste and distinction (de Solier, 2005).

Typical of the kind of everyday lifestyle programming popular with Chinese-Singaporeans are cheap, down-home formats like *Good Food Fun Cook* (GFFC), aired in 2008 on Friday at 8pm on Channel 8. Targeted at housewives and showcasing the talents of celebrity chef Sam Leong, 'the idol in the cooking world',[9] and Quan Yi Feng, one of Singapore's top TV hosts, GFFC brings 'the kitchen out to the public', with episodes featuring Sam cooking in an open-air kitchen, haggling with vendors and mingling with locals at street markets.

As in *Home Décor Survivor* and *Energy Savers*, the mode of address on the show is informal and zany again aiming for a feeling of *renao*. The show has a highly populist agenda reflected in the way in which Sam and Quan Yi Feng interact with the ordinary members of the public who gather to watch and learn as they cook—people whose very ordinariness is framed to reflect the 'aunties' that are the show's target audience. At the same time, like many Singaporean shows, GFFC combines an entertainment oriented approach to lifestyle with an educational agenda. Sam on the one hand is positioned as a man of the people—struggling with a very stilted Cantonese-inflected Mandarin, in distinction to the fluency of the Taiwan-born Quan Yi Feng[10]—but at the same time, he is there to teach the audience about practical recipes, quality food, and style and aesthetics. As the show's Senior Executive Producer, Tay Lay Tin notes, 'on GFFC, it's the first time we are educating the audience to say you can do this five star cuisine at home. The food is very simple but the presentation is upper class. Sam Leong is famous for this'.

The show also teaches the audience about healthy food, with each episode focusing on one of thirteen themes, such as how to manage hair loss, how to look youthful, how to keep fit, etc. The show's research team thus includes a Chinese physician who helps choose healthy ingredients for the show's dishes and a research writer who, as Tay Lay Tin notes, makes 'these issues simple, lighter… more approachable for a general audience.' GFFC then combines a focus on taste and aesthetics (similar to *Home Décor Survivor*) with the kind of public educational focus apparent in shows like *Energy Savers*. The show initiates ordinary citizens into cosmopolitan forms of taste while at the same time addressing them as good healthy Chinese-Singaporean citizens.

In varied ways the three Chinese-Singaporean productions discussed can all be seen to position local audiences as reflexive cosmopolitan consumer-citizens negotiating western, regionalist Chinese and local models of lifestyle consumption and social identity.[11] Through transforming the home *Home Décor Survivor* performs a complex hybridised cultural

---

9 Interview with Tay Lay Tin, Senior Executive Producer, Chinese Entertainment Productions, Singapore, January 2008.

10 Thanks to Fran Martin for this observation.

[11] While Singaporean TV might be seen to legitimate global middle class lifestyles there are limits to the kinds of cosmopolitan lifestyles it will portray. For instance, featuring queer identified actors or hosts is a no go zone for Singapore TV (although it can feature camp hosts who are 'read' by the audience as gay) as evidenced by the recent case of a home makeover show that was fined for featuring a gay couple who wanted to transform their game room into a new nursery for their adopted baby. 'Singapore government fines TV station for gay show' http://www.boston.com/news/nation/articles/2008/04/24/singapore_fines_tv_station_for_gay_show/

aesthetics tied to both global and Asian taste cultures but framed largely in consumerist terms. *Energy Savers* likewise speaks to both global and national-governmental concerns around thrift, responsible consumption and self-regulating modes of citizenship. *Good Food Fun Cook* (GFFC), meanwhile, is a particularly localised and ordinary mode of lifestyle television—tied to local people and places, overtly addressed to 'ordinary' (middle-aged, working class, female) audience members, and offering practical how-to advice on simple, everyday home cooking. But here also we see a degree of cosmopolitan aspirationalism on display (again framed in highly localised ways)—played out in this instance through its concern with teaching audiences how to appreciate 'five star cuisine' and the show's healthy agenda.

GFFC's consumer message (which ties aspirational taste to healthy lifestyles) can easily be read as affirming the simple spread of a global, neoliberal agenda of enterprising selfhood. But as I have suggested such modes of lifestyle consumption need to also be understood in relation to local and regional Chinese cultural values (for instance on GFFC Chinese traditions of medicinal food is an important focus of the show). Likewise the healthy, responsible model of selfhood promoted on such formats is articulated to a rather localised form of neoliberalism, here paradoxically reflecting the close regulatory relationship between the media and the Singapore government, which has been actively pushing a healthy lifestyle campaign through media sites such as television. While makeover shows like *Home Décor Survivor* and *Energy Savers* and lifestyle programs like GFFC then all speak to a certain extent to the globalizing rubric of western 'lifestyled' forms of identity, they nevertheless do so in ways that complicate universalistic models of lifestyle and modernity.

CONCLUSION

This essay has discussed the global mobility of the reality TV format through the specific lens of the lifestyle makeover show, examining the way in which this form of programming has been shaped by both globalising forces and domestic concerns and contexts. As we have seen, from Australia to Asia, TV markets around the world have been marked by the emergence of programming concerned with modelling and promoting particular modes of identity, consumption, taste and lifestyle. On lifestyle makeover television, in particular, the homes and lives of ordinary people (and at times celebrities) are paraded as examples of ideal (or in the case of makeover shows, not-so-ideal) models of selfhood while lifestyle experts provide us with rules and guidelines for managing increasingly complex lives. What does this global embrace of makeover formats tell us however about local TV markets and cultures? Does the transnational mobility of the makeover show mean that Anglo-American models of enterprising individualism and self-improvement are becoming hegemonic?

The ready acceptance of the makeover ethos within the Australian market suggests a certain universality (at least within Anglo-American TV territories) of late capitalist concerns around the promotion of flexible lives and selves and the centrality of style, aesthetics and 'good' forms of consumption to optimising one's lifestyle. At the same time, Australian makeover shows have a significantly different cultural 'feel' to them compared with US and UK shows, reflecting the fact that Australia's reception and reworking of these formats has been shaped by distinctive cultural concerns and televisual traditions. The familiar ordinariness of Australian lifestyle TV, played out in programmes like *Backyard Blitz*, points to the role of culturally-inflected values and mythologies, in particular concerns with communality, social egalitarianism and 'mateship'. While articulated within the rubric of 'lifestyle' on makeover TV, such concerns can be seen to complicate notions of a unitary, homogeneous and seamlessly global 'western' lifestyle culture.

Likewise the Singapore case study presented here suggests that the notion of 'lifestyle' needs to be understood not only in relation to global shifts in identity around consumer culture and late modernity but also articulated to specific geo-cultural contexts and local/regional modernities. Reality programming, while on the one hand seemingly 'selling' global models of lifestyle, taste and consumption, tends to be also tied to the familiar and the domestic. The double-edged nature of these forms of programming is evident in Singaporean makeover shows, where the examples discussed display a complex and varied blend of local embeddedness and nostalgia for local and regional Chinese traditions with a global sensibility. On these shows, an emphasis on Chinese medicine and the health-giving properties of food sits cheek by jowl with the modelling of cosmopolitan middle class taste and 'five star cuisine'; advice on adapting Peranakan traditions next to a focus on modernist aesthetics.

Both case studies then—'western' and 'Asian'—point to the limits of assumptions that the forms of selfhood emerging through reality TV can be read as merely 'western' or even 'westernized' in any simple sense. Instead, these forms of subjectivity and identification— like the French cuisine or the pop art design taught by the programs—are themselves likely to be significantly indigenised and 'made-over' in their uptake in these specific cultural contexts.

ACKNOWLEDGMENTS

I'd like to thank Fran Martin for her valuable critical insights into television in Asia and her English translations of the Chinese Singaporean shows discussed in this chapter.

REFERENCES

Andrejevic, M. (2004). *Reality TV: The work of being watched*. Lanham, Md: Rowman and Littlefield.

Bignell, J. (2005). *Big Brother: Reality TV in the twenty-first century*. Basingstoke, UK: Palgrave Macmillan.

Biressi, A., and Nunn, H. (2005). *Reality TV: Realism and revelation*. London: Wallflower.

Bondebjerg, I. (1996). Public discourse/private fascination: hybridization in 'true-life-story' genres. *Media Culture and Society, 18*(1), 27-45.

Bonner, F. (2003). *Ordinary television: Analyzing popular TV*. London: Sage.

Bratich, J. Z. (2007). Programming reality: Control societies, new subjects, and the powers of transformation. In D. A. Heller (Ed.), *Makeover television: Realities remodeled* (pp. 6-22). London: I.B. Tauris.

Brunsdon, C. (2003). Lifestyling Britain: The 8-9 slot on British television. *International Journal of Cultural Studies, 6*(1), 5-23.

Brunsdon, C., Johnson, C., Moseley, R., and Wheatley, H. (2001). Factual entertainment on British television: The midland's TV research group's 8-9 project. *European Journal of Cultural Studies, 4*(1), 29-62.

Chua, B. H. (2000). *Consumption in Asia: Lifestyles and identities*. New York: Routledge.

Corner, J. (2004). Afterword: Framing the new. In S. Holmes and D. Jermyn (Eds.), *Understanding reality television* (pp. 290-299). London: Routledge.

Crawley, M. (2006). Making Over the New Adam. In D. Heller (Ed.), *The Great American Makeover: Television, History, Nation* (pp. 51-64). New York: Palgrave Macmillan.

de Solier, I. (2005). TV dinners: Culinary television, education and distinction. *Continuum: Journal of Media and Cultural Studies, 19*(4), 465-481.

Ellis, J. (2000). *Seeing things: Television in the age of uncertainty*. London: I.B. Tauris.

Everett, A. (2004). Trading private and public spaces @ HGTV and TLC: On new genre formations in transformation TV. *Journal of Visual Culture, 3*(2), 157-181.

Harvey, D. (2005). *A brief history of neoliberalism*. Oxford, UK: Oxford University Press.

Heller, D. (Ed.) (2006). *The great American makeover: Television, history, nation*. New York: Palgrave Macmillan.

Hetsroni, A. (2005). Rule Britannia! Britannia rules the waves: A cross-cultural study of five English-speaking versions of British quiz show format. *Communications: The European Journal of Communication Research, 30*(2), 129-153.

Hill, A. (2005). *Reality TV: Audiences and popular factual television*: Routledge.

Kavka, M. (2006). Changing Properties: The Makeover Show Crosses the Atlantic. In D. Heller (Ed.), *The great American makeover: Television, history, nation* (pp. 211-229). New York: Palgrave Macmillan.

King, V. T. (2008). The middle class in southeast Asia: Diversities, identities, comparisons and the Vietnamese case. *IJAPS, 4*(2), 73-109.

Lewis, T. (2008a). *Smart living: Lifestyle media and popular expertise*. New York: Peter Lang.

Lewis, T. (2008b). Transforming citizens: Green politics and ethical consumption on lifestyle television. *Continuum 2*(2), 227-240.

Lewis, T. (2009). Changing Rooms, Biggest Losers and Backyard Blitzes: A history of makeover television in the UK, US and Australia. In T. Lewis (Ed.), *TV transformations: Revealing the makeover show* (pp. 7-18). London: Routledge.

Magder, T. (2004). The end of TV 101: Reality television, formats and the new business of TV. In S. Murray and L. Ouellette (Eds.), *Reality TV: Remaking television culture* pp. 137-156). New York: New York University Press.

McCracken, E. (1993). *Decoding women's magazines*. New York: Palgrave Macmillan.

McRobbie, A. (2004). Notes on 'What Not To Wear' and post-feminist symbolic violence. *Sociological Review, 52*(2), 97-109.

Miller, T. (2007). *Cultural citizenship: Cosmopolitanism, consumerism and television in a neoliberal age*. Philadelphia: Temple University Press.

Moran, A. (1998). *Copycat television: Globalisation, program formats and cultural identity*. Luton: University of Luton Press.

Moran, A., and Keane, M. (2006). Cultural power in international TV format markets. *Continuum, 20*(1), 71-86.

Moseley, R. (2000). Makeover takeover on British television. *Screen, 41*(3), 299-314.

Moseley, R. (2003). The 1990s: Quality or dumbing down? In M. Hilmes (Ed.), *The television history book* (pp. 103-106). London: British Film Institute.

Ouellette, L., and Hay, J. (2008). *Better living through television*: London: Blackwell.

Palmer, G. (2004). 'The New You': Class and transformation in lifestyle television. In S. Holmes and D. Jermyn (Eds.), *Understanding reality television* (pp. 173-190). London: Routledge.

Redden, G. (2007). Makeover morality and consumer culture. In D. Heller (Ed.), *Reading makeover television: Realities remodeled* (pp. 150-164). London: I.B. Tauris.

Rose, N. (1989). *Governing the soul: The shaping of the private self.* London: Routledge.

Rose, N. (1996). *Inventing ourselves: Psychology, power, and personhood.* Cambridge, UK: Cambridge University Press.

Skeggs, B., and Wood, H. (2009). The labour of transformation and circuits of value 'around' reality television. in T. Lewis (Ed.), *TV transformations: Revealing the makeover show* (pp. 119-132). London: Routledge.

Taylor, L. (2002). From ways of life to lifestyle: The 'Ordinari-ization' of British gardening lifestyle television. *European Journal of Communication, 17*(4), 479-493.

Thussu, D. K. (2007). *News as entertainment: The rise of global infotainment.* London: Sage.

Waisbord, S. (2004). McTV: Understanding the Global Popularity of Television Formats. *Television and New Media, 5*(4), 359-383.

Watts, A. (2006). Queen for a Day: Remaking consumer culture, one woman at a time. In D. Heller (Ed.), *The great American makeover: Television, history, nation* (pp. 141-157). New York: Palgrave Macmillan.

Wood, H., and Skeggs, B. (2004). Notes on ethical scenarios of self on British reality TV. *Feminist Media Studies, 4*(2), 205-208.

Xu, J. H. (2007). Brand-new lifestyle: consumer-oriented programs on Chinese television. *Media Culture and Society, 29*(3), 363-376.

In: Reality Television-Merging the Global and the Local
Editor: Amir Hetsroni, pp. 211-258

ISBN 978-1-62100-068-6
© 2010 Nova Science Publishers, Inc.

Chapter 13

FROM REALITY TV TO COACHING TV: ELEMENTS OF THEORY AND EMPIRICAL FINDINGS TOWARDS UNDERSTANDING THE GENRE

Jürgen Grimm
University of Vienna, Austria

This chapter [1] has a twofold goal: theoretical clarification of the reality TV genre and testing the theoretical essentials by empirical findings referring to the *Supernanny* formats in five countries. Part 1 tries to define a basic framework of reality TV and aims at answering the questions which social developments add to the popularity of this truly global television genre and which criteria form its smallest common denominator. Based on Alfred Schutz' sociological phenomenology of everyday life (called *"lifeworld"* theory)[2], reality TV is analyzed with regards to its specific contributions to the recipient's everyday living environment. Especially the latest trends towards coaching TV (lifestyle, upbringing) show clearer than before what makes up reality TV's extra-medial reference and what lines of development its sub-genres follow. The second part presents results of a comparative international study of *Supernanny* programs in England, Germany, Austria, Spain and Brazil gained by a research project at the University of Vienna. Together with the survival and celebrity shows, the *Supernanny* format marks the most dynamic area of development in the post-*Big Brother*-era. The interpretation of results on British edutainment television is governed by the question whether and if so, to what degree global marketing and stable formats work well with the adaptation to the respective countries' parenting traditions and which different upbringing problems on the one hand are visible and what different parenting recommendations are being given on the other hand in the various countries. The data of the content analysis are supplemented by results from a survey of 1611 *Supernanny* viewers in Germany and Austria that allow to double-check some of the theoretical essentials on reality

---

[1] For helping with the English version of the chapter the author thanks Oliver Hoffmann and Petra Schwarzweller.
[2] On the use and spelling of the term "lifeworld", cf. Ihde (1990).

TV concerning the audience's motives. Finally, I rely on the findings to evaluate and forecast the future perspectives of reality TV.

PART ONE: TOWARDS A THEORY OF REALITY TV

"The cinema can be defined as a medium particularly equipped to promote the redemption of physical reality. Its imagery permits us, for the first time, to take away with us the objects and occurrences that comprise the flow of material life." (Kracauer, 1960, p.300)

Redemption of Reality?

50 years ago, the sociologist and culture theorist Siegfried Kracauer emphatically saluted the era of the cinema as the "redemption of physical reality". In an adaption to an abstraction caused by ideologies, science and literature he saw the main reason for the alienation from the basic facts of life. Cinema was supposed to change that because unlike paintings, novels or theatre it works directly with reality particles, or more exactly with "the material world with its psychophysical correspondence" (Kracauer, 1960, p.300). Though this did not create perfect naturalism, it created a special form of art that's able to teach the perception of reality 'bottom up'. Today, the situation has changed drastically: We are surrounded by moving pictures that have long since lost their innocence as "psychophysical correspondence" because their orchestrated character is all too clear. Therefore, there is no reason to hope for pseudo-religious redemption – on the contrary: the media images seem to rather blot out than to reveal "reality". Media users react to the loss of cognitive unambiguousness by searching for remains of objectivity lost in the quicksand of ephemeral media choices. That is the hour of birth of "reality TV" that has been motivating program innovations time and again since the 1980s – and that, in a sort of double frontline position, protests against the ideological top-down techniques of educationalization and against the naïve equalization of appearance and reality.

The problem with exact definitions of reality TV is almost proverbial (Kilborn, 1994; Andrejevic, 2004; Hill, 2006; Krakowiak et al., 2008; Ouellette and Hay, 2008). Program experiments span from formats focusing on catastrophes and crime to dating and talent shows and hybrid programs like *Big Brother* and *Survivor*. The phenomenon is heterogeneous and global: *Big Brother* is broadcast in dozens of countries (Mathijs and Jones, 2004; Frau-Meigs, 2006). Even in China (Keane et al., 2007) and the Arabic countries (Kraidy, 2008), reality TV formats are immensely popular. Diffusion and change happen so quickly that the theorists are hard-pressed to adapt their attempts on definition and ascription of cultural functions in time. Hill compares reality TV research to a *moving target*: "Just as you get your bearing on the latest reality format, another format steps in, and you have to change direction." (Hill 2006, p.192). Still, there is a lowest common denominator shared by all reality TV programs that is schematically as follows:

Reality TV programs show
1) ordinary people on the screen
2) in positive and negative everyday situations

3) with an appeal of authenticity (lay actors, clumsy show-offs, members of the lower classes etc.)
4) in a way that interacts with the show participants' everyday life.

Reality programs are centered
5) around crucial events in life that destroy the ordinary everyday routine and mark a challenge.

The participants' efforts
6) aim at solving a problem (e. g. relationship conflicts) or are part of an attempt to upgrade social positions (e.g. gain fame).

The viewer
7) can draw conclusions from the participants' success or failure.

These seven basic elements of the reality TV definition apply to various implementations of the genre. They are tied together by a *culture of announcement* in which the private life becomes available to the general public for watching, noticing and assessing, be it out of an existential or everyday emergency or for reasons of culmination of attention (Franck, 1998). In a neutral, proto-moral sense, it is all about confessions of everyday people who hope to gain something by presenting themselves to the general public. The publication of the private is often fueled by a personal problem and offers the audience a chance to optimize their own crisis management by comparing their own everyday situations and coping strategies to the ones featured on the screen. The confrontation with others' reality experiences helps working on one's own model of reality.

My thesis postulates that reality TV reacts to the reality crisis of the electronic media whose founding in reality has become brittle. It is at the same time an embodiment of this crisis and an attempt at solving it. Reality TV doesn't bring the redemption of the physical reality as a whole, but it restitutes everyday life as a basic reality and thus offers to each individual a shelter from systemic affronts by the economy, politics and pedagogics. Earlier educational TV programs saw the audience as "a gullible mass that needed guidance in the liberal art to participate in the rituals of public democracy. Today's popular reality TV addresses the viewer differently. The citizen is now conceived as an individual whose most pressing obligation to society is to empower her or himself privately." (Ouellette and Hay, 2008, p.3).

Characteristic for this is the "savvy viewer" (Andrejevic, 2004; 2008), who responds to the strategies of deception with equally strategic unveiling techniques by trying to see through the media orchestration (Hill, 2000; 2004; 2006). For the British *Big Brother* viewers, Hill shows clearly that they perceive the "true' self" of the show participants behind their performing techniques and thus get a satisfying light bulb moment. The participants' mutually thwarting profiling attempts, especially failed razzle-dazzles, allow conclusions regarding the characters behind them (Grimm, 2001). Thus, the "savvy viewers'" distance from the depicted media reality is a built-in fact of reality TV.

The Voyeur's Change

Alternately, the viewer can unveil her or himself and change – as Mark Andrejevic puts it – from "voyeur" to "exhibitionist". Indeed, one of reality TV's defining characteristics is the flexible role permeability between the shows' viewers and participants. But it seems arguable if the connection of voyeurism and exhibitionism must actually be interpreted with what Andrejevic (2008, p.333) – following Sigmund Freud - calls *scopophilia*. In a fit of ardor, this alludes to St. Augustine's "concupiscentia oculorum" (397 A. D.). St. Augustine generally suspects all forms of lust of the eyes in the late Roman Empire (theatre, fine arts and of course gladiatorial fights) of being sinful. The problem is the lack of specificity in his judgment. In an era of ubiquitous optical media, voyeurism becomes ever-present and thus meaningless. It does not explain the affinity to any particular program, as it – in for a penny, in for a pound – concerns all forms of reception of audio-visual media. Kracauer is very modern with regards to that, stating neutrally: "However, the supreme virtue of the camera consists precisely in acting the voyeur." (Kracauer, 1960, p.44) As the goal of 'voyeurism', he regards the "change of the agitated viewer", be it by seeing the Medusa's head in Achilles' shield and confronting his or her fears or by everyday's banality showing itself in full detail and vacuity. Thus, we are not talking sheer lust of the eyes; rather, the viewing is functionally embedded into working on oneself and one's own life.

Here, Kracauer comes very close to the reality TV phenomenon. For *Big Brother* with its principle of observation is not the ultimate reality TV (and neither is it the habituation to a surveillance society), but the performative presentation of everyday situations in the media to make them controllable by the everyday lifeworld subjects. Reality TV is "everyday life in a state of emergency" (Grimm, 1995) by and for ordinary people. The man on the street's self-empowerment of course does not stop the agents of the media system from using and exploiting the mass audience's needs to raise the quota and/or pedagogic or political elites from superimposing their own goals to the new forms of media communication. But – and that is the core thesis – basically, reality TV is a form of reality affirmation *operationally included into everyone's everyday life*. The viewer's sense of reality (and thus, reality TV's global popularity) does stem neither from the immediate form of watching nor from its functionality for political and economical goals, but from the connection of everyday experiences to the media events and the consequences for the viewers' lives that can be derived from them. This is the "practical value" of the commodity reality TV without which no "exchange value" (and no media economical gain) can be realized – and without which no political campaigns brokered by reality TV will work. It is interesting that American attempts at reality TV from Iraq showing heroic soldiers flopped, while reality programs on the consequences of the war in Iraq broadcast by the station Al-Sharqiya (The Eastern) run by Saddam Hussein's former secretary of propaganda, Saad al-Bazzaz, amongst them the format *Labor Pulse Material* showing debris clearings after bombings, were extremely popular (McMurrie, 2008, p.194). In the former case, there is no everyday surplus value for the Iraqi audience (that feels abased by the occupation), while in the latter the scenario coincides with the urge to not only survive, but to cope with an extremely violent everyday life.

The "world of everyday life", marked by routines (repeating actions, situations and interactions) and crucial changes (birth, marriage, sickness, death) is, according to Alfred Schutz (Schutz and Luckmann, 1973; 1983) paramount reality that we take for granted, within which our routines develop, where we communicate with others, meet with resistance and with which we interact actively. Schutz opines an acentrical world view centered on the everyday lifeworld subject. "The knowledge about the world is in two ways, socially and biographically based on experience; i. e. the subject is affected by its predecessors' and teachers' interpretations of the world as well as by its own experiences which in the form of 'knowledge at hand' function as a scheme of reference". (Schutz, 1945, p.533).

Intersubjectivity with others ensures that the worlds of socially close contemporaries converge to a certain degree, even if they constantly need to be compared anew and, if necessary, renegotiated. The *world of everyday life* forms the master pattern from which other worlds or related "finite provinces of meaning" like e. g. religion, art and science, but also dreams, are created and is the measure for all of them.

In the *world of everyday life*, we are players and voyeurs at once. Our environment is kept under surveillance for dangers or problems that could endanger our routines. The automatic chain of actions and experiences is broken once the unexpected happens. Enter the voyeur who is driven by either curiosity or fear. Psychologists call this a dissonance of cognitive patterns (Berlyne, 1960; Schank and Abelson, 1977; Anderson, 1995), which disrupts ongoing actions and triggers *orientation reactions* (Sokolov, 1965). The exploration of our field of vision helps us identify potential dangers and look for possible solutions. If watching is pleasant on top, all the better.

In any case, the voyeur's change while watching reality TV marks a learning process. A heated public discussion in Germany accused the watchers of the reality format *Notruf* (German version of *Rescue 911*) showing accidents and rescue missions of "catastrophe voyeurism". However, an experimental study showed that people were more ready to help others after watching the show. It is also proven that the *Red Cross* received more applications for first aid courses, when this show was on the air. Watching catastrophes in *Notruf* turned a lot of people into dedicated rescue helpers, and even those that did not develop any interest in professional knowledge showed a certain interest in catastrophes while watching *Notruf*, but only if fire brigades and police could effectively quell them (Grimm, 1995). It seems telling that the test resulted in under-average values for "sensation seeking" for heavy viewers of *Notruf*. There was no trace of sensationalist motives for their interest. The satisfaction does not primarily spring from watching, but from gaining a better orientation regarding a problematic situation. Watching reality TV can be fun, but the intentional reference to the everyday living environment remains the main motivating variable endowing the watching with meaning. At least, this is true for regular viewers of reality formats. "Better living through reality TV" is Ouellette's and Hay's (2008) short formula. The goal is to better control one's own living conditions, something which Rotter (1966) calls "locus of control" and Bandura (1994) terms "self-efficacy".

Everyday Life and Reality TV

Media scholars like Tamar Liebes and Sonja Livingstone (1994, 1998) show through the example of Daily Soaps how viewers relate fictional stories to their everyday situation by relating the complications on screen to their own experiences and state similarities or differences. Analogically, reality TV users do not simply transfer the situations and actions one-to-one into their own world view or repertoire of practical activities, but critically question the media scenario. Errors when checking for everyday suitability inevitably lead to a loss of reality that can become pathological in extreme cases (e. g. daydreaming, social isolation). The controlled shifting between "areas of meaning", though, is part of a "normal" media use in which the suitability check of fictional scenarios (minimum condition is the understanding that everything is fictional) protects from uncritical transfers from the symbolical world of entertainment into our everyday living environment and from unwanted shocks when switching worlds.

The necessity for reality checks using earlier or recent experiences goes beyond media reception into the immediate experiences in our living environment. In the "natural attitude" of everyday life, individuals according to Schutz (1945) tend to "idealized assumptions" to be able to continue their routine forever (*I can do it again and again*); also, experiences are basically automatically projected into the future (*everything remains as is*). This tendency, highly functional for automatized processes in everyday life, becomes a problem when the unexpected happens or a changed situation demands re-orientation. For example, a recent experience in a partnership conflict cannot be transferred to all potential future partnerships. If we disregard this, we risk practical failures in life. The lifeworld subject thus is forced to check recent and past experiences for reality suitability and compare them with "new" information from the social environment and from the media world. To understand reality TV's inner logic, one must consider the ubiquity of the reality check and the permanent nature of experience-related comparisons, both postulated not only by the lifeworld concept, but also by more recent memory theories.

According to Markowitsch and Welzer, a twofold adjustment process between a person and their concrete lifeworld environment on one hand and between the person and various concepts of meaning and action on the other hand forms the "autobiographical memory" that keeps clearly self-related early memories with an emotional index "autonoetically available" (Markowitsch and Welzer 2005, p.11). By remembering the "good-old-days" (e. g. past happy partnership situations), mechanisms of stimulus and reaction in the current perceived to be problematic everyday routine (e.g. the long-term partnership is in a crisis) are put in a reflective context, also with regards to the input of external information (e. g. Internet makes it possible to choose partners or to develop sexual promiscuity). This "mental time travel" helps avoiding *dysfunctional lifeworld idealization*, "governing all conduct in the natural sphere, namely, that I may continue to act as I have acted so far and that I may again and again recommence the same action under the same conditions." (Schutz, 1945, p.547). To remain sustainable, the "explicit" and the "implicit" memories (Schacter, 1987) are constantly checked and rearranged in some aspects, modified in others with regards to their place within the structure of relevance and activity routines (Schutz and Luckmann, 1973, p.186). This memory rearrangement is embedded into a continuous *cognition and emotion management* (Grimm, 1999b) by individuals and groups that creates the relevant definitions of situations, motivations and readiness to act in the reference frame of changing lifeworld conditions. If one soccer team scores a goal, the opponent must re-motivate themselves for a counter attack; they will look for a chance to regain lost confidence (e. g. safe short passing game in the own half of the field) and will if necessary knock the tactical concept on the head. A stressed employee may watch an action movie on TV after work to forget abasements by his boss and to notify his tired body that there are also encouraging examples of individual strength and assertiveness. In both cases, the lifeworld subject creates a specific "informational environment" (by acting accordingly or through symbolic communication) that enables the subject to mobilize resources for problem solving and everyday coping. Brenda Dervin (1976; 1980; 1986) calls this "gap bridging" in the reference frame of her "sense making" idea. The "gaps" resulting from everyday problems lead to information seeking within and without media submissions whereas the information sought for bridges the gap to problem solving.

Schutz points out the importance of the fear of death within the individual relevance system. "I know that I shall die and I fear to die. This basic experience we suggest calling the fundamental anxiety. It is the primordial anticipation from which all the others originate. From the fundamental anxiety spring the many interrelated systems of hopes and fears, of wants and satisfactions, of chances and risks which incite man with the natural attitude to attempt the mastery of the world, to overcome obstacles, to draft projects, and to realize them." (Schutz, 1945, p.550). Ernest Becker (1973) thinks along similar lines, when he defines conquering the fear of death as the everyday human core task. Following Paul Tillich's theory of existential anxiety (see Weems et al., 2004), existential and social psychology have long since opened up an own field of research "terror management" which goes well with Schutz' *lifeworld* theory. Terror management can explain techniques of coping with fear that heighten the self-worth (Greenberg et al., 1986; Hart et al., 2005), at the same time empowering individuals to act in their everyday life ("Yes, I can") and cater to the interior of the lifeworld subject as a reference for *emotion management* ("I'm controlling myself"). The goal of *terror management* as part of *emotion management* (regulation of affects like fear, sorrow and anger, cf. Grimm, 2006b) is not primarily to be happy but to be able to act adequately in a given situation. Too much existential anxiety or too much fear in everyday life situations undermine the individuals' competence for social behavior and problem solving. Thus, *terror management* is linked to self-efficacy as well as to the need for safety which forms a firm reference parameter for shaping everyday life, be it with regards to parental attachments that guarantee safety (Cox et al., 2008), be it via religious ideas promising an afterlife (Jonas and Fischer, 2006). The two sides of *terror management* correspond with two varieties of coping with fear that media in general support immensely (Vitouch, 1993): The coping strategy of "sensitizers" is aimed at confronting provocative situations (bungee jumping, dangerous sports, horror movies) to experience the success of survival and to increase the feeling of controlling and power. On the contrary, "repressers" prefer harmonic environments (candle light dinner, romantic movies) which give them a kind of tranquillizer and satisfy their need for safety.

Fear-mitigating motives of increased self-efficacy have, by the way, also been proven for the identification with heroic characters. in movies and for computer game use, especially first-person games with weapons (Geyer, 2006). The media users partially transfer the assertiveness experienced in fiction and virtual reality into the world of everyday life. It is clear to see that reality shows like "Survivor" directly cater to claims of coping with fear, self-empowerment and self-efficacy. Reality TV's everyday roots obviously do not stem from a similarity to everyday situations only, but are also and primarily operative – by brokering situational options of intervention and by enabling the lifeworld subject to act.

Now, what makes reality TV special is the fact that the story's inner logic and probability are not the only criteria in the evaluation of everyday suitability (that would be true for fictional entertainment), but that it additionally aims at strategies of revealing realities beyond the performers' scenical intentions. As "savvy viewer", the reality TV watcher knows that we all "act" as we have to present ourselves in social interactions (Goffman, 1959; Perinbanayagam, 2000). Nevertheless – and because of this – we try to see behind the others' curtains and see through each other. That is a pretty rational behavior that can (but not necessarily will) strengthen the foundation of social behavior by balancing it from a critical distance. This does not stop anti-deception maneuvers from becoming a sort of control mania (e. g. checking the partner's e-mail account or the mobile numbers he has saved) that step by step undermines social trust. That is what suitability check techniques within and without the media are for that are not restricted to unmasking and unveiling alone, but must prove their potential in the reference frame of a critical and reflected – and, if you want to put it that way, "phenomenologically enlightened" everyday pragmatism.

The "savvy viewer" of reality TV adds to his love for unveiling strategies a critical attitude towards the sort of pedagogical or propaganda communication intentions (Hill, 2006; Biressi and Nunn, 2005) attributed to political statements by party functionaries in the media as well as expert-guided advisory broadcasts and sometimes fictional entertainment if they are overly obtrusive geared to deliver a specific moral or political message or political/pedagogical indoctrination. That is also part of the rationality of a mediatized everyday living environment to resist recognized persuasive intentions. The living environment subject's only problem, therefore, is that the authorities cannot simply be trusted as they are unfamiliar with the subject's personal situation. Insofar, skepticism towards authorities in everyday matters is necessary to keep the individual sovereign and adaptable to changing situations. The lack of direct pedagogical intentions in many reality TV programs gives viewers a chance to get their own picture of things and exactly thus is regarded as a gain in "realism" offering an ambigious set of positive and – to a higher degree – negative role models without exaggerated guidance and patronizing. The price for realistic gain and openness for free interpretation, however, is paid by embarrassed show participants, the celebrity wannabes who are very publicly duped or explicitly serve as a dissuasive example right from the beginning.

Neutrally speaking, reality TV implicates for its viewers a dominance of *negative learning* ("I see how it does not work "; "I realize what I shouldn't do", cf. Grimm, 1999b) as opposed to didactics via positive models and exemplary action. The communication's credibility is in so far strengthened through negative examples as the recipients' freedom in interpreting the media scenarios is respected and thus, a possible "reactance" (resistance against persuasion attempts, cf. Brehm, 1966; McGuire, 1999) can be avoided. The feeling of sublimity towards the failed reality TV "stars" is doubly gratifying: It can diminish the recipients' existential fears in the sense of *terror management* and at the same time heighten the belief in *self-efficacy* in practical living conditions. The situation of the embarrassed show participants on the other hand is precarious. While they are publicly noticed, this attention is diminished by a certain degree of contempt. This is the most seductive aspect of reality TV promising attention to average people (who miss exactly this attention in their "normal" lives) and at the same time offering to the audience Mr Nobody's currying for prominence as a chance for contempt that increases the feeling of self-worth. The viewers' *negative learning* corresponds to the negative attention for the viewed.

Reliability of Orientation

The appearance of participants that call for differentiation and hyper-critical downward comparisons, which is typical of casting shows, daily talks and *Big Brother*, strengthens the "voyeur's" position as a moral judge and creates orientation, especially with regards to avoidance. Pfau and colleagues have found in the frame of their "inoculation approach" that negative imagery of smokers can become effective in an anti-smoking campaign as arguments against addictive habits can provoke a protest behavior of the forcefully indoctrinated smokers (Pfau et al., 2003; Yin and Pfau, 2003; Pfau, van Bockern and Kang, 1992). Since within entertainment settings, there is no resistance against persuasion attempts (if the viewers do not assume such intents on the producers' part), all resistance focuses on the negatively portrayed personnel on screen. The public embarrassment of semi-stars in talent and casting shows surely offers a certain protection against the "addiction" to publicity that many reality TV formats fuel. But reality TV producers try to defuse the participants' depreciation by humor (sarcastic off comments, parodist exaggeration, self-irony of the participants) to keep the "exhibitionists" from the ranks of the "voyeurs" coming. Under the aspect of everyday orientation, the discredited (and often ridiculed) reality TV participants "inoculate" the viewers against an adoption of models of action into their lives. Thus, the "savvy viewers" of a docu-soap that shows teenage mothers despairing from everyday problems know exactly that after the show, the monotony of changing diapers starts anew, not to mention the sadness of impossible trips to the disco. They mainly remember their aversion against this lifestyle.

One problem of *negative learning* consists in the fact that the criticized or humorously depreciated models are uncertain with regards to what ways of action promise success. Additionally, the repeated, sort of ritual self-empowerment of reality TV viewers with regards to effective orientation remains incomplete if it is only based on gloating or laughing about those already weak and society's socially disadvantaged groups (the unemployed, alcoholics, the mentally ill). It stimulates dissociation attitudes and a critical mindset, but it does not tell the viewers what to do and how. The gain of sovereign judgment and error awareness is balanced by a loss of authoritative orientation. Especially intense reality TV users could thus be faced with new problems for their everyday orientation, because if you concern yourself too much with foolish or weird characters, damaged people and problematic behavior on TV, this abets a relativist "all the same" attitude. In other words, a chaotic situation with many or too many dubious options the viewers try to dissociate themselves from, potentially leads to a state of negation of orientation – something which Paul Virilio (1999) calls "polar inertia": the state of clueless torpor or a morally acquitted voluntarism in which any behavior can be justified or aggressively judged. The negation of orientation is best represented by the motto "each to his own!" that is repeated indefinitely in daily talks. That is definitely progressive in a dogmatically highly regularized, repressive culture. It constitutes reality TV's explosive force in autocratic systems like China and Saudi Arabia (Keane et al., 2007; Khatib, 2005; Kraidy, 2008). But the same principle becomes a liability when the mutual relativization of lifeworld orientation leads to a general arbitrariness in the end.

Developments of the Genre

Not all media researchers count daytime talk shows explicitly as reality TV: Neither Hill (2006) nor Murray and Ouellette (2008) mention them in their lists of subgenres. But they are definitely pioneers (Tally, 2008) that feature elements that are typically found in reality TV like helping *ordinary people* to screen presence and establish *problems of everyday life* (e. g. partnership conflicts, unemployment and alcoholism, outfit and body, lifestyle) and use these problems as topics for social media communication. Daily talks present *everyday life in a state of emergency* (s. definition at the beginning of this chapter), as they focus on and scandalize crisis and deviant people and behavior (e. g. promiscuous sexuality, violent hooligans). There are also moments of *intervention* when the talk show presents everyday problems as "live life drama" (Grimm, 1999a). Should the confession talk or the shouting contest not be enough to recognize the "truth" and untie knots of interaction, lie detectors or paternity tests bring additional light to the darkness of private conflict and deceit.

The daily talks' strength is *pluralism of situation* with a wide range of everyday life's trouble, their weakness is *relativism of orientation* caused by contradictory experiences of the show participants, the arbitrariness of audience comments and the multiplicity of possible consequences for viewers. In addition, the number of washed-up people and action models boost the uncertainty to judge. In the multi-optional scenarios of daily talks with their focus on *negative learning,* there is uncertainty with regards to

(a) positive behavioral alternatives,
(b) successful coping strategies and
(c) reliable co-judges who intersubjectively collateralize the own ability to judge.

Newer developments within reality TV programs contain answers to all three problems. The daily talks' success that dominated the daytime program of many European private networks in the 1990s (e. g. in Germany, Italy and France) was followed by more affirmative formats that signal a return of authorities into reality TV. For reasons of everyday rationality, the self-relativizing relativism enforces a limited recourse to people who have, by office or by education, superior knowledge and definite power of judgment. The comprehensive launch of court TV in Germany ended the daily talk boom. The first show in that genre was the format *Richterin Barbara Salesch* (after the US format *Judge Judy*) where a "real" judge heard first real and after October 2000 fictional cases, played by lay actors. It is remarkable that attempts to replace talk show participants by actors to revive the declining genre have failed before. The non-professional talk guests guaranteed the degree of unpredictability and surprise that the *Big Brother* audience digs as well and that adds a more "realistic" sincerity to the comparison to one's own everyday experiences. Viewers disliked actors in talk shows because their opinions seemed planned, artificial and manipulative. Here, the reality accent lays in the authenticity of everyday people's experiences and opinions that cannot be fictionalized without losing quota. In court TV, on the other hand, actors and scripts were no problem. The question of guilt first remains unanswered which heightens the viewers' suspense (and is thus a challenge to the "savvy viewer"), but was scripted beforehand just like in a regular court motion picture. Here, the definition criteria of Ouellette and Murray (2008, p.3) who deem the "non-scripted access to 'real' people in ordinary and extraordinary situations" as constituent for reality TV fail. Court TV can still be construed as reality TV, though, since the moment of authenticity switches from ordinary people to the judge as a person of authority who guarantees the quality of orientation (her decisions are based on written law and follow acknowledged principles of justice). Additionally, the orderly game of accusation and defense brings structure and rationality into the unveiling technique that for example Andrejevic (2004) criticizes in the MTV format *One Bad Trip* because it practices and propagates the unregulated spying on interaction partners within the family or a circle of friends which causes a "climate for conspiracy" in private relationships and does enormous damage. Those who want to avoid such extremities of the unveiling mania but do not want to completely stop looking for reality behind the facades of self-presentation need a reliable compass like court TV that establishes the unveiling techniques on a constitutional level and certifies them via a professional judge.

A similar shift in the reality focus happens in the format *Zwei bei Kallwass* (a German format with a professional psychologist counseling clients with relationship problems played by lay actors on coping with their problems and conflicts in a sort of systemic family therapy in front of a live audience). Again, it is no problem here that the cases are fictional and scripted: what counts is that the psychologist is "real"! The proof of authenticity is situated right where it is needed for an orientation in the world of everyday life: with the authority and credibility of the recommended solution that had become brittle in other reality TV formats. In a way, the program development reacts twofold, to a changed need of the everyday lifeworld (psychological counseling is a mega trend) and to dynamical needs that the development of reality TV itself has caused.

Along this line comes the latest trend in coaching TV. The prototype here is the British edutainment format *Supernanny* in which hobby pedagogue Jo Frost has been visiting families with parent-child-conflicts in need of counseling at home since 2004 to heighten the lifeworld subjects' ability to control their immediate life situation by direct intervention (admonishing the children, counseling the parents) and via recommendations on restructuring their everyday life (time schedules, systems of gratification/punishment, restructuring of the living area). The motive of control has a triple relevance here: *First,* the format is about the control of children avoiding parental influences. *Second,* the parents' precarious self-control that adds to the "family chaos" due to social factors (double stress of job and household, unemployment etc.) and own negative childhood experiences (biographical "wounds") is to be restituted. *Third,* the intervention aims at helping the children to largely control themselves so that direct parental intervention becomes less and less necessary. The program has meanwhile been licensed to more than a dozen countries, among them Germany, Austria, the US, Brazil, Spain, France, Israel and China. Symptomatically, in Germany hobby pedagogue Jo Frost's role is taken by the certified social pedagogue Katja Saalfrank and in Austria by the couple Sabine Eder (social pedagogue) and Sandra Velásquez (psychologist), both also professionals. Like in court TV and psycho coaching, the orientation achievement is "authoritatively" certified in edutainment TV. While Frost refers back to her past experiences as a "real" nanny in British families, in Germany and Austria (also in Spain and Brazil), the counseling claim is academically backed by university degrees. The supernannies' tips are published in parenting guidebooks (e. g, Frost, 2005a; Saalfrank, 2005) outside the show and discussed in Internet forums. The professionalized TV discourse thus interacts with everyday life discourses. As opposed to court TV and psycho coaching shows, the *Supernanny* format's participants are "real". This increases the authentic parts and heightens the sincerity with which viewers spread the recommendations, which in turn intensifies the mandatory critical everyday suitability check (s. above), but also increases the probability of a transfer to the viewers' life.

The *Supernanny* format answers to all three weaknesses of reliability and the concreteness of the everyday orientation mentioned above that daily talks share with the totalitarian surveillance show *Big Brother* (where chance and intrigue rule) and the survival show *Survivor* aiming at abstract self-empowerment. In the *Supernanny* programs, *positive alternatives of acting* are phrased prominently by the protagonist in the form of concrete recommendations, and the *efficiency of problem solving* is guaranteed by direct intervention and counseling for the participating families. For the recipient, that means an impressive demonstration of everyday suitability. Furthermore, the supernanny persona represents a *reliable co-judge* who "authoritatively" accompanies and pre-structures the viewers' processing of what they saw. *Supernanny* programs definitely know no problem of orientation negation like talk shows (or *Big Brother*). Thus, Hill is wrong when she says: "Whereas in traditional reality formats the relationship between information and entertainment was fairly explicit ..., in contemporary reality formats the relationship is more implicit." (Hill 2006, p.179) While it is true that in an early phase of reality TV focussing on emergency service and crime centred formats the "*problem solving*" (regarding first aid for accident victims and crime fighting) was more important than in *Big Brother* or *Survivor*, the evolution of the genre has long since passed by the noncommittal attitude and artificiality of game situations in a container or on an adventure island; at least, the reality genre has differentiated itself towards commitment and concreteness of everyday orientation. When it first aired, *Big Brother* could be construed to be an answer to *Rescue 911* and *Top Cops* – in the sense of both liberation from systematic patronizing by police and fire fighters and of opening the setting towards "ordinary" everyday accidents – but the most recent coaching formats offering not only upbringing tips, but for example also cooking recipes and ideas for debt reduction react to the brittleness of the orientation transfer resulting from the artificial situation in the *Big Brother* house (or far-off islands and jungle camps) into the viewers' world of everyday life. Those who are suffering from the relativist dogma should turn to *Supernanny;* the "savvy viewing" of egocentrics and tricksters will not be enough for them in the long run.

With regards to the evolution of the genre we see that the current development is towards differentiated functions for various areas of the everyday lifeworld. In a way, post-modern reality TV stressing pluralism and openness has moved on into a post-post-modern phase reducing the arbitrariness of orientation. Of course, that does not exclude that the genre will recur again to more open and relativist formats when the authoritative recommendations and professional solution approaches are less trusted again. Obviously, there are two ideal forms of reality TV: *Variety A* is geared towards unveiling reality and acknowledges a high degree of freedom when it comes to everyday orientation; *variety B* is more tied to an operative coping with concrete everyday situations and combines addressing problems with their potential solution (that should be as authoritative as possible). In both cases, we find a predominance of everyday life that reality virtually raises a monument to by claiming public attention for the lifeworld subjects' sorrows and hardships beyond political systems of relevance. Reality TV is situated in the "political antechamber" but it remains open to connectable political appropriations that are compatible with everyday life. In casting ordinary people in a stark bright light that catches the most remote corner of their private existence, the camera's eye can potentially increase not only the individual reflexivity in the world of everyday life, but also the chance for governance and social control by institutions of the political and economic system.

In the next – and last – paragraph of theoretical approaches to reality TV, I will analyze more thoroughly the conditions of popularity and the socio-political implications of the *Supernanny* format and include findings of the Vienna research project that will be more thoroughly and more systematically described in the chapter's second part to empirically support theoretical essentials. With the *Supernanny* format, the evolution of reality TV as a genre reaches an interlocutory climax by strongly altering the question for a "real" reality on behalf of the intervention in the everyday lifeworld. What does the operative focusing of coaching TV mean for the genre's popularity and worldwide marketing? What consequences do surveillance cameras, interventionism and globalization have with regards to the self-efficacy claims of the lifeworld subjects?

Glocalization of the Supernanny

Austrian Supernanny: "Of course it is strange at first when there's a camera present. But because it is there for so long, parents and children get used to it quickly. I even have the impression that the camera makes one aware of certain things even more ... One must not forget that the families we had are not used to reflecting on themselves at great length. They also don't have the social environment to give them feedback on their social behavior. This means that they really see themselves for the first time on film: "OK, this is how I really am!" Or, they see themselves for the first time in the video analysis and are taken aback. This has always been a strong experience for the family." (Edinger, own interview 2005)

The camera in the kids' room radicalizes the watching eye's authority far beyond normal "voyeurism". It does not deliver simple eye candy but offers a powerful instrument of intervention to the lifeworld subject. It aims at holding the mirror to stressed out parents allowing them to perceive their interaction with their children from a distanced watcher's position. This opportunity of self-watching opens the parents for the nanny's intervention and recommendations. This media-brokered connection between self-reflection and indoctrination is highly effective: the majority of cases are "successfully" solved within the intervention phase of about a week. While more than half of the Austrian participating families we questioned showed to be in need of counseling even after participating in the *Supernanny* experiment, their progress consists of understanding and more readily accepting counseling offers. A problem of professional governmental institutions offering help (Federal Ministry for Family Affairs, Senior Citizens, Women and Youth; Youth Welfare Office) is that their clients – especially those from lower classes – mistrust them because the help is linked to governmental power that uses the forceful removal of children (e. g. commitment to a youth institution) as an ultimate means. Those who mistreat their children and let them become "problem kids" are in danger of losing them to governmental institutions. The reality format *Supernanny* easily conquers the barriers this erects between the system's agents and the lifeworld subjects because TV not only has no sanctioning function, but also promises public attention as gratification. While there is the risk of public embarrassment, the public attention offers an additional incentive for the participants to solve their own family problems. As opposed to casting shows operating by a competitive system of exclusion and willingly accepting the defeat of the many to create one "star", the participants of *Supernanny* programs decide upon their success themselves. Their practical test does not consist in eliminating competitors, but is only assessed by their everyday life. In the context of home and family that forms the inner core of it, the lifeworld subject potentially becomes a "star among stars" if it can master the practical challenges of upbringing by reflective techniques (camera, talking, therapeutic games), guided by the nanny.

A central factor of successfully globally marketing reality TV formats (Moran, 1998; Moran and Malbon, 2006) is their focus on the world of everyday life that has similar basic structures and is experienced similarly in all countries and cultures: "impact zones" and "zones of potential accessibility" (Schutz & Luckmann, 1973; 1983) do exist in every society; all lifeworld subjects share lifeworld idealizations of "I can do this again and again" and schematic everyday routines. The individuals' reflexivity also belongs to the general consequences of modern times (Giddens, 1990; 1991; Archer, 2007) that create a need for reflective support. Finally, the double structure of society as system and lifeworld that

Juergen Habermas (1985a; 1985b) calls "uncoupled" is a universal thing. Indeed, the dictum of "colonization of the lifeworld by the system" is dubious when it comes to reality TV since here, systemic resources (e. g. police and fire fighters or psychologists and pedagogues as agents of the welfare state) usually serve the lifeworld subjects. In any case, the relative independence of lifeworld structures that the uncoupling thesis correctly describes means a connection point for globalized formats that the everyday life primacy is programmatically inscribed to and that the systemic level only incorporates according to the principle of subsidiarity. Finally, the similarity of parenting problems around the world gives Jo Frost's tips and recipes a worldwide popularity. Equally surprised and proud, the British edutainment pioneer stated in an interview with the Australian newspaper *The Sun-Herald*:

> "It's universal. You could put every nationality in a room, and it's the same thing going on. I've got parents with children who have taken control, who don't want to eat, who rule the roost, purely because they have been given too much choice." (cited by Teutsch and Browne, 2005)

Frost sees the emotional involvement into the parents-child-conflict as a reason for business myopia and adds: "Objectively, I'm able to observe. For parents, it's difficult to see when they are so emotionally involved." (Frost, ibid)

Optimistically speaking, for the participating parents the *Supernanny* programs mean "self-therapy" which Rachel Dubrofsky (2007) calls a basic trait of reality TV, exemplified by dating shows like *The Bachelor* and *The Bachelorette*. The author sees reality TV as part of a "culture of therapy", whereas "therapy" does not really refer to pathological states, but signifies a sort of identity work aiming at personality changes and heightening the control ability. The description of the camera situation within the family by the Austrian supernanny Edinger (cf. the opening quote of this paragraph) hints at the potential for self-change within self-adulation. This effect is further enhanced by the awareness of being watched by many people, provided the proof motive is stronger than the fear of embarrassment. The participant can trust the supernanny's power to, as Melissa Lenos (cit. in Tally, 2006) ironically remarks, ride in like a "gunslinger" in a wild west movie and help the parents in distress – happy ending of course guaranteed. In continuation of George Herbert Mead (1934), the authoritatively accompanying camera could be seen to represent the "generalized other" that first of all represents a moral instance for the child that it keeps internalizing more and more during the socialization process. In adults, recursions to what "the people" think make the self-control of the impulsive "I" by the reflective "Me" more effective. The "Me" stores outside expectations that the subject uses to control maladjusted parts of his personality. In religious terms, this could be described as a public stimulation of conscience in order to gain moral self-discipline. Is *Supernanny* thus aiding our reflective and moral maturation like the spiritual exercises once were that Michel Foucault (2009) sees as a specifically Christian tradition of individualization and self-evidence?

Seen as a whole, admittedly, Foucault's analysis of power conveys a critical understanding of self-control that he analyzes as an alternative or rather an add-on to classical forms of ruling with power and punishment, i.e. as a part of modern "gouvernmentality" where control from without and from within are fundamentally interlocking. The term sounds like the French word *"gouvernante"* (a distant relative to the British nanny model, responsible for etiquette and good behavior). As a stereotype, the "gouvernante", famous for her patronizing style of upbringing that tries to indoctrinate the pupil's self-control by rigid forms of morality – is basically condescending. In that sense, the parents' self-empowerment through *Supernanny* could mean the children's incapacitation that is linked to punishment and force according to Foucault (1995). But "gouvernmentality" goes deeper and incorporates the parents' position. Even if it does not look like heteronomy of the parenting role, strengthening the parents' position by the supernanny's recommendations could in the end mean an illusion of freedom that only implements the rationality of systemic governance coming from the political sphere of society. In that case, the receivers of *Supernanny* counseling would not have become more competent in handling their everyday challenges, but had only reproduced the state's job, be it governing its citizens, be it with regards to the social services' help, on an individual basis. Tally (2006) complains that *Supernanny* programs have a surrogate function: Instead of providing nurseries and full-time schools, the upbringing responsibility is completely left to the parents whose problems caused by the lack of pedagogical institutions and resources can hardly be solved via *Supernanny*. According to Ouellette and Hay (2008), reality TV with its claim to help manage everyday life matches well the political "outsourcing" common in the UK since Thatcher. Nikolas Rose (1999) thinks similarly in his critical impetus, when he perpetuates Foucault's theory and does not see more freedom in neo-liberal politicians' call for self-responsibility and self-control but rather a change in the form of control that only provides a better mask for governance and political domination.

Now, in the face of the worldwide financial and economical crisis, one can doubt the success of self-regulation in the sense of neo-liberal politics, but a growth in affect control and a shift of control functions towards the individual is incontestable these days. Already in the 1930s, Norbert Elias (1939) diagnosed the decrease of direct governmental power and the increasing importance of self-regulating individuals throughout the civilization process. The individuals' ability for self-control may sometimes include suffering from freedom, but is definitely to be preferred to a dictatorship's lack of freedom. Those who tend to see liberal systems close to fascist ones cannot appreciate the freedom of the lifeworld with family and home at its core: there and only there, anti-dictatorial resistance systems can form. Bakardjieva uses Afro-Americans as an example of people who saw their homeplace as a place to fall back to, to regenerate and to develop solidarity "in the context of the black liberation struggle" (Bakardjieva, 2005, p.73). Robert Silverstone (1994) ascribes a more ambiguous role to the "domestic sphere" that generally gains more importance through TV and Internet. Also, the neo-Marxist everyday theorist Henri Lefebvre (1947) stresses the family everyday world's "dual potential" because it can be used for the individuals' authoritarian domestication as well as for their emancipation.

Thus, we need to ask precisely and decide empirically how much the British *Supernanny* format tends towards authoritarian control or prefers emancipatory and democratic styles of controlling and parenting children within the process of global diffusion. Following Roland Robertson (1994) we act on the assumption that globalization really means "glocalization" because the successful implementation of the original form in other countries as a rule needs cultural modifications. In the interest of global marketing, the "identity" of the brand should be kept, though. Therefore, we will compare the original British program with the adaptations produced in Germany, Austria, Spain and Brazil and recognize the amount of *glocalization* by the format's differentiation of the authoritarian and democratic potential.

PART TWO: INTERNATIONAL COMPARISON OF EDUTAINMENT PROGRAMS

Parenting Styles and Cultural Diversity

The image of British parenting in continental Europe is authoritarian. Victorian pedagogic principles are seen as strict, rules-oriented and set on obedience. At least in the past British upbringing included corporal punishment as well. While it has long since been officially banished in UK schools, the term "English education" still connotes to values like consequence and discipline. The nanny system practiced in English middle to upper class families adds to this (a third person next to father and mother is responsible for supervision and educational duties within the family's house for a certain period of time), for it formalizes the pedagogic relationship to a certain degree and helps the parents. The ideal nanny sets up authoritative rules (e. g. for doing homework or going to bed) to avoid long discussions with the pupil. Of course, the nanny enforces the rule, if necessary, with the authority of something like a "constitutional entity". The advantage is clear: Fraught arguments between "stressed-out" parents and "contumacious" kids about homework and bedtimes do not take place as there is no effective way of protesting against a "higher jurisdiction". Also, the motive of negative attention that a neglected child could count as a personal benefit has no grounds anymore. On the other hand, socially deprived parents (e. g. single mothers, unemployed people) have no chance to compensate the frustration they suffer from in a personal power struggle against the child. This could constitute a first factor for the popularity of British *edutainment TV*: It makes a former socially exclusive right to get help and relief for the parents accessible to poorer classes (directly for the participants, for the viewers by taking part in a symbolic form of order).

The TV supernanny is a symbol of domestic order, of aiding parents and of strengthening pedagogical assertiveness that for various reasons have become precarious in our society where the individualization of lifestyles and the relativization of traditional norms and values in the permanent process of modernization (Giddens, 1990) add to parental uncertainty: Which adaptations do children have to manage in an enormously fast world? Which parenting style is correct? Some parents that are faced with this situation abdicate the arduous business of upbringing and leave the children to their own devices or to electronic "baby sitters" like TV and play station.

In the 1960s, the US education expert Diana Baumrind defined three parenting styles: *"permissive"* (compliance with all the child's impulses, imposing no rules and no punishments), *"authoritarian"* (operating sanctions to break the child's will; setting rules which are mostly theologically motivated) or *"authoritative"* (employing justified rules, both demanding and responsive to the child) (Baumrind, 1966). Later, the *"uninvolved"* parenting style was added, describing "rejecting" and "neglecting" parental ways of conduct. Of course, Baumrind had advocated the "authoritative style" and latter studies showed indeed that this style strengthens success at school and minimizes drug abuse (Baumrind, 1991). Baumrind sees an upbringing crisis in the US that, according to her, started in the 1940s and has to do with the influence of psycho-analysis (especially with Sigmund Freud's positive take on child sexuality) and the German neo-Marxist "Frankfurt School" or was at least intensified by both. Together with US colleagues, the emigrants from the "Frankfurt Institute for Social Research" had done the famous study "The Authoritarian Personality" (Adorno et al., 1950) in the 1940s on behalf of the US government in which Theodor Adorno and Max Horkheimer attributed part of the responsibility for the NS rule in Germany and the readiness of many of their compatriots to contribute to the hunt for the Jews to the "authoritarian personality" (repressive control of urges, fixation on scapegoats, uncritical slavish obedience). In the eyes of the anti-fascist US public, this discredited authoritarian parenting methods for all times. Except for notorious fascists, no one wants to be responsible for educating future concentration camp guards or to lay the groundwork for anti-Semitic pogroms. In the late 1960s, this idea was re-actualized in the US and even more in Germany and became popular throughout the world. Freud and Adorno gained the support of the British educational reformist Alexander S. Neill (founder of Summerhill) as warrantor for the "anti-authoritarian" student protests against fascist relics in their societies and against the war in Vietnam. The Summerhill project (Neill, 1960), practicing optional school attendance and to this day based on the pupils' democratic autonomy and "self-regulation", aims to be an alternative to the UK's classical authoritarian upbringing model. For Baumrind, Neill is expression and catalyzer of a "philosophy of permissive and child-centered attitudes" (Baumrind, 1966, p.888) that falls prey to the illusion of the children's "self-regulation" and opens the floodgate to a "permissive" parenting. Baumrind – and many other critics of the 68ers' ideals – see the concept of "anti-authoritarian" (or in the original calling by Neill "anti-coercive") parenting as an overreaction to a misguided authority that disavows any form of parental control. The results are maladjusted children: without frontiers, without discipline, easily gullible, tending to addictions and without societal success.

Those who see a permissive attitude as responsible for the recent upbringing crisis and include the paternal loss of control in the definition of the problem of delimitated difficult children will find secondary virtues like discipline and respect for authorities attractive again. Thus, the second assumption on the popularity of British *edutainment TV* is: *Supernanny* is part of a trend towards restituting authority and the family's controlling ability that have been undermined by modernization and globalization processes and responds to anti-authoritarian *zeitgeist* phenomena and generally to the relativist crisis of values in education and upbringing. In this context, the supernanny's look tells us a lot: The "authoritarian touch" transported by Jo Frost's conservative outfit and strictly coiffed hair reminds us of parenting methods from back in the day. At the same time, the protagonist's ever-present umbrella not only transports the label "British", but also lends her sort of a Mary Poppins charm that glosses over the authoritarian moment congenially. Is this only old "authoritarian" wine in new "entertaining" skins? Does the authoritarian model become acceptable through the backdoor of humor? Or does the British model of edutainment actually mean a new form of parenting by strengthening the clearness of parents' behavior and of parental control without repeating the old mistakes of a repressive authoritarianism?

Baumrind had been forced into a heated public discussion a few years ago, when she declared at the American Psychological Association's conference that certain forms of corporal punishment are not necessarily detrimental in certain situations (Baumrind, 2001). This brought her criticism from colleagues who called it a return to anti-humanist methods (Gershoff, 2002). At Berkeley university, psychologists and pedagogues signed a public statement against Baumrind objecting all forms of corporal sanctions. The debate still goes on and has damaged the differentiation between "authoritarian" and "authoritative" parenting styles. Additionally, international comparative studies showed that the equations "authoritative parenting means success at school" and "authoritarian parenting means exaggerated submission, lack of self-esteem and failing at school" do not seem to be valid in all cultures. According to Baumrind's criteria catalogue, children from Asian families (China, Japan, Korea) with an "authoritarian" parenting style, e. g., are by far the most successful at secondary schools (Chao, 1995; Chao and Tseng, 2002). They also exhibit less maladjusted behavior like illicit drug use and delinquency. In stark contrast, studies in Spain showed that of all things, children with "indulgent parents" (which conform to Baumrind's permissive category) develop a greater self-esteem than those from families with "authoritative" parents, not to mention those with "authoritarian" parents (Musitu and García, 2005; Martinez and García, 2007). In Spain, indulgence obviously is especially beneficial for the children's development and performance at school – maybe because of the repercussions of the repressive Franco regime that undermine the differentiation between different concepts of authority. In Brazil, surveys on the upbringing situation are ambiguous. Like in Spain, Brazilian kids with indulgent parents on average have a greater self-esteem than those from families with "neglecting" and "authoritarian" parents (Martinez et al., 2007). But in "authoritative" families, children have equally high degrees of self-esteem, and their system of values is much less "conservative" than the one of children with "indulgent" parents.

The findings from international comparisons show the highest variance with regards to the difference between "authoritarian" and "authoritative" – subtle nuances of language in English as well. In a newer survey from the UK, we see the expected relations between "authoritative" parenting and positive outcome for personality development (Chan and Koo, 2008) that partially blur outside the Anglo-Saxon culture. Rudy and Grusec (2001) generally find that authority (in whatever form) has a different basic function in "individualist cultures" than in "collectivist" ones. Among Canadians hailing from Egypt, who are basically seen as more "authoritarian" and "collectivist" than Anglo-Canadians, the degree of collectivist values was the best predictor for authoritarianism. Authoritarian ways of upbringing were best predicted in Anglo-Canadians by collectivist values (on a lower base level), but now connected to a "lack of warmth". In the light of these findings, the positive results of "authoritarian" parenting style in China can be explained by Confucianism that teaches parents to combine a strict supervision and punishment of children with a loving, warm family climate. The positive role of indulgence in the Spanish and Latin American cultures becomes plausible, when we realize that the individual values are ideologically and socially rooted here but are also constantly threatened by structures of violence and power. The execution of power goes along with social cold, while the individual finds attention that embeds his spontaneity socially and emotionally in networks of mutual cooperativeness.

The problems of analyzing authoritarian ways of upbringing in international comparison suggest that we should review the technique of control in relation to the social climate like Kurt Lewin does in his *field theory* of leadership styles. Since moreover, Baumrind's remarks on corporal punishment are likely to associate authority with violence and to neglect the intra-family climate, I will from now on replace Baumrind's categories by Lewin's when analyzing TV parenting styles. In the 1930s and 1940s, Lewin and colleagues (Lewin, Lippitt and White, 1939; Lewin, 1948) conducted a series of experiments with ten-year-old school boys who organized themselves in clubs in which an adult educator varied his leadership style over the course of several weeks either authoritarian-autocratically (condescending, decreeing) or democratically by using explanations, discussions and help for the weak. Lewin's third category is the laissez-faire concept correlating to Baumrind's "permissive" category. The researchers working with Lewin were especially interested in the consequences of leadership styles on group dynamics, aggression and the acceptance of the leader. They found out that authoritarian-autocratic leadership causes open and/or latent aggression and undermines the leader's popularity. Based on a seemingly strong control, the group's controllability is undermined in the long run. The result is what parents fear most: the "quiet of the grave", sometimes longer, sometimes shorter, followed by unbridled fits of rage by the children and rebellion.

Rudolf Dreikurs and Eva Dreikurs Ferguson (Dreikurs, Cassel and Dreikurs-Ferguson, 2004; Dreikurs-Ferguson et al., 2006) as well as Eva Schenk-Danzinger (1992) developed these theories further which finally led to a pattern of 4 parenting styles that the analysis of *Supernanny* formats in the next paragraph is based on.

Authoritarian/autocratic parenting style:	authoritarian, strict, undemocratic.
Democratic parenting style:	fair, arguing, active.
Permissive/Laissez-faire parenting style:	indulgent, passive, anti-coercive.
Negating parenting style:	passive, lazy, uninterested.

Alexander S. Neill, the figurehead of "anti-coercive upbringing", would not have been placed fully in the laissez-faire category with the attributes "indulgent" and "passive", but at the border to a "democratic parenting style", because Summerhill's organizational norms are namely "fair", "arguing" and "active". Like Lewin, Neill advocates training of various forms of pupils' self-regulation (discussions, votes) which the mainstream of Western school psychologists still celebrates as the pre-school of democracy. So why not also trust the coaching format *Supernanny* to go the way from bottom to top: from democratic control in the small area, the family, to exercising political control on the systemic level? In that case, Jo Frost would, like a good democratic leader should, help people to help themselves to heighten problem families' ability to govern themselves. Or is she still just being dominatrix who as a socially unrelated interventional heroine establishes a "governmental order" at the expense of the lifeworld subjects?

Democratic Ranking of TV Nannies in Five Countries

All examined supernannies categorically refuse violence in upbringing. Here, the TV pedagogues have a clearer position than Baumrind. Their renunciation of violence is remarkable since superheroes in Hollywood movies usually are less picky when it comes to their means of choice (Jewett and Lawrence, 1988; Lawrence and Jewett, 2002) and the critical remark on the "gunslinger" narrative hinted at more tolerance towards violence. The edutainment nannies are completely restricted to non-violent demonstrations of assertiveness – basically, they are *supercalifragi-pacifistic*. Sometimes, though, participating parents can use force but are criticized by the nanny during the rest of the show. Instead of slaps in the face, be it as punishment or out of desperation, the nanny recommends a restructuring of the parent-child-relationship or of behavioral patterns and daily routines that further aggression. In a violent climate, the social trust that is the foundation for interactions between people with a shared lifeworld cannot grow. With the invention of the state's monopoly on violence, private individuals have sworn to renounce violence anyway. Whoever breaks that vow becomes a criminal. But the private home can very well become a place where violence can bloom and will be used in the dark recesses of seclusion. The TV publication means the end for this sort of privacy. The trite lament over the growing dissolution of privacy in an increasingly mediatized society ignores an important aspect: the gain of control in the face of abuse, rape and death brought on by the publication of private niches of violence. Nanny TV builds efficient dams against destructive forms of inner domestication.

Two weak points of the notorious media chastisement accompanying any form of reality TV are a generalized criticism of its voyeurism and a lack of esthetical differentiation, i.e. the inability to properly abstract effective components of the show and discern them from each other (such as the depiction of "positive" and "negative" action models). The child protection agencies in Germany and Austria have protested against the nanny TV in public because they think it (a) exposes children to the viewers' uninhibited voyeurism and (b) propagates authoritarian means of parenting: obedience drill and repressive oppression.

We have already spoken about the voyeur's change in reality TV, and the argument of exposure does not apply to children and adolescents threatened by abuse. Many of the children participating in Supernanny programs are marked as "out of control" which implies a clear appeal to parents, school and society to regain control. Before the publication of the control problem, the children were alone and under the arbitrary control of their parents who tend to reproduce their suffering from low self-efficacy in a domestic context in endless chains of conflict – at the expense of the family's weakest part. Thus, the children as a rule are already threatened before becoming exposed to the intrusion of public eyes. This contains a certain risk of embarrassment and hurt feelings, but is also a prerequisite for solving the control problem according to therapeutic and morally reflective considerations and thus for surmounting the state of unchecked parental interventions (or the refusal of parental responsibility through neglect) that are dangerous and damaging for the child. The relation between potential damage and gain is thus a question of an actual evaluation of the higher legal interest, not of a general refusal of public attention for the child's privacy zone that is supposedly to be protected always and everywhere. The dungeon of Natascha Kampusch who had been at the mercy of her tormentor for years in a spectacular Austrian case of kidnapping and captivity and the case of Josef Fritzl (continuous sexual abuse of his own children in the family house's basement over a period of 24 years) are eloquent examples for the limits of a dogmatic privacy policy with regards to the interaction with children. Publicity for a house's intimate zones is an essential instrument of democratic control to correct domestic life in certain situations of intensified conflict. Reality TV's focus on these areas replaces or adds to forms of immediate neighborhood control common in village communities but eroding more and more in anonymous cities and even more in the *Internet Society* (Bakardjieva, 2005). In any case, the publication of children's "private problems" in nanny TV does not make the protagonists anti-democratic at all, but heightens the attention for the children's problems.

Of course, that is only advantageous for the children if the focus on the problem also mobilizes resources to solve it. Within the show's dramatic structure, the participating family represents the problem, whereas the nanny represents the solution. Thus, we need to discern exactly between the parents' parenting style (before the nanny's intervention) and the one recommended by the supernannies. The child protection agencies' second accusation quoted above with regards to authoritarian parenting detrimental to the children is only true when the TV pedagogues rate the incriminated methods as exemplary or apply them themselves. In the case of a representation of authoritarianism on the problem description level (e.g. in the form of failed disciplinary attempts by desperate parents), it would become the object of "negative learning" through which the viewers distance themselves from behaviors that are unsuitable under typical daily circumstances. To check (as opposed to confound) the communication cautionary and appellative structure, the Vienna research project "TV Supernanny" surveyed the nannies' and the parents' behavior separately in a systematic content analysis.

So far, the project group,[3] under my supervision, has examined a total of 48 shows of the format in five countries (Austria, Germany, UK, Brazil, Spain) that included 2170 sequences. In the case of Britain, Austria and Germany the sample consisted of 25 random episodes from the show's first season (2004). Additionally, five episodes of the British original from the most recent season (2008) were added to test the format's stability. The Spanish and Brazilian *Supernanny* episodes (16 all in all) were also randomly selected from their first season (2006). The coding unit was the *sequence*, defined as a scene which keeps up a coherence of place and time and /or creates a greater, meaningful frame of interaction, counseling or reflection. *Interactive* sequences focus on the parents' and children's everyday behavior. *Counseling* sequences are dominated by the nanny's or the editorial staff's tips (often provided through voice-over comments and assessments). In *reflection* sequences, the parents consider the biographic modalities of their current behavior, often aided by the nanny through quasi-therapeutic techniques of discourse and autobiographical memory. Within the sequence, we recorded the parents' upbringing practice and the nannies' recommendations in various aspects ("admonish the child", "encourage the child", "order the child to have some downtime", corporal and non-corporal "punishment of the child" etc.) and analyzed the parenting style. Based on synthetic codes, the parenting styles can be allocated to acting groups (father, mother, parents as a couple, grandmother, nanny) and thus yield differentiated parenting profiles, e.g. for the nannies and parents. Additionally, the way of addressing the upbringing practice was analyzed with regards to intentional content; here, the coders searched the scenic context for indicators of expounding the problems of the upbringing practice or its quality of recommendation. Thus, precise differences between positive and negative models, which offer totally different ways of "positive" and "negative" learning to the viewers, can be noted.

Overall, the message system analysis consisted of 249 variables. After an intensive coders training, the inter-coder reliability consistently reached a satisfying level from R=0.82 (for *interactive* sequences), through R=0.88 (for *reflection* sequences) and up to R=0.97 (in sequences with a dominant counseling reference). These figures are in line with what can be expected from a reliable analysis of complex content (see Holsti, 1969; Neuendorf, 2002). The complete method is extensively described in Grimm (2006a).

Table 1 contains the content analysis results with regards to the parenting style before the supernanny's arrival that was rated problematic through its presentation or explicit comments. The focus here is on the parenting style that starts conflicts with the kids and added to the motivation to ask the TV nanny for help. Code 0 was used if in a sequence related to problems, the respective parenting style was not used; the upper extreme of 5 marks the problematic parenting style as "very much applied". The in-between codes were chances for the coder to gradate. Only sequences expounding the problems of some parenting behavior were coded; for sequences unrelated to problems, the respective variables were marked as missing data and excluded from further analysis. This means that the values on the scale show

---

[3] The content analysis was done by Gabriele Tatzl, Nora Sells, Kim Sztrakati, Manuela Brandstaetter and Christiane Grill whom I hereby thank for their research work. I am indebted to Nora Sells who also kindly organized the online survey, see below.

the amount of the respective parenting style being used to define problems in relation to the other parenting styles.

Across the board, the participating parents often show authoritarian techniques when the *Supernanny* program defined the problem, e.g. "shouting at the child" or "condescending behavior towards the child" and sometimes used light forms of corporal punishment. In three countries (Austria, Spain, Germany), authoritarian parenting is the predominant behavior that at the same time calls for the most criticism. The *permissive parenting style*, criticized by pedagogues, adds a lot to domestic problems in dealing with children in nanny TV (especially in Spain and England). Added up, authoritarianism and permissiveness represent the predominant pattern of problematic parenting styles in all countries.

Table 1. Parenting styles expounded on (pre-intervention)

Mean, N=1542, N(Brit)=231 N(Germ)=258 N(Austria)=223 N(Spain)=432 N(Braz)=398 Scale: 0-5 Parenting Style Practised by Parents	Supernanny Programs						
	Britain	Germany	Austria	Spain	Brazil	F-Test	Total
Authoritarian	1.50	1.71	1.88	1.83	.75	***	1.49
Democratic	.44	.26	.61	1.27	.31	***	.64
Permissive	1.61	.77	.53	1.80	.85	***	1.17
Negating	.60	.82	.58	1.21	.33	***	.74
Sample period: 2004 - 2008. Scale: 0=not practised at all; 5=strongly practised. ***=highly significant difference between Supernanny programs, p<0.01; **=significant difference, p<0.05.							

Authoritarian parenting style: The *authoritarian-autocratic parenting style* assumes the necessity of exerting authority on children. Therefore, the children's own initiative is suppressed, and their opinion is deemed worthless and they have very few chances to develop freely. *Democratic parenting style:* The *democratic parenting style* sees children and adolescents as serious dialogue partners who hold their own opinion. The older the offspring, the more independent and self dependent they are expected to act. Still, parental instructions and help are deemed necessary. Democratic parents are open towards their children and give them a feeling of safety and acceptability. *Permissive/laissez-fair parenting style:* Permissive parents are rather reluctant to educate. Therefore, the children and adolescents have to be proactive when it comes to personal decisions. A laissez-faire upbringing knows no firm rules; everyone is left to their own devices. When personal decisions need to be made, children and adolescents are, as a rule, more active than their parents. *Negating parenting style:* In a *negating parenting style*, the parents do not influence the children's behavior at all. They are not interested in taking part in the child's development.

If you add to that the lack of interest in active parenting in the "negating style" which correlates highly significantly (r=0.43) with the "permissive style" and has a negative correlation with the "authoritarian" and "democratic" parenting styles, the *Supernanny* programs reveal a double crisis of upbringing: problems and the need for counseling arise, on one hand, through too heavy-handed parental demonstrations of power barely camouflaging the actual powerlessness. On the other hand, the lack of assertiveness is manifesting as a passive non-interventionist attitude. Thus, authoritarianism within the problem defining sequences of nanny TV, indicates an upbringing practice with parents' bias towards over-controlling. The Supernanny problem families suffer from a lack of a "democratic" parenting style, according to which a parental responsibility and a positive attitude towards the child would have to be connected to non-repressive and non-argumentative behavior. This parenting style is on the last position in all countries with regards to problem definition and – as I am going to show – also with regards to independent problem solving attempts by the parents. Apparently, there is a threefoldly graduated reflection offered to the *Supernanny* audience: 1. A lack of effective and enduring control on the parental behavior level (which is shaped by ineffective authoritarian control attempts or the relinquishing of control); 2. The ease of successfully practiced democratic forms of upbringing; 3. A value assessment of the *Supernanny* producers that ultimately favor a democratic style. Table 2, containing the parenting style recommended by the nannies, shows how wrong the accusation of an anti-democratic concept of the show actually is. The average program is dominated by the *"democratic parenting style"* most nannies favor. At the same time, "permissive" and "negating" parenting styles basically play no role in their recommendations. Nanny TV's appellative structure thus is directed against laissez-faire and advocates a fair, arguing and active treatment of children.

Table 2. Nannies' recommended parenting style (while intervention)

Mean, N=768, N(Brit)=129 N(Germ)=159 N(Austria)=150 N(Spain)=160 N(Braz)=170							
Scale: 0-5	**Supernanny Programs**						
Parenting Style Propagated by Nanny	Britain	Germany	Austria	Spain	Brazil	F-test	Total
Authoritarian	2.74	1.09	.14	1.00	.45	***	1.02
Democratic	1.40	2.57	3.93	3.68	3.72	***	3.13
Permissive	.00	.00	.00	.08	.26	***	.08
Negating	.00	.00	.00	.09	.00	***	.02
Sample period: 2004 - 2008. Scale: 0=not propagated at all; 5=highly propagated. ***=highly significant difference between Supernanny programs, p<0.01; **=significant difference, p<0.05.							

Only in the UK, the authoritarian concept is more popular with the nannies than the democratic one. Is the British original format thus at least purposefully "authoritarian" as the English pedagogic cliché suggests? In an international comparison, Jo Frost's authoritarian elements are actually stronger than those of other nannies – for several reasons. In British edutainment, the permissive parents are especially numerous and are deemed the cause of a lot of problems. Of course, considering this, a harder hand seems to promise more effective solutions. On the other hand, the democratic component is especially strong in Austrian supernannies who, as we know, have to deal with exceptionally authoritarian parents. In both cases, we see a logical relation between the definition of the problem and a potential solution. Aside from nation-specific peculiarities regarding the problems of everyday upbringing practices causing a variety in the TV supernannies' recommendations, any interpretation of the findings must take into account the fact that the British nanny's characteristic style is ironically authoritarian and serves as a symbol of strength for all parents after losing control. A common trait of the too-authoritarian and the too-indulgent parents is that their "children get out of control", as the producers point out often in the trailer. Without strength and assertiveness, the crisis can hardly be managed. Now, it is part of the special dialectic of challenge that in the past, authoritarian parental interventions often failed and worsened the conflict. Thus, it is all about a more significant meta-control that includes not only the child, but also the parents who cause the problems. This form of control is unattainable by "classical" authoritarian parenting. It may be achieved by an ironically transformed variety of British authoritarianism, Jo Frost style, but only if it is adapted to the respective countries' conditions and fulfills the postulate of glocalization.

In Germany, Austria, Spain and Brazil, the supernannies copied Frost's conservative look and wear outfits that suggest strictness and authority. In their recommendations, though, they avoid the air of authoritarianism more than the British original, probably to avoid giving the wrong hints to parents with authoritarian tendencies. The important contrast to England, though, lies in the different traditions of upbringing that ignore the subtle difference between "authoritarian" and "authoritative". If you do not want to be authoritarian outside the Anglo-Saxon territory, you have to show it clearly, e.g. through decidedly democratic actions. In England, there is still the in-between position of a democratically mitigated authority concept. It is not entirely out of the question that even my well-trained Austrian coders sometimes coded the British model as "authoritarian" when a different coding decision in the sense of authoritative-democratic could have been reached. To understand the glocalization process, it is definitely helpful to know that in comparison to the original program, all follow-up formats mitigate the authoritarian components and strengthen the democratic moment that forms the original nucleus of the appellative structure and ranks foremost amongst the nanny's recommendations even in the British format.

The democratic parenting style is most predominant in the Austrian *Supernanny* program that works more than all others with reflective methods of parent counseling and avoids directly intervening with the child. It is part of its self-concept to keep a distance to the English and – especially – German authoritarianism, as the words of the Austrian Nanny Sandra Velásquez (who was born in Mexico) prove: "I see Austria as much more Romanic than Germany; in the sense of a more casual and improvising lifestyle. (...) Germany is rather 'snappy' and looking for structures. Austria doesn't work like that." (Velásquez, own interview 2005). Spain as a "Romanic country" and Brazil as a member of the Romanic language family are pretty close to Romanic Austria with its subjective tendencies when it comes to democratic parenting. Spain (and with some reservations Brazil) show, as reported above, a positive correlation between non-authoritarian parenting styles and success at school as a specialty of their educational culture that is not present in Anglo-Saxon territory. This additionally boosts democratic to permissive parenting recommendations (the latter are much less frequent, though). The Spanish nanny Rocio Ramos-Paul has studied clinical psychology. Authoritarian educational methods are well known to her, but her main interest is the children's welfare that she tries to secure by procuring room for development and democratic coaching. Permissive elements come into play too. This applies even more to Cris Poli, an Argentinean, who is the current Brazilian supernanny and like her Spanish and Austrian colleagues values highly psychological counseling and reflective work with the parents' self.

Additionally, the Brazilian format is the least didactic. In problem definition, as well as in coaching, it focuses on showing educational realities that make a clear allocation to parental styles and direct solutions which are difficult for the parents due to the complexity of everyday situations. Compared to other formats, it gives less interpretational help which certainly is detrimental for a clear orientation. However, this does not diminish the popularity of the show among Brazilian viewers. Brazil is the country where most families apply for participation (over 30.000 according to the producers) with the best prime time quota. Poli and the Brazilian *Supernanny* team rarely assess anything explicitly; an exception is made when it comes to basic principles of democratic parenting that she advocates almost as often as the Austrian nannies. Thus, she can sharpen her democratic profile without using Jo Frost's directive coaching methods. Extreme fans of unambiguousness may not like this, but Brazil's "savvy viewers" who value their autonomy like Poli all the more for it. A similar concept was much less successful in Austria. Here, the *Supernanny* program was cancelled after only three seasons and was replaced by the coaching format "Bauer sucht Frau" (*Farmer wants a Wife*). Even disregarding the fact that in Austria's rural alpine climate the farmers' relationship problems seem to be better related to a typical national reality (and thus potentially more appealing to Austrian viewers) compared to the everyday problems of domestic upbringing (something which possibly inspired the broadcasters to make that programming switch), *Supernanny's* marketing problems could also have been caused by cultural implementation difficulties. By distancing themselves from Frost and Saalfrank and through their well-developed criticism of the "authoritarian model", the Austrian nannies might have overtaxed the audience doubly: with regards to the wide acceptance of authoritarian parenting methods and to the relatively strongly developed need for guidance and unambiguous orientation. What supports this interpretation hypothesis is the fact that authoritarian parenting is emphasized the most in the Austrian *Supernanny* format. At the same time, the Austrian version of the format reduces the participants' authoritarian parenting more than in all the other countries. What must be assessed as a desirable gain in democratic educational culture from a pedagogical-psychological perspective led in Austria to a loss of viewers whose cognition management was unable to bridge the emerging dissonances only by learning consistently from the nanny.

The German *Supernanny* program with its medium authoritarian and democratic tendencies is (in Velásquez' words) situated in between the "Romanic" and the "Anglo-Saxon" pole. The German nanny Katja Saalfrank phrases a clear "on the one hand – on the other hand" by saying "according to my experiences, especially with the families in the *Supernanny* TV series, children need strong parents with (...) clear points of view." In her guidebook for parents, she writes: "To control children in any way goes against my basic ideas of how to deal with human beings." (Saalfrank, 2006, p.6) The German show that mixes authoritarianism and democracy is among the most successful and most persistent *Supernanny* formats in the world and has a strong standing on the German market. According to our content analysis, the format is editorially more clearly structured and less ambiguous with regards to the orientation performance than the Austrian program. This emphasizes the finding that the relative openness of exchange and the low level of authoritarianism that brings popularity in Brazil cannot be transferred to the German speaking central European area without a loss of attractiveness for the audience.

Different Learning

Table 3 contains the solution oriented parental style that the participating parents practiced after the nanny's arrival under her direct tuition or autonomously. The higher the value, the more the parents used the respective parental style when trying to resolve conflicts or to cope with the child's disciplinary problems and other pedagogical challenges.

Again, the British format shows the highest degree of authoritarianism. When comparing this finding with results reported on Table 1 and Table 2 it becomes clear that the value of solution oriented parental style lies right between the initial situation posing the problem and the nanny recommendations. Obviously, the parental practice has come closer to the propagated parenting standard through the nanny intervention. To revise this interpretation, Table 4 systematically shows the differences between the old problem defining parenting practices and the new post-intervention parenting style.

Table 3. Solution oriented parenting style (post-intervention)

Mean, N=1096, Chan=218 RTL=222 ATV=207 Cuatro=223 SBT=226							
Scale: 0-5	Supernanny Programs						
Parenting Style Practised by Parents	Britain	Germany	Austria	Spain	Brazil	F-Test	Total
Authoritarian	2.25	1.74	1.33	1.47	.93	***	1.54
Democratic	.95	1.21	1.92	3.36	3.31	***	2.17
Permissive	.63	.45	.40	.58	.61		.54
Negating	.51	.30	.26	.51	.19	***	.35

Sample period: 2004 - 2008. Scale: 0=not practised at all; 5=strongly practised. ***=highly significant difference between Supernanny programs, p<0.01; **=significant difference, p<0.05.

Table 4. Difference between the new parenting style and the old upbringing practices

Diff Mean						
Scale: -5 - 5	Supernanny Programs					
Parenting Style Practised by Parents	Britain	Germany	Austria	Spain	Brazil	Gesamt
Authoritarian	,75	,03	-,55	-,36	,18	,05
Democratic	,51	,95	1,30	2,08	3,00	1,54
Permissive	-,98	-,32	-,14	-1,23	-,24	-,64
Negating	-,09	-,53	-,32	-,70	-,14	-,38

Sample period: 2004 - 2008. Scale: 0=not practised at all; 5=strongly practised. Diff Mean: Problem solving minus trouble making parenting style practised by parents.

The British format, the only one in which, as we know, the nanny recommendations supersede the authoritarian level of the problem definition shows the highest increase of authoritarian parenting styles. In the German format, the authoritarian behavior remains more or less the same, while it decreases the strongest in Austria. This can be interpreted as England and Austria showing effects that are nanny compliant, even if they have different directions. Jo Frost with her relatively authoritarian points of view knows how to successfully convince permissive parents, while the Austrian nannies Velásquez and Edinger who are, in comparison to Frost, more interested in reducing authoritarian parenting prefer democratic recommendations, and are also successful.

On the unchanging authoritarianism level of German nanny TV we can assume that the protagonists do not always apply the counseling advice they receive from the nanny.

Saalfrank (like all nannies) recommends a much more democratic parenting style than the participants normally use. Still, the parents' upbringing behavior remains quite stable.

In Brazil, the nanny communication's paradoxical dialectic went so far as to turn the show's democratic, anti-authoritarian intention upside down in a way. Although the parents' authoritarian tendencies are relatively weak and although Cris Poli is even beneath this weak level and massively advertises democracy, the participating parents' readiness to use authoritarian measures increases after the nanny's intervention. This can either be explained by reactance (resistance against persuasive pressure) or as an epi-phenomenon of self-empowerment brought on by participating in the supernanny's "therapy sessions". Remember, the lifeworld subjects' self-empowerment is a central motive of their cognition and emotion management in general. It means that individuals plagued by experiences of powerlessness use any chance for psychic armament and motivational consolidation of their assertiveness (e. g. symbolic offers of TV action heroes). One does not need a lot of imagination to assess the circumstances in Brazil as relatively unstable from a lifeworld point of view: high everyday violence rate, lots of crime, strong differences between the rich and the poor. If you add the relative openness of communication in the *Supernanny* program, reactance as explanation seems highly unlikely. In Brazil's case, a relatively strong need for the bolstering of self-efficacy and internal locus of control in the face of a wide variety of threats rather hints at the increase of the show's participants' authoritarianism levels being due to a spontaneous increase of self-empowerment. In this case, reactance is not to be expected because the slight increase of parental authoritarianism has the highest plausibility. This doesn't mean that the participating Brazilian parents refuse the predominant democratic counseling altogether, but they re-shape the main tendency by authoritarian techniques motivated by self-efficacy.

In general, the parents' "authoritarian" insistence seems to contain a certain degree of symbolic self-empowerment if the country's parenting culture does not especially gratify alternative parenting styles. The latter could explain why participating Spanish parents (just like the Austrians) reduced their authoritarianism the most. On the other hand, permissive practices helping Spanish pupils' success at school according to empirical studies (Musitu and García, 2005; Martinez and García, 2007) were reduced at the same time. A closer look at the difference in values between old and new parenting style separately for the sequences with direct nanny tuition and those with spontaneous parental solution attempts seems useful here. With regards to the authoritarian style, in the case of direct nanny guidance we get Diff =-0.48 in the Spanish example; without Nanny guidance, the difference between parental behavior focused on solutions and the problem definition beforehand is in the positive area: Diff=0.71. The values for the permissive style are: Diff (directly Nanny guided)=-1.71; Diff(without Nanny guidance)=-0.20. My interpretation of these findings suggests that even the Spanish parents feel the well-known insistence against the nanny's anti-authoritarian recommendations; only when she is personally present, do the parents reduce their authoritarianism compliant to the norm. Otherwise, it grows in the course of the general self-empowerment through nanny TV. With regards to the socially successful pattern of indulgence and permissiveness common in the Spanish parenting culture, the decrease in the case of direct nanny guidance is much higher than in the case of spontaneous parental behavior focused on solutions. This, too, can be seen as the participants' tendency to resist learning – obviously, they did not want to give up the successful pattern to the degree the nanny wanted. This can also be phrased in a way that criticizes the media: The Spanish nanny follows the British model in fundamentally criticizing permissive parenting styles without keeping in mind the Spanish culture's specific conditions; the result is a loss of learning success. At the same time, this finding can be used as proof of the concept of a permanent reality check by the lifeworld subjects checking incoming information for its everyday suitability to adapt their capability of orientation to changing life conditions that I developed in the theory part. Thus, differences between learning demands and actual learning success are generally no surprise.

Three things are noteworthy for the analysis:

1) Persuasive intention does not necessarily mean persuasive effect on the recipient's side.
2) Message recipients prove to be especially resistant when the coaxing persuasive pressure penetrates and does not conform to one's own needs and views.
3) Only when the glocalization offer is massively disregarded or when the intended effect conflicts with other effects, the direction of the effect can be reversed (as in the Brazilian example).

In most cases, though, the result of the show's immanent learning success accords to the way shown by the nannies, at least in the direction of effect. The parents' post-interventionist upbringing practices with and without nanny guidance differ gradually, but as a rule not in principle from the recommended basic direction. This is especially true for the main effects of a solution-focused change of parenting style towards an increase of democratic parenting techniques that we found and that promise a double gain of control – for the parents and for the child alike. No other dimension of parental upbringing practice has changed more under the influence of nanny TV (see Diff=1.54, summary column of Table 4). Across all nanny editions that we examined the democracy gain represented the best measure for the capacity for solution with regards to the problems of the twofold upbringing crisis of the parents intervening too much or too little. In the case of direct nanny guidance the democratic success is highest: Diff(directly nanny guided)=3.00. In the case of missing nanny guidance, the gain still remains in the positive range: Diff(without nanny guidance)=0.27. The learning pattern with regards to the permissive parenting style also conforms to the direction the nannies intend across all the nanny editions in our analysis: Diff(directly nanny guided)=-1.13; Diff(without nanny guidance)=-0.21. We also find this pattern of a gradual decrease with regards to the second low control parenting style, the negating one: Diff(directly nanny guided)=-0.68; Diff(without nanny guidance)=-0.11. As I said, the authoritarianism dimension proves to be the exception to the rule, where the difference values vary atypically. All in all, the parent's authoritarianism remains almost unchanged (Diff=0.05). While the authoritarian dimension significantly decreases in the nanny guided upbringing practice (Diff=-0.47), it increases when the parents act outside the nanny's line of sight (Diff=0.53).

We explained this difference above by motives of self-efficacy and self-empowerment of the lifeworld subjects. This difference cannot be ignored, not even if you insinuate a pedagogical intention on the part of the producers in the nanny guided sequences. No one could have stopped the editorial teams from perfectly staging the non-nanny guided sequences for the audience, too. Obviously, a certain reflection of extra-media reality remains in the film analysis data – a reality whose redemption, as we know from Kracauer, is the most distinguished job of reproduced images and one that we try to filter by means of statistical comparison as "savvy researchers". Otherwise, the participating parents' motive of self-empowerment is compliant to the decrease of permissive and negating parenting that we

From Reality TV To Coaching TV 239

found. All in all, the parental learning successes presented in nanny TV can be described as an increase in parental control and, at the same time, a gain in democratic containment of parental control excesses over the child.

Parenting Success

Table 5 shows that the self-assured demonstration of parental educational strength ("Self-confident agency/stand up to the child") represent with N=282 applications the most commonly practiced and in about 50% of all cases immediately successful parenting measure.

Table 5. General parenting measures successfully practiced

%, N=1281	Supernanny Programs							
Upbringing Action	Britain	Germany	Austria	Spain	Brazil	Sign	Total	N
Structuring the daily routine / regular play times	53,8	84,2	100,0	13,3	9,4	***	40,0	85
Reflecting the parenting practice	100,0	100,0	100,0	12,5	12,9	***	24,7	73
Self-confident agency/stand up to the child	25,6	78,5	71,0	65,6	25,0	***	51,8	282
Communication at the child's eyebrow level / eye contact	50,0	73,3	76,9	71,4	38,5		60,5	81
Use of calm deep voice / short clear instructions	43,8	96,6	97,5	63,8	27,3	***	68,2	192
Parents act in concert	44,4	71,4	76,9	10,0	,0	***	38,5	52
Familiy activities suitable for children	80,0	83,3	77,8	44,4	,0		69,0	42
Play with the child / have fun	100,0	90,5	91,7	57,1	26,3	***	76,0	129
Loving interaction with the child / show the child affection	83,9	94,7	93,7	60,0	36,4	***	81,7	175
Set up rules / draw clear lines	33,3	45,7	69,2	56,5	26,3	**	46,5	170
Mean	61,5	81,8	85,5	45,5	20,2		55,7	128

Sample period: 2004 - 2008. Percentage of sucessfull executed upbringing action measured by the number of execution of respective action. ***=highly significant difference between Supernanny programs, p<0.01; **=significant difference, p<0.05.

The motive of control is also reflected quantitatively in the attempts to solve educational problems in nanny TV. It is remarkable that the success quota is by far the highest in non-autocratic, respectful and loving manners: "Loving interaction with the child", followed by "Play with the child" and "Family activities suitable for children". The attributes of an empathy-based relationship recognizing the children's rights as human rights that are decisive for a democratic upbringing are thus not only preferably advertized by the nannies and accepted by the parents as a learning result, but are also virulent on the level of parenting measures and can be independently deducted from the scenic development beyond persuasive verbal strategies by the viewers.

The most successful loving interaction with children is shown in the German and Austrian *Supernanny* formats. In Austria, though, this upbringing technique is much more frequently on the screen: While here "Loving interaction" and "Show the child affection" form 35% of the general parenting measures, it is only 8% in the German *Supernanny* program (see Table 6). Decisive for the audience's learning success in a sense of role model learning is, of course, the success factor that is even higher in German nanny TV than in the Austrian format. "Self-confident agency" is only successful with 25% in the British *Supernanny* format, but is practiced all the more intensely. More than a third of the general parenting measures in the British *Supernanny* format are part of this central nanny TV component that is so important to the participants' self-empowerment and the identification of the recipients interested in symbolic self-efficacy. The relatively low success quota shows

240 Jürgen Grimm

that the "authoritarism" in the British *Supernanny* format is not only broken by Jo Frost's tongue-in-cheek attitude, but also by an openly shown defeat of authoritarian parenting actions. What may be detrimental for the audience's role model learning is a gain with regards to *negative learning* because the viewers can make their own decisions based on success *and* defeat of what is right and has a chance for a differentiated reality check. But the low quantitative share of "Set up rules" in the British *Supernanny* format hints at Jo Frost most often acting as a "gunslinger" or encouraging her clients to intervene directly instead of setting up new rules for all family members to create a better framework for the parent-kid-interactions. Except for their value for the emotion management of the participant feeling strong and the recipient riding along on symbolic horses, interventionist practices do not lead to a lasting success in positive *and* negative learning.

Table 6. General parenting measures practised[4]

%, N=1203, N(Brit)=259 N(Germ)=242 N(Austria)=225 N(Spain)=232 N(Braz)=245 Supernanny Programs							
Upbringing Action	Britain	Germany	Austria	Spain	Brazil	Sign	Total
Structuring the daily routine / regular play times	5,0	7,9	2,7	6,5	13,1	***	7,1
Reflecting the parenting practice	1,2	1,2	1,8	13,8	12,7	***	6,1
Self-confident agency/stand up to the child	33,2	26,9	13,8	27,6	14,7	***	23,4
Communication at the child's eyebrow level / eye contact	10,0	6,2	5,8	6,0	5,3		6,7
Use of calm deep voice / short clear instructions	12,4	12,0	17,8	29,7	9,0	***	16,0
Parents act in concert	3,5	2,9	5,8	4,3	5,3		4,3
Familiy activities suitable for children	3,9	5,0	4,0	3,9	,8		3,5
Play with the child / have fun	9,7	8,7	16,0	12,1	7,8	**	10,7
Loving interaction with the child / show the child affection	12,0	7,9	35,1	15,1	4,5	***	14,5
Set up rules / draw clear lines	5,8	14,5	5,8	29,5	15,5	***	14,1
Sample period: 2004 - 2008. Percentage of respective upbringing action measured by the number of sequences with any solution oriented action. ***=highly significant difference between Supernanny programs, p<0.01; **=significant difference, p<0.05.							

With approximately 30% of the general parenting measures practiced, "Set up rules" and "Draw clear lines" is especially characteristic for the Spanish nanny TV of which we already know that the decrease of the parents´ permissive views counts among the show's greatest learning results. At least, the desired effect of new rules and clear lines actually happens in more than 50% of all cases. Only the Austrian protagonists' success quota is a little higher based on a very small quantitative base of 5.8%. In Brazil, the low success of loving manners (only 4.5% of the cases) and the less than average attainment of "self-confident agency" (25% of the cases) indicate that the Brazilian *Supernanny* program is construed as less strongly pedagogic and not exceptionally clear in aiming at giving orientation. Here, too, it must of course be considered that clearness is only one factor for reality TV's popularity and that it depends on the amount of suffering from the multi-optional society, in a given cultural context. Obviously, the Brazilian audience's readiness to tolerate stronger ambiguities and use this as a gain for their own aplomb of judgment is especially high. In any case, the Brazilian format offers to the viewers relatively open ways of interpretation that have not led to any popularity loss so far. The Austrian and the German format show the strongest success orientation for the general parenting measures practiced: 85.8% resp. 81.8% of the upbringing

---

[4] The varying case numbers in Table 5 and Table 6 can be explained by the different percent base: the first case is based on the sequences (all sequences =100%), the second on parenting measures practised (=100%). Also cf. Table footnotes.

action immediately lead to a success here, while the sample average is just 55.7%. On the other end of the scale, we find the Brazilian nanny TV with 22.2% success. England and Spain are situated right in the middle of the range of successful parenting demonstration.

The cultural diversification of the *Supernanny* formats clearly shows the different ways of coping with parenting problems and the range of linking the relation between open ways of addressing problems to more or less unambiguous forms of problem solving in a culturally specific offer of orientation. The presentation's degrees of unambiguousness vary just like the participants and the supernannies' emphasis on the parents' self-assured demeanor and the loving interaction with the children.

Gender Bias and Parental Divide

Who are the problem solvers in nanny TV? Can their identity assist us in consolidating criteria for the socio-political quality of edutainment? Table 7 lists the family role of the protagonists who act as advice executants and problem solvers in *Supernanny* programs in the five examined countries. We see that across the board mothers are most frequently the ones who fill this position. They are five times more often than men directly guided or indirectly encouraged by the nannies to use new strategies in coping with educational problems. Also, mothers alone are being addressed twice as often as father and mother together, which can be valuated as a deficit in joint parental action. On the one hand, this reflects the social reality of upbringing conditions (where women still carry the main load of educational work); on the other hand, nanny TV can preserve or intensify an existing gender bias in the exertion of upbringing tasks. The mothers' nanny guided island position thus contains an element of social inequity since along with the control options, the mothers take on a higher degree of responsibility for the reproduction of the world of everyday life and for society's supply of socialized lifeworld subjects (including potential dysfunctions like deviance and crime). By taking over upbringing tasks, notoriously overstrained mothers lose control in other areas. The price for nanny TV's one-sided addressing of the mothers is an increase of the parental divide that enables fathers to duck their responsibilities or – phrased negatively – excludes them from the upbringing business and thus keeps an important resource of lifeworld happiness and social power out of their reach. Grandmothers and other family members play almost no role in upbringing practice. The gender bias is complemented by a focus on the core family that in the asymptotic perspective leads towards a model where mothers sail the ship of upbringing alone.

With smaller variations between the countries – the most in England, the least in Brazil – the supernannies more or less count on *supermummies* that are even explicitly in the name of a competitor to the *Supernanny* program in Germany (*Supermamas*, RTL2). In Brazil, the coaching of both parents together reaches a considerable value of 41%, but the value for fathers as individual problem solvers drops to a marginal level of 6%. Here, the fathers as individuals disappear almost completely and rarely stand out, being marginal background figures in their sometimes seemingly forced duet with the mothers.

Still, *gender bias* and *parental divide* seem the least grave in Brazilian edutainment TV. Compared to Cris Poli, Jo Frost addresses mothers as competent upbringers much more often, and she rarely addresses the parents together. Even though fathers' inclusion is remarkable, Jo Frost has the overall lowest value of gender equalization and co-parental advice.

Table 7. Problem solvers as advice executants

Column%, N=1165, N(Brit)=133 N(Germ)=230 N(Austria)=215 N(Spain)= N(Braz)=							
Problem Solver / Advice Executant	Supernanny Programs						
	Britain	Germany	Austria	Spain	Brazil	Chi^2	Total
Nanny	8.6	3.1	1.9	4.3	7.4	***	5.2
Mother	60.2	59.6	53.0	59.8	45.5		55.6
Father	19.3	18.9	18.1	15.0	6.1		15.4
Parents	11.5	14.9	27.0	20.9	41.0		23.1
Grandmother	.4	3.5	.0	.0	.0		.8
Sample period: 2004-2009. All problem solving persons= 100%. ***=highly significant difference between Supernanny programs, p<0.01; **=significant difference, p<0.05.							

Another characteristic of the British original is the relatively high value of direct nanny intervention (cf. Table 7, line 1). Here, the nanny is supervisor and problem solver at once and directly intervenes in the family's upbringing business. More than her colleagues, Jo Frost prefers direct contact with the child she wants to set on the right path by insistent speeches and admonishing – without taking the detour of working with the parents.

On the other side of cultural diversification is the Austrian nanny model with the least direct influence on the child and the most psychological counseling style of the Supernanny using psycho-therapeutic techniques of working with the parents. The fixation on the mother is slightly below average here. The viewing quota significantly dropped when Sandra Velásquez emphasized the need to work with the fathers more in the season's second half after the author's personal critique. The reasons for that are the circumstances of TV as means of mass communication that is in the case of *Supernanny* programs highly affected by a female audience. Of course, female viewers claim the emotional gain of the heightened self-efficacy for themselves after carrying the burden of educational challenges. Here, we face a basic limit of social effectiveness of reality TV programs with a lifeworld focus which we will discuss later on. Without satisfying the need for emotion and cognition management in the context of the viewers' everyday lifeworld, no social transformation can be engineered, no matter how desirable it is. Reality TV in general and especially coaching TV are structurally conservative with regards to this.

What gain does the completely feminized nanny concept offer to women? Table 8 discloses that mothers set the tone, but do not act more successfully as problem solvers than their male counterparts. Their success rate (38.8%) is actually even lower than the success rate of fathers (44.7%). Although men are less often shown throughout the show, they are more successful, when they show up. Whether realistic or not, this success can at least give nanny TV credit for encouraging a more participatory parenting style on the part of fathers.

The values in Table 8 show how often the sequences with problem solving attempts lead to an immediate success, subject to the problem solver types. The balance of success and failure entails general parenting practices reported above and also specific parenting measures such as "Ignore the child for a while", "Admonish the child", "Give the child timeout" and "Hit the child". Because the success rates of specific upbringing measures are lower than of general measures reported in Table 6 the total score in Table 8 is lower too, but more

From Reality TV To Coaching TV   243

representative for parenting success in *Supernanny* overall and also more reliable for evaluation of the problem solvers' effectual upbringing. The table proves that the Austrian and Spanish formats feature the largest share of success of nanny action. In these two countries even those mothers who are relatively unsuccessful reach at least an above-average score in international comparison. The best men's balance is again found in the Spanish format, followed by the Austrian one. The least successful men reside in Germany, where the nanny, Katja Saalfrank (a single mother herself), may know about the limitations of male upbringing from her own experience.

Table 8. Problem Solvers' Balance of Success

%, N=1111, N(Brit)=243 N(Germ)=223 N(Austria)=210 N(Spain)=219 N(Braz)=216							
Supernanny Programs							
Specific and General Upbringing Action	Britain	Germany	Austria	Spain	Brazil	Sign	Total
Success rate: Nanny	53,9	51,8	75,0	77,8	40,0		55,4
Success rate: Mother	28,7	37,6	48,1	46,5	34,9	***	38,8
Success rate: Father	36,7	33,0	50,0	67,2	36,4	**	44,7
Success rate: Parents	51,9	50,6	58,9	26,7	15,6	***	35,5
Sucess rate: All	*34,2*	*40,4*	*52,9*	*47,1*	*26,3*	***	*40,0*
Sample period: 2004 - 2008. Success rate (%): percentage of successful action measured to all action for solution. ***=highly sign. difference between Supernanny programs, p<0.01; **=sign. difference, p<0.05.							

The most noticeable finding in Table 8 is the failure of the joint parental action. As far as success is concerned, it is ranked lower than that of fathers and mothers and fails exceptionally in Brazil. While the Brazilian format is by far leading, when it comes to the instances of parents as a couple being addressed, the success rate of joint parental action in this South American country is very low (barely 16%). In the other Latin country, Spain, the discrepancy between the frequency of couple adressing and the success at solving is not as high as in Brazil, but a success rate of 26% is weak enough to suggest that the Spanish viewers take the failure of joint parental action in the *Supernanny* program as a reason for their own avoidance of engaging in this activity. In Austria, on the other hand, the tradition of joint parental responsibility is very strong. Reflecting that, the Austrian programs contain the highest share of successful joint parental problem solving (just a bit less than 60% of the cases). This is especially meaningful as it is joint parental action where problem solvers use the democratic parenting style the most often. On a scale of 0-5, the average practicing intensity of the "democratic parenting style" for couples' upbringing action focusing on solutions is 3.41. Even the nanny's score (M=3.30) is lower than that. The least democratic are the completely overstrained mothers whose average score is 1.62; but in this case, not only the women, but also the democratically-pedagogically undernourished society pays for the unjust assignment of educational tasks. Of course, the men's corresponding democracy value is only slightly higher (M=1.92). The mothers may as well find solace in the fact that in the gender competition, the men clearly take the buck of leading in the category of authoritarism (M=1.95). The mothers' average score (M=1.62) is still higher than the nannies'

(M=1.30) on the authoritarism scale and far ahead the one of the married couples who are well situated with regards to this, too (M=1.05).

To sum up, partners as double heads of the family are underrepresented in the nanny TV families and show a great variety of success and failure. Socio-politically and pedagogically these underrepresented couples represent the most desirable educational model, for only under these circumstances, the families' democratic potential will be stimulated at maximum level to conquer educational crises effectively and permanently.

Viewers' Motives in Germany and Austria

Now, I finally turn to the viewers' motives which are founded in their concrete everyday lifeworlds, just like those of the shows' participants. The lifeworld subjects are faced with numerous everyday problems and emergencies that need to be quickly resolved. The solutions need not conform to the demands on the systemic level (i.e. in the interest of developing the parenting culture as a whole) nor must they set a public example. The participating families' interest is focused on solving their very personal parenting problems; the viewers' interest is a paradigmatic TV demonstration of how to use their own everyday experiences as free and as effectively as possible for their own orientation. The gain from watching *Supernanny* thus is not a direct solution for a specific everyday problem, but lies in the symbolic confrontation with certain situations. It is very much part of the logical comparison process to ask if the TV situation or a similar one took place in the recipient's past or could take place in his future. The foundation of the para-social interaction between participants and recipients is the congruency of their thematic interests to think and talk about.

Thus, it is no wonder that the online survey of 1611 TV viewers in Germany and Austria, which took place between Oct. 17th and Dec. 29th 2005 showed a specific socio-demographic profile with matching results for both countries.[5] The probability to become a *Supernanny* viewer or even heavy viewer increases, if the person is female and has no higher education. Having a child and feeling involved in the situation increases the reception. Less relevant are retrospective biographic influences (such as one's own problems during childhood) and the wish to have a child in the future. One certainly cannot assume that the findings would hold in any country, but a certain generalizability can be justified by the fact that the pattern found is compatible with the basic structures of the world of everyday life according to Alfred Schutz that we deem universal. Schutz assumes that the lifeworld subjects develop specific *thematic and motivational relevancies* based on their everyday experiences that guide their attention and determine the orientation towards certain informational milieus of the social environment and – as we add – the media world. As long as women are more or less the only ones entrusted with the upbringing business, their upbringing relevancies are stronger than men's – especially when they have children themselves. Then, the probability of an orientation towards *Supernanny* programs rises highly significantly according to the survey results, no matter what subjective motives are crucial in detail. The moment relevant for the orientation is a so-called "because motive" (Schutz and Luckmann, 1983, p.215f) "objectively" embedded in the individuals' biographical and lifeworld context and "causing" the inviduals' behavior.

---

[5] On the procedures of the online survey and the various questions asked cf. Grimm (2006a).

The "uses and gratifications" approach can also be used to check the viewers' needs and conscious motives for having a strong orientation towards the program. These are called intentional "in-order-to motives" in the terminology of lifeworld theory. The most common *in-order-to motives* according to the survey were "Learn about things that might be important for me", "See how others deal with their problems" and "Can see people like you and me". When assembling the single motives into four groups, "cognitive-reflective" motives (with an element of comparing one's own everyday reality to the media scenario) ranked first, followed by "cognitive-stimulating" motives that are based on curiosity and expecting something extraordinary. Ranked third and fourth respectively are "para-social" (sort of intimate link to TV personas like the supernanny) and "emotional-reflective" motives (working with emotions in the sense of emotion management). Last but not least ranked "emotional-stimulating" motives referring to intense emotions and states of arousal. On a scale from 1 (when the motive does not exist) to 5 (when the motive strongly exists), we find the following mean values in the survey of *Supernanny* viewers:

Motives of Cognitive Reflexion	3.0
Motives of Cognitive Stimulation	2.7
Motives of Para-social Interaction	2.3
Motives of Emotional Reflexion	2.1
Motives of Emotional Stimulation	2.1

Obviously, the interest in the shows' sensationalism that would be expressed by high values in the motives of emotional stimulation is limited; those values are low. Rather, the possibilities of social comparison that nanny TV brings are important to the viewers because they cater best to their predominant reflexive needs. Here, the motives oriented towards everyday life clearly outweigh escapist attitudes. Motives rooted in the everyday life reach an average of m=2.7, while escapist motives with m=1.9 are far beneath this level. This means that the majority of nanny TV viewers do not want to forget their own sorrows; rather, the *Supernanny* viewers want to position themselves in relation to behavioral models of the participating families. This result conforms to what Hill (2000; 2004) detected about the British *Big Brother* audience and to Ouellette's and Hay's (2008) generalized thesis on coping with some everyday life problems through viewing reality TV.

When asked why they watch *Supernanny*, Mrs A (1 child, aged 2 ½ years old) who took part in one of our group discussions with *Supernanny* viewers, answered: .

> "To see that other parents' problems are bigger than mine." (All laughing.)
> *Supervisor:* "Is that a motive for watching?"
> *Mrs A:* "Well, this experience is sort of satisfying." (All laughing.)
> *Supervisor:* "What do you think when you see a raging child completely going wild?"
> *Mrs A:* "Well, you react kind of amused because you know that from your own child as well. Maybe not in that way, but that simply amuses. It is actually entertainment then."

Mrs A associates the "raging child" on the screen to her own offspring who is of course not as bad as the one on TV. This satisfies her which she directly links to "entertainment". This example shows how the comparison processes related to everyday life work when dealing with entertainment. In this case, the gratifying amusement is clearly at the cost of the show's participants the "savvy viewer" can laugh about heartily without considering the consequences for those laughed about. The coaching format Supernanny poses the same problem as *Big Brother* and the casting shows, namely that show participants and viewers do not act in concert in the zero sum game of reality TV. The more the viewers can feel strong and sublime, the more the participants lose their added gratification in the realm of public attention. Instead, they become infamous as negative paradigm in the public area. The viewers' orientation gain remains untouched by that. The more dramatically the participants fail (or the more dramatic their problems are at least), the more effective the audience's *negative learning* becomes. So while the participants try to solve their everyday problems, the viewers wait for mistakes and failure. Safe in their TV armchairs, watching others' problems doesn't only transmit a feeling of safety and superiority but also clarifies the difference between their own behavior and that of the problem families on nanny TV. Obviously, *negative learning* is not only an aspect often neglected when analyzing media effect processes, but also a rather underestimated dimension of fun with regards to the attractiveness of reality TV programs, where amusement, increase of self-worth and affirmative orientation mix inseparably with regards to precarious behavioral routines. That can in extreme cases lead to denying the learning aspect altogether.

We do have fundamental doubts as to how much the actual impetus of *Supernanny* use can be measured by motive catalogues that refer only to the conscious aspect of program use, are prone to social desirability, and can easily miss subconscious parts. Seeing reality TV use as a substitute for an adult education center and outing oneself as an information oriented user thus may seem socially captious and potentially reputation-damaging to some test persons because no one wants to be seen as an "idiot" or "naïve person". To overcome these shortcomings, we measured the psycho-social attributes of *Supernanny* viewers and non-viewers with the help of standardized personality tests – similar to the ones used by Reiss and Wiltz (2004) (only with a different test procedure) to discover relevant personality structures. The personality tests measured "locus of control" (Rotter, 1966; Rost-Schaude, 1978), "empathy" (Davis, 1980; 1983), "sensation seeking" (Zuckerman, 1979), "disaster sensitivity" (Grimm, 1999b) and "conflict behavior" (Grimm, 2006a). In Table 9, psycho-social dispositions were sorted by groups defined by the intensity of *Supernanny* program use. The higher the value, the more the attribute in question is true for the group. All the test values were projected to a scale ranging from 1 to 100, so that the acceptance percentage of the test items can be interpreted in relation to the highest possible acceptance. The asterisks in the table show that in the personality dimension in question, there is a significant difference between the groups hinting at a hidden *because motive* of nanny TV attraction in the sense of Alfred Schutz. If the highest value is in the heavy viewer group, the personality dimension has a non-accidental connection to the program use.

Supernanny viewers have significantly more external locus of control in politics than non-viewers. This means that the viewers feel heteronomous in the social and political area. What, then, makes more sense than to vie for the control missing in the social and political area within the family by all means? Supernanny viewers are, to the highest degree, interested in harmonic intra-family relations. Their family tolerance (readiness to cut back on one's own interests in family conflicts) is highly developed, just as the focus on solutions in arguments. The participating families' upbringing problems are thus seen as a threat demanding crisis intervention and conquering the loss of control. Thus, they prefer the show's participants facing a happy ending. Then, everyone is satisfied, and the potential conflict between participants and recipients remains within a dramaturgy that guarantees solutions – assuming the producers know this audience disposition and do not take their viewers as negativists keen on sensations by mistake.

From Reality TV To Coaching TV

Table 9. Psycho-social dispositions of Supernanny viewers

Affirmation%, N=1611, N(G1)=78 N(G2)=288 N(G3)=939 N(G4)=299

Psycho-Social Traits	Supernanny-Viewing Groups[1]					Total
	Not at All (0)	Rarely (1)	Regular (2-4)	A Lot (5+)		
Locus of Control for Everyday Life	56,4	59,7	56,4	57,7		57,2
Locus of Control for Politics/Society	50,9	48,6	37,8	34,2	***	39,7
Thrill and Adventure	54,9	49,1	44,3	38,5	***	44,6
Experience	41,7	36,3	29,4	26,7	***	30,6
Disinhibition	26,8	24,4	20,0	17,4	**	20,6
Boredom Susceptibility	36,8	36,6	34,2	32,0		34,3
Sensation Seeking (total)	40,2	36,4	32,1	29,0	***	32,7
Empathic Concern	70,6	71,6	73,0	72,9		72,6
Personal Distress	44,9	48,5	50,2	52,1	***	50,0
Disaster Affinity	64,3	64,7	76,2	76,8	***	73,7
Negative Realism	53,8	42,6	46,5	54,3	**	47,6
Negativity Intolerance	58,0	57,6	52,0	49,3		52,8
Conflict Avoidance	35,5	36,1	38,0	39,2		37,8
Problem-Solving	57,6	57,2	59,3	61,9	**	59,3
Readiness to Fight	24,3	23,8	25,1	27,2		25,2

Online survey in Austria and Germany (10/17-12/29/2005). [1] In brackets see the number of shows watched by viewing groups per month. Given are percentages of affirmative answers to trait constructs, scales: 0-100. ***=highly significant difference between groups, $p<0.01$; **=significant Difference, $p<0.05$

Supernanny viewers are the opposite of "High Sensation Seekers". They are neither interested in risks nor in adding to their experience or in losing inhibitions. Nanny fans avoid intense experiences: they exhibit no trace of sensationalism. Their empathy and sensitivity are above average. Pain and wounds of victims cause them physical stress. Thus, it is implausible to assume that joy in others' suffering was the main reason for watching *Supernanny*.

Still, disaster sensitivity (the attitude of considering disaster news as more important than other news) and negative realism (the opinion that negative media reports reflect reality) are stronger among *Supernanny* viewers than they are among non-viewers. We know from earlier studies (Grimm and Sells, 2006) that disaster sensitive viewers are also sensitive "Low Sensation Seekers" who turn to bad and negative things primarily out of emotional distress to make their emotion management more efficient. Thus, the nanny audience's disaster sensitivity is no proof for a perverted lust for disasters as the voyeurism critics keep insinuating. In both cases it is not about a general fascination of the horrible, but about fear of the bad things that you face via TV. The goal of this confrontation is to heighten the own ability to control emotions and to train coping with difficult situations.

Kracauer recommended standing firm under the Medusa's gaze in the era of film. We need to keep standing firm under the gaze of reality TV that already in the early days presented itself as disaster TV with redemptionist perspectives. In the more advanced state of genre evolution, the basic motive for attending to this form of mass communication remains the continuous battle for control of many emergencies that everyday life is full of.

DISCUSSION AND CONCLUSION

The British pediatrician Edward Christophersen praised the *Supernanny* program and the publication of Jo Frost's parenting guidebook euphorically in the renowned trade journal *Pediatrics*.

> "Her book and her television show may be one of the best resources available today for dealing with common behavior problems in children. Her suggestions for dealing with temper tantrums, sibling rivalry, mealtime problems, sleep problems, and toilet training are probably as good as anything currently available. Her television show does a good job of demonstrating the manner in which she recommends dealing with compliance issues in children." (Christophersen, 2005, p.1768).

The parents that I have talked with found Frost to be practical and down to earth, giving recommendations that they feel they can implement. Not all the pedagogues reacted quite so positively. A very critically minded nursery pedagogue from Vienna, who took part in one of our group discussions with upbringing experts, made no secret of her dislike of the Supernanny program. She regarded the concepts of upbringing the supernannies stand for as "outdated and antiquated", their outfit alone reminded her of "the good old authoritarian days" when it was still common to oppress and to hit kids. This is why she bluntly disapproved of edutainment TV and opposed it in general.

The programs' content analysis showed clearly that the supernannies are far from exculpating parental violence. The claim that they oppress and deal repressively with kids hardly corresponds with our data. Rather, stressed-out parents with a tendency towards extreme control and towards a loss of control at the same time act that way and turn to the *Supernanny* team in their distress. Then, the counseling aims to change the show participants' parenting style. This usually happens towards democratic principles of parenting as described by Kurt Lewin and advocated by a majority of pedagogues. Thus, it is surprising how excited and radical in her judgment the aforementioned nursery pedagogue was. Maybe the job-related need to discern herself from the "disdainful" taste of the masses that Pierre Bourdieu (1986) described as a general trait of intellectuals played a role here. What everyone likes cannot be used to construe special identities or job-related claims of supremacy. Against this backdrop, Christophersen's remarks actually seem heroic, as he risks ostracism by his colleagues. But his extremely positive judgment is also only partially compatible with the various empirical findings.

While there is a predominant democratic tendency in nanny TV, the international comparison of the programs also showed a certain range of representation and of limited approval for authoritarian-autocratic ways of upbringing. This was partly due to the parents' motives of self-empowerment, but it is also in parts supported by some of the supernannies. The British show, the only format where the nanny gave more authoritarian recommendations than democratic ones, proved to be the most "authoritarian". When assessing this, one must keep in mind that Jo Frost is faced with "permissive" parents more often than other nannies and tries more or less successfully to teach them self-assured action towards the child. Furthermore, her "authoritarianism" has a funny British touch which defuses the repressive parts. The Austrian format has the strongest democratic profile which of course had a tendency to overstrain the audience and suffered from popularity losses. The Spanish *Supernanny* program is also very democratically inclined and even more successful than the Austrian one in teaching the respective parenting attitudes to parents. The latter is also true for the Brazilian example where the democratic intention has the highest impact on parents. All in all, Brazilian nanny TV is very open and little didactic in presentation. This show leaves a lot to the audience's interpretation. This should be considered a hint that the need for authoritative guidance depends on the respective culture.

In most cases, the parents that appeared on the shows have followed the direction of the nanny's recommendations, but not to the intended degree. In some cases, the change in parenting style can even reverse itself like in the Brazilian nanny TV edition with regards to the authoritarian parenting style. Although the Brazilian supernanny avoids authoritarian practices like no other and although the parents, who seek guidance, exhibit a low level of authoritarian concepts, during the show the authoritarian trend in the parental actions increases. This can be celebrated as democratic resistance against the supernanny's "dictatorial" intentions or can be criticized as democratic deficiency of orientation aid; from the parents' point of view, all of this is not about a systemic quality of democratic culture, but about action in a concrete situation of their everyday life. In this context, self-assured, assertive action means a gain of freedom for the lifeworld subject.

All participating parents more or less suffer from a deficiency of control related to their interaction with their children *and* to their own helplessness, frustration and anger. Those who cannot control their emotions in the heat of battle with unwilling kids will be unable to convince a child: neither to do homework nor to go to bed in the evening. A minimum of emotion management thus is a prerequisite for constructively shaping the parent-child-relationship and at the same time for a liberal form of social interaction that is free from emotional pressure and in that sense, controlled. Recently, psychological approaches of "self-regulation" emphasizing the importance of emotion management and other techniques of control over one's own mind and body have become more important (Forgas, Baumeister and Tice, 2009). This leads to a shift of the problem perspective in many areas, e.g. with regard to addictions and aggression management, that are not seen any longer primarily under the aspect of model learning, but under the aspect of control over impulsive behavior. When the individual's ability to choose from various options of acting is restricted due to manic and uncontrollable behavioral tendencies, the lifeworld subject's sovereignty is damaged in exactly this sense. Now, the participants of *Supernanny* programs try to avoid the threatening loss of control, guided by the nanny. Camera and audience additionally support this quasi-therapeutic process because the attention of others at the same time heightens the ability for self-regulation. The TV performance generates a supportive climate of critical self-reflection and of wanting to prove oneself that heightens the chances for change and development of control abilities.

Thus, control in the world of everyday life means something totally different than at the systemic level. In the world of everyday life, the lifeworld subject's freedom, whose only limit is the freedom of others in the social environment, grows along with its control abilities. Systemically, control means domination over the mass of the populace which thus loses degrees of freedom. It seems barely plausible to construe a fundamental equation between self-control and governmental rule and generally accuse the individuals' freedom of "gouvernmentality" like Rose (1999) suggests based on Foucault's power analysis; Ouellette and Hay (2008) assume this as reality TV's fundamental tendency. In the double perspective of system and lifeworld, simple equations are not justifiable, not even Habermas' thesis of a "colonization of the lifeworld" by systemic powers. Rather, reality TV tries to configure the system to lifeworld concerns, to appoint it a subsidiary, supporting role. This is not easy; to a certain degree, indissoluble frictions between system and lifeworld are to be expected. Thus, dubious nanny interventions that do not take into consideration the local conditions have counter-productive effects. For example, the Spanish format noncritically adopts the British criticism of liberal parenting without considering the positive role of "permissive" parent attitudes for success at school among Spanish children.

The globalization of the *Supernanny* format is a source of systemic lifeworld dysfunctions, when the format does not adapt well enough to local conditions. Roger Silverstone felicitously phrases the ambivalence of interventions into the world of everyday life by media as a part of the socio-system: "… the media are players, shifting expectations, both tools and troubles in the management of lifeworld." (Silverstone, 2007, p.111). The *Supernanny* programs are not systemically neutral at all, nor do they automatically lead to democratic conditions, only because the mother communicates with her child on eye level. The analysis of the parents' solution attempts showed a marked gender bias that was increased by the fact that the nanny preferred to address mothers and consolidates the uneven allocation of parenting responsibility between the sexes. One of the paradox results of nanny TV is that the men – mostly having no part in the upbringing business – act more successfully than the quantitatively predominant women when they appear on TV. The mothers get more public attention, but their parenting style is portrayed less positively. For mothers who take part in the show, the risk of public embarrassment adds to the affirmation of uneven allocation.

There is a tension between reality TV participants and viewers as the viewers obtain their orientation gains in the sense of *negative learning* at the cost of publicly stigmatized people and behaviors. This is basically also true for the *Supernanny* programs that minimize the conflict by adding a dramatic happy ending. In general, *Supernanny* viewers are not interested in failure, but just like the show's participants they try to enhance their everyday control ability. *Supernanny* viewers feel an above-the-average social and political external locus of control and thus want to prove their control ability within the close area of their family. This can be construed as a special quality of reality TV: to keep the frictions between system and lifeworld bearable by emphasizing the everyday control ability. The image of the sensationalist or social voyeur watching nanny TV out of sheer curiosity does not go together well with this study's data. The majority of the *Supernanny* audience is anti-sensationalist and is only interested in disasters because of the expected rescue. Already in the 1990s, reality TV audience at the heyday of *Rescue 911* proved to be "Low-Sensation-Seekers" (Grimm, 1999). Reality TV viewers thus are not "classical voyeurs" but at best "socially changed voyeurs". The observing of the lifeworlds of others does not take place for its own sake, but is functionally embedded in the viewers' social environmental conditions and their attempts to solve problems. The *Supernanny* audience is highly empathetic, interested in harmony and to a high degree interested in the solution of intra-family conflicts. The "average", ideally constructed *Supernanny* fan is female, under 30, has a relatively low income, no secondary school diploma and at least one child. The low social barriers and the lack of educational prerequisites must be seen as a social profit for watching *Supernanny* since they tear down barriers that could make the use of official counseling recommendations more difficult.

Even the very critically minded nursery pedagogue (mentioned earlier in this section) regards the principle of assistance on site as a plus of nanny TV. She thinks, though, that it can be barely adapted to the job practice. She also sees the principle of mobile family counseling as a failure.

> "At the risk of becoming a doomsayer, up until now I was always against it (against the mobile intensive care for families, - J.G.). I've worked a lot with difficult kids and difficult parents. The intensive care for families was poorly received here. I think that there is a fundamental difference: we were forced onto people and there – with the supernanny – the family calls the person. When I say 'help please' I listen, but when someone comes and says 'this is the way it should be done' I close my ears. That is my experience."

Perhaps without intention, our great worrier brings up a strong pro-nanny argument. The voluntariness of attention that the client families of edutainment TV exhibit opens them up for the nanny's interventions. Why not, then, acknowledge the motivational work in the run-up to professional counseling and use nanny TV to further own professional intention? After the *Supernanny* program had been caught in the crossfire of criticism of pedagogical experts in the beginning, there seems to be a mood swing now. The federal congress of the professional association of German social pedagogues called its closing event in 2009 "Parenting Help from TV?" Even as a question, that would not have been thinkable a few years ago, when the protests of the pedagogues ruled the public discussion on parenting TV. Finally, the Austrian example shows that *transcendences of reality TV* can bring surprising results in the extra-media world. The former supernanny Velásquez was hired by Vienna's magistrate department 11 that is responsible for educational issues and psychological services to fight the crisis in the youth welfare office's mobile family counseling and the communal social pedagogy. The idea was to heighten the clients' counseling readiness and get better intervention results through reflective techniques (amongst others, via cameras and other means borrowed from the TV *Supernanny* repertoire). With the aid of this concept (FIT) it was possible to reduce the institutionalization of children by 30%. Meanwhile, FIT is being discussed as a possible addition to traditional ideas of social services and professional family consulting on a European level.

Such a positive transformation of reality TV within and for society cannot hide the fact that coaching TV has its basic limits, e.g. in the monotony of unchanging recipes. The success of FIT in Vienna followed the cancelling of the *Supernanny* program on TV. Here, we find hints that the culture of announcement that spends itself in the media and makes the publication of the private a topic of social learning can contradict the intentions of the public confession. The sheer intention of a quasi-therapy via the general public does not even begin to guarantee success, neither for the participants nor for the viewers. Maybe the side-effects are even graver than the benefit of a gained change. Thus, FIT's clients are purposefully not delivered to a mass of viewers; the camera is only there for the participants to watch themselves reflectively and for a qualified professional audience to watch them. Like on the Internet (cf. e. g. the Facebook debate), in the area of coaching TV, the insight grows that unchecked publication frenzy may cause uncontrollable damage. Obviously, even a media saturated society needs reservations of unwatchedness to engage fully in reflexion.

The question now is what is about to come after coaching TV? I have four assumptions on that.

1) The desire for a constantly increasing approximation of media programs to the lifeworld - one could say: "lifeworldization" - in the end creates a lust for "real" reality not limited to this or that simulation of reality. The development of reality TV tends to transcend into the extra-media world supporting a consciousness of media deficits.

2) In the program segment of coaching TV, this trend manifests in the fact that the authoritative monitoring of everyday problems promotes the restitution of the "private" in professional areas that the audience cannot reach. This does not implicate a complete retreat into the private life or even a monadic seclusion that would be neither desirable nor possible, but the relativization of an unchecked and uncritical variety of public profiling, no matter what the cost.

3) On the other hand, professional life counselling needs an exonerating, preparing and motivating addition by media orientation the way reality and coaching TV perform under the conditions of a gratified voluntary attention of a huge public every day without overtaxing the limited capacities for professional counselling. In developing optimized interfaces and accepting the mutual performance limits lies the future of professional counselling and of a higher quality coaching TV.

4) The suffering from the relativism of orientation that marked the development from reality TV to coaching TV will boost a re-popularization of reality formats to the same degree to which the doubts against authoritative counseling concepts grow. In a situation of dogmatic torpor the relativism of orientation will be necessary for managing every day life just as increasing reliance is in demand if plural agency options produce decision uncertainty.

In general perspective the conclusive presumption seems plausible that the development of reality TV is being determined by balancing between openness and reliability of orientation functions referring to the changing audiences' needs. Thus, after a period of preference of authoritative counseling the "savvy viewers" will probably return to where they feel best: to fishing for various authentic reality particles beyond orchestration and forgery, be it in the media world or beyond, in the social world. In both areas, though, the reality particles are no reality cores that keep the innermost reaches of reality together, but fractal resistors in the social area – metaphorically speaking "reality splints" or "reality fragments" that the subjects are confronted with on their journey through everyday and media worlds whenever orchestrated appearance and reality are drifting too far apart. They bring a light bulb moment to the "savvy viewer" that challenges him to check behavioral routines and assessment schemata. In the best case, this prises open worn-out thought patterns and creates an incentive to optimize and readjust everyday action. But the duality of simulation and reality is eternal. That is why this is not about exposing reality in the light of the 'naked truth', but about the reflexive potential of *simulacra* that only operate without becoming pathological as long as they prove themselves in the framework of the world of everyday life.

REFERENCES

Adorno, T. W., Frenkel-Brundswik, E., Levinson, D. J., and Sanford, N. (1950). *The authoritarian personality: Studies in prejudice*. New York: Norton.

Anderson, J. R. (1995). *Cognitive psychology and its implications* (4th edition). New York: W.H. Freeman and Company.

Andrejevic, M. (2004). *Reality TV. The work of being watched*. New York: Rowman and Littlefield.

Andrejevic, M. (2008). Reality TV, savvy viewers, and auto-spies. In: S. Murray and La. Ouellette (Eds.), *Reality TV: Remaking television culture* (2nd edition) (pp.321-342). New York: New York University Press.

Archer, M. S. (2007). *Making our way through the world: Human reflectivity and social mobility*. Cambridge, UK: Cambridge University Press.

Augustinus, A. (1950/397). *Bekenntnisse* (Confessiones, first edition - 397 AD). Zürich: Artemis.

Bakardjieva, M. (2005). *Internet society: The internet in everyday life*. New Delhi: Sage.

Baumrind, D. (1966). Effects of authoritative parental control on child behavior. *Child Development, 37*(4), 887-907.

Baumrind, D. (1991). The influence of parenting style on adolescent competence and substance use. *Journal of Early Adolescence, 11*(1), 56-95.

Baumrind, D. (2001, August). *Does causally relevant research support a blanket injunction against disciplinary spanking by parents?* Paper presented at the 109th annual convention of the American Psychological Association.

Becker, E. (1973). *The denial of death*. New York: The Free Press.

Berlyne, D. E. (1960). *Conflict, arousal and curiosity*. New York: McGraw-Hill Inc.

Biltereyst, D. (2004). Big Brother and its moral guidance: Reappraising the role of intellectuals in the Big Brother panic. In E. Mathijs, and J. Jones (Eds.), *Big Brother international: Formats, critics and publics* (pp.9-15). London: Wallflower Press.

Biressi, A., Nunn, H. (2005). *Reality TV: Realism and revelation*. London: Wallflower Press.

Bourdieu, P. (1986). *Distinction: A social critique of the judgement of taste*. London, New York: Routledge.

Brehm, J. W. (1966). *Theory of psychological reactance*. New York: Academic Press.

Chan, T. W., Koo, A. (2008). Parenting style and youth outcome in the UK. *Oxford Sociology Working Paper, 6*(11).

Chao, R. K. (1995, August). *Beyond authoritarianism: A cultural perspective on Asian American parenting practices*. Paper presented at the 103rd Annual Meeting of the American Psychological Association, New York.

Chao, R., Tseng, V. (2002). Parenting of Asians. In: M. H. Bornstein (Ed.), *Handbook of parenting, volume 4: Social conditions and applied parenting* (pp.59-89). Hillsdale, NJ: Lawrence Erlbaum.

Christophersen, E. R. (2005). Super Nanny. *Pediatrics*, 115, 1768-1769.

Cox, C. R., Arndt, J., Pyszczynski, T., Greenberg, J., Abdollahi, A., and Solomon, S. (2008). Terror management and adults' attachment to their parents: The safe haven remains. *Journal of Personality and Social Psychology, 94*(4), 696-717.

Davis, M. H. (1980). A multidimensional approach to individual differences in empathy. *JSAS Catalog of Selected Documents in Psychology, 10*, 85.

Davis, M. H. (1983). Measuring individual differences in empathy: Evidence for a multidimensional approach. *Journal of Personality and Social Psychology, 44*, 113-126.

Dervin, B. (1976). The everyday information needs of the average citizen. In: H. Kochen, and J. C. Donohue (Eds.), *Information for the community* (pp.19-38). Chicago: Amercian Library Association.

Dervin, B. (1980). Communication gaps and inequities: Moving toward a reconceptualization. In: B. Dervin, M. Voigt (Ed.), *Progress in communication sciences, Vol.2* (pp.73-112). Norwood N.J.: Ablex Publishing Corporation.

Dervin, B. (1989). Audience as listener and learner, teacher and confidante: The sense-making approach. In: R. Rice, and C. K. Atkin (Eds.), *Public communication campaigns* (pp.67-86). Thousands Oaks, CA: Sage.

Dreikurs Ferguson, E., Hagaman, J., Grice, J. W., and Peng, K. (2006). From leadership to parenthood: The applicability of leadership styles to parenting styles. *Group Dynamics: Theory, Research, and Practice, 10*(1), 43-56.

Dreikurs, R., Cassel, P., and Dreikurs Ferguson, E. (2004). *Discipline without tears: How to reduce conflict and establish cooperation*. New York: Wiley.

Dubrofsky, R. E. (2007). Therapeutics of the self. *Television and New Media, 8*(4), 263-284.

Edinger, S. (2005). Interview of myself with the Austrian supernanny. Vienna: Author.

Elias, N. (2000/1939). *The civilizing process* (revised edition). Malden, MA: Blackwell.

Fink, G. R., Markowitsch, H. J., Reinkemeier, M., Bruckbauer, T., Kessler, J., and Heiss, W. D. (1996). Cerebral representation of one's own past: Neural networks involved in autobiographical memory. *Journal of Neuroscience, 16*(13), 4275-4282.

Forgas, J., Baumeister, R. F., and Tice, D. M. (2009). *Psychology of self-regulation: Cognitive, affective, and motivational processes*. London: Psychology Press – from Taylor and Francis Group.

Foucault, M. (1995). *Discipline and punish: The Birth of the prison*. New York: Vintage Book.

Foucault, M. (2009). *Security, territory, population: Lectures at the College de France, 1977-1978*. New York: Picador.

Franck, G. (1998). *Ökonomie der Aufmerksamkeit. Ein Entwurf* (The economy of attention). München: Carl Hanser.

Frau-Meigs, D. (2006). Big Brother and reality TV in Europe: Towards a theory of situated acculturation by the media. *European Journal of Communication, 21*(1), 33-56.

Frost, J. (2005a). *Supernanny: How to get the best from your children*. London: Hodder and Stoughton.

Gershoff Thompson, E. (2002). Corporal punishment by parents and associated child behaviors and experiences: A meta-analytic and theoretical review. *Psychological Bulletin, 128*(4), 539–579.

Geyer, S. (2006). *Computerspiel, Gewalt und Terror Management. Grundlagen-Theorie-Praxis* (Computer game, violence and terror management. Basics-theory-practice). Saarbrücken: VDM Verlag.

Giddens, A. (1990). *The consequences of modernity*. Oxford: Polity Press.

Giddens, A. (1991). *Modernity and self-identity. Self and society in the late modern age*. Cambridge: Polity Press.

Goffman, E. (1959). *The presentation of self in everyday life.* Garden City, NY: Doubleday/Anchor Books.

Greenberg, J., Pyszcynski, T., and Solomon, S. (1986). The causes and consequences of a need for self-esteem: A terror management theory. In: R. F. Baumester (Ed.), *Public self and private self.* New York: Springer.

Grimm, J. (1995). Wirklichkeit als Programm? Zuwendungsattraktivität und Wirkung von Reality TV (Reality as program? Attraction and impact of 'reality TV' - German). In: G. Hallenberger (Ed.), *Neue Sendeformen im Fernsehen. Ästhetische, juristische und ökonomische Aspekte. Arbeitshefte Bildschirmmedien 54.* University of Siegen: DFG-Sonderforschungsbereich 240.

Grimm, J. (1999a). *Fernsehgewalt. Zuwendungsattraktivität – Erregungsverläufe – sozialer Effekt. Zur Begründung und praktischen Anwendung eines kognitiv-physiologischen Ansatzes der Medienrezeptionsforschung am Beispiel von Gewaltdarstellungen* (TV violence: Attraction, arousal, social effect: Reasoning and application of the cognitive-physiological approach for study media reception on the example of violence imagery - German). Opladen, Wiesbaden: Westdeutscher Verlag.

Grimm, J. (1999b). Talkshows – aus Sicht der Rezipienten (Talk shows – from the viewpoint of the recipients - German). *tv diskurs, 7,* 66-79.

Grimm, J. (2001). Wirklichkeitssplitter im Container. Ergebnisse eines Forschungsprojekts zu 'Big Brother' (Fragments of reality within the container: Results of a research project on Big Brother - German). *Medienheft, 15,* 41-56.

Grimm, J. (2006a). *Super Nannys. Ein TV-Format und sein Publikum* (Supernanny. A TV format and its audience - German). Konstanz: UVK.

Grimm, J. (2006b). Vom Umgang mit Gefühlen beim Fernsehen: Theoretische Modelle und empirische Befunde (emotion management during TV viewing: Theoretical models and empirical results). In: B. Krause and U. Scheck (Eds.), *Gefühle und kultureller Wandel* (Emotions and Cultural Change - German) (pp.279-299). Tübingen: Stauffenburg-Verlag.

Grimm, J., and Sells, N. (2006). *Vom Guten des Schlechten. Mediale Wirkungen der Tsunami-Katastrophe* (About the good in the bad: Media effects of the tsunami catastrophe - German). *tv diskurs, 35*(1), 46-51.

Habermas, J. (1985a). *The theory of communicative action, vol.1: Reason and the rationalization of society.* Boston: Beacon.

Habermas, J. (1985b). *The theory of communicative action, vol.2: Lifeworld and systems, a critique of functionalist reason.* Boston: Beacon.

Hart, J., Shaver, P. R., and Goldenberg, J. L. (2005). Attachment, self-esteem, worldviews, and terror management: Evidence for a tripartite security system. *Journal of Personality and Social Psychology, 88*(6), 999-1013.

Hill, A. (2000). Fearful and safe: Audience response to British reality programming. *Television and New Media, 1*(2), 193-213.

Hill, A. (2004). Watching Big Brother UK. In: E. Mathijs and J. Jones (Eds.), *Big Brother international: Formats, critics and publics* (pp.25-39). London: Wallflower Press.

Hill, A. (2006). Reality TV: *Audiences and popular factual television.* London: Routledge.

Holsti, O. R. (1969). *Content analysis for the social science and humanities.* Reading, MA.: Addison-Wesley.

Ihde, D. (1990). *Technology and the lifeworld. From Garden to earth.* Bloomington, Indiana: Indiana University Press.

Jewett, R., and Lawrence, J. S. (1988). *The American monomyth, (2*nd edition). New York: University Press of America.

Jonas, E., and Fischer, P. (2006). Terror management and religion: Evidence that intrinsic religiousness mitigates worldview defence following mortality salience. *Journal of Personality and Social Psychology, 91*(3), 553-567.

Keane, M., Fung, A. Y. H. , Moran, A. (2007). *New Television, globalisation, and the East Asian cultural imagination.* Aberdeen, Hong Kong: Hong Kong University Press.

Khatib, L. (2005, May). *Language, nationalism and power: The case of reality television in the Arab world.* Paper presented at the 55th annual meeting of the International Communication Association, New York.

Kilborn, R. (1994). How real can you get? Recent developments in 'reality' television. *European Journal of Communication, 9*(4), 421-439.

Kracauer, S. (1997/1960). *Theory of film: The redemption of physical reality.* Princeton, NJ: Princeton University Press.

Kraidy, M. M. (2008). Reality TV and multiple Arab modernities: A theoretical exploration. *Middle East Journal of Culture and Communication, 1*(1), 49-59.

Krakowiak, K. M., Kleck, C., and Tsay, M. (2006). *Redefining reality TV: Exploring viewers' perceptions of nine subgenres.* Paper presented at the 56th annual meeting of the International Communication Association.

Lawrence, J. S., and Jewett, R. (2002). *The myth of American superhero.* Cambridge, UK: William B. Erdmans Publishing Company.

Lefebvre, H. (2008/1947). *Critique of everyday life, volume 1: Introduction.* New York: Verso.

Lewin, K. (1948). *Resolving social conflicts.* New York: Harper and Row.

Lewin, K., Lippitt, R., and White, R. K. (1939). Patterns of aggressive behavior in experimentally created 'social climates.' *Journal of Social Psychology, 10*, 271–299.

Liebes, T., and Livingstone, S. M. (1994). The structure of family and romantic ties in the soap opera: An ethnographic approach. *Communication Research, 21*(6), 717-741.

Liebes, T., and Livingstone, S. M. (1998). European soap operas. The diversification of a genre. *European Journal of Communication, 13*(2), 147-180.

Lundy, L. K., Ruth, A. M., and Park, T. D. (2008). Simply irresistible: Reality TV consumption patterns. *Communication Quarterly, 56*(2), 208-225.

Markowitsch, H. J., and Welzer, H. (2005). *Das autobiographische Gedächtnis: Hirnorganische Grundlagen und biosoziale Entwicklung* (Biographical memory: Brain bases and bio-social development - German). Stuttgart: Klett-Cotta.

Martinez, I., and García, J. F. (2007). Impact of parenting styles on adolescents' self-esteem and internalization of value in Spain. *Spanish Journal of Psycholoy, 10*(2), 338-348.

Martinez, I., García, J. F., and Yubero, S. (2007). Parenting styles and adolescents' self-esteem in Brazil. *Psychological Reports, 100*, 731-745.

Mathijs, E., and Jones, J. (2004). Big Brother international. Introduction. In: E. Mathijs and J. Jones (Eds.), *Big Brother international: Formats, critics and publics* (pp.1-8). London: Wallflower Press.

McGuire, W. J. (1999). *Constructing social psychology: Creative and critical processes.* Cambridge, UK: Cambridge University Press.

McMurria, J. (2008). Global TV Realities: International markets, geopolitics, and the transcultural context of reality TV. In: Susan Murray and Laurie Ouellette (Eds.), *Reality*

TV: Remaking television culture (2nd edition) (pp.179-202). New York: New York University Press.

Mead, G. H. (1934). *Mind, self and society from the standpoint of a social behaviourist.* Chicago: Chicago University Press.

Moran, A. (1998). *Copycat TV: Globalisation, program formats and cultural identity.* Luton: University of Luton Press.

Moran, A., and Malbon, J. (2006): *Understanding the global TV format.* Bristol, UK: Intellect.

Musitu, P., and García, J. F. (2005): Consequences of family socialization in the Spanish culture. *Psychology in Spain, 9*(1), 34-40.

Neill, A. S. (1960). *Summerhill: A radical approach to child rearing.* New York: Hart Publishing.

Neuendorf, K. A. (2002). *The content analysis guidebook.* Thousand Oaks, CA: Sage.

Ouellette, L., and Hay, J. (2008). *Better living through reality TV: Television and post-welfare citizenship.* Malden, MA: Blackwell.

Ouellette, L., and Murray, S. (2008). Introduction. In: S. Murray and L. Ouellette (Eds.), *Reality TV: Remaking television culture* (2nd edition) (pp.1-22). New York: New York University Press.

Perinbanayagam, R. S. (2000). *The presence of the self.* New York: Rowman and Littlefield.

Pfau, M., Bockern, S., and van, Kang, J. G. (1992). Use of inoculation to promote resistance to smoking initiation among adolescents. *Communication Monographs, 59*, 213-230.

Pfau, M., Roskos-Ewoldsen, D. R., Wood, M., Yin, S., Cho, J., Lu, K. H., and Shen, L. (2003). Attitude accessibility as an alternative explanation for how inoculation confers resistance. *Communication Monographs, 70*(1), 39-52.

Reiss, S., and Wiltz, J. (2004). Why people watch reality TV. *Media Psychology, 6*(4), 363-379.

Robertson, R. (1994). Globalisation or Glocalisation? *Journal of International Communication 1*(1), 33-52.

Rose, N. (1999). *Powers of freedom: Reframing political thought.* Cambridge, UK: Cambridge University Press.

Rost-Schaude, E., Kumpf, M., and Frey, D. (1978). Untersuchungen zu einer Deutschen Fassung der 'Internal-External-Control'-Skala von Rotter (Investigating the German version of the 'internal-external-control'-scale by Rotter). In: *Bericht über den 29. Kongreß der Deutschen Gesellschaft für Psychologie* (pp.327-329). Zürich: Hogrefe.

Rotter, J. B. (1966). Generalized expectancies for internal versus external control of reinforcement. *Psychological Monographs, 1.*

Rudy, D., and Grusec, J. E. (2001). Correlates of authoritarian parenting in individualist and collectivist cultures and implications for understanding the transmission of values. *Journal of Cross-Cultural Psychology, 32*(2), 202-212.

Saalfrank, K. (2005). *Die Super Nanny. Glückliche Kinder brauchen starke Eltern* (Supernanny. Happy children need strong parents.) München: Goldmann.

Schacter, D. L. (1987). Implicit memory: history and current status. *Journal of Experimental Psychology, 13*(3), 501-518.

Schank, R. C., Abelson, R. P. (1977). *Scripts, plans, goals and understanding.* Hillsdale, NJ: Lawrence Erlbaum.

Schenk-Danzinger, L. (1992). *Entwicklung, Sozialisation, Erziehung: von der Geburt bis zur Schulfähigkeit* (Development, socialisation, education: From birth to school age- German). Vienna: Österreichischer Bundesverlag.

Schutz, A. (1945). On multiple realities. *Philosophy and Phenomenological Research*, 5(4), 533-576.

Schutz, A., and Luckmann, T. (1973). *The structures of the life-world, volume I.* Evanston, Illinois: Northwestern University Press.

Schutz, A., and Luckmann, T. (1983). *The structures of the life-world, volume II.* Evanston, Illinois: Northwestern University Press.

Silverstone, R. (1994). *Television and everyday life.* London: Routledge.

Silverstone, R. (2007). *Media and morality. On the rise of mediapolis.* Cambridge, UK: Polity.

Sokolov, E. N. (1965). The orienting reflex, its structure and mechanism. In: L. G. Voronin, N. Leontiev, A. R. Luria, E. N. Sokolov, and S. Vinogradova (Eds.), *Orienting reflex and exploratory behaviour* (pp.17-24). Baltimore, MD: Garamond Pridemark Press.

Teutsch D., and Browne, R. (2005, May 1). Super nanny backs her instinct over expert critics: An interview with J. Frost. *The Sun-Herald.*

Virilio, P. (1999). *Polar Inertia.* Thousand Oaks, CA: Sage.

Vitouch, P. (1993). *Fernsehen und Angstbewältigung. Zur Typologie des Zuschauerverhaltens* (Television and coping with fear. About a typology of viewers' behaviour- German). Opladen: Westdeutscher Verlag.

Velásquez, S. (2005). Interview of myself with the Austrian supernanny. Vienna: Author.

Weems, C. F., Costa, N. M., Dehon, C., Berman, S. L. (2004). Paul Tillich's theory of existential anxiety: A preliminary conceptual and empirical examination. *Anxiety, Stress, and Coping*, 17(4), 383-399.

Yin, S., Pfau, M. (2003, May). *Priming in the inoculation process: An alternative route of resistance to persuasion.* Paper presented at the 53rd annual meeting of the International Communication Association, San Diego, CA.

Zuckerman, M. (1979). *Sensation seeking. Beyond the optimal level of arousal.* Hillsdale, N.J.: Lawrence Erlbaum.

In: Reality Television-Merging the Global and the Local
Editor: Amir Hetsroni, pp. 259-277

ISBN 978-1-62100-068-6
© 2010 Nova Science Publishers, Inc.

Chapter 14

REALITY NATIONS: AN INTERNATIONAL COMPARISON OF THE HISTORICAL REALITY GENRE

Emily West
University of Massachusetts Amherst, USA

When 1900 House (Hoppe, 2000) premiered in the UK in 2000, a hybrid television form was born that would spawn spin-offs and imitators over the next several years in several other countries. These series place people in historical settings, asking them to leave their 21st century lives behind, and live within the material and social constraints of the past for a period of three or four months. Part historical documentary, part re-enactment, part gamedoc - like Survivor, and part observational reality show or docusoap - like The Real World, the new historical reality genre drew upon a number of formulae. From the historical documentary tradition it inherited the pedagogical mission of addressing historical ignorance and shoring up national collective memory; from reality genres it drew emphases on entertainment and putting "real people" in visually and emotionally interesting situations.

Historical reality programs have been border-crossers not only in terms of genre, but literally, across national boundaries. One of the prominent features of reality television in general is the part it plays in the increasingly global flows of television concepts (Bignell, 2005). The success of 1900 House led to spin-off House series not only in the UK but in the USA, Australia, New Zealand, and Spain, and closely related imitator series in Canada, Australia, and Germany (Gardam, 2003; Outright Distribution, N.D.). Many of these series also aired across national boundaries, such as when the US-UK co-productions aired in both countries, and when Australia and New Zealand broadcasters imported the American, British, and Canadian reality series.

An international comparison of the historical reality genre provides a new case study of the global circulation of a reality format. However, it also allows an international comparison of discourse about national identity and its perceived reality, taking the programs themselves as sites for that discourse as well as the discourse (highly mediated of course) of the people who volunteer to take part in them. This chapter takes national sentiment and identity as its primary analytic focus. The historical reality genre might be somewhat globalized, but its orientation is resolutely national, something it has in common with most other genres of reality TV. While much social theory at this historical moment is rightly focused on the transnational, the global, and the cosmopolitan, we cannot lose sight of the dominance of the nation in delimiting the boundaries of the public sphere for many, if not most, people (Schlesinger, 2008). Further, the nation still serves as a dominant context for feelings of group identity and belongingness. Reality TV has played its part in the examination and re-articulation of national identity and belongingness, across numerous national contexts. These include some of the most successful global exports, such as Big Brother and Pop Idol, always linked to national or regional contexts. These programs ask who is the best American or French singer, or what will happen when a cross-section of national or continental (as in Big Brother Africa) subjects live together under constant surveillance. In addition to these global reality formats there are examples of national self-examinations through the reality genre unique to specific nations and regions – such as Macedonia's That's Me, a Big-Brother-like reality show featuring representatives of different parts of the former Yugoslavia (Volcic, 2008); or George ka Pakistan (Georges' Pakistan), in which an Anglo, British volunteer tries to earn Pakistani citizenship, through a popular vote, by proving that he can "make it" living life in Pakistan.[1] Reality TV, while a global trend that often features the export and circulation of global formats, tends to focus on questions of local, and particularly national, interest.

Historical reality TV, in particular, falls into the category of media memory projects that posit national collective memory, and by extension national identity, as a problem or lack that should be repaired. The settings for these programs are hardly random. They feature times and places with symbolic resonance in national mythologies. Whether turn of the century middle-class Brits, pioneers on the American frontier, or Newfoundlanders eking out an existence in a remote fishing village, these programs purport to revisit the past through the experiences of the reality participants, in order to shed light on the origins of the nation today. The subtext, and sometimes the explicit text, is that the programs seek to recapture the essence of the nation, an essence that is currently elusive. The relation between contemporary national subjects and the historical ones whose experiences are to be re-enacted is presented as problematic. Are contemporary national subjects worthy - are they as hardy and courageous as their national forebears? These questions provide the premise for historical reality television.

For this chapter I examine a sample of seven historical reality mini-series that aired between 2000 and 2005 in English-speaking countries, ranging from four to eight episodes each. They were chosen based on the accessibility of the program DVDs, how well they were received, as well as on my familiarity with the national contexts in which they are set. These series are mainly from the British Commonwealth, and in many cases, are set in a colonial context. The programs examined here include 1900 House (Hoppe, 2000), 1940s House (Graham, 2001), and Edwardian Country House (Graham and Willis, 2002) (retitled Manor House when it aired in the United States), meant to represent 1905-1914 in a big country manor house – all from Great Britain; American series Frontier House (Hoppe, 2002), set in 1880s Montana and Colonial House (Hoppe, 2004), a re-creation of the Plymouth Colony in 1628; from Canada Quest for the Bay (Brown, 2001), which re-creates an 1840 voyage of "Yorkmen" who carried goods and furs for the Hudson Bay Company, by boat, from Winnipeg to Hudson Bay, and Quest for the Sea (Brown, 2004), which depicts a remote

---

[1] Thanks to Dr. Salman Hameed for bringing this program to my attention.

outport fishing village in Newfoundland in 1939; and Australia's Outback House (Burum, 2005) – set on an 1861 sheep station in New South Wales, and The Colony (Hilton, 2005) – set in a remote valley near Sydney in the early 19th century. The British series were joint productions by Channel Four – a British public service broadcaster that, unlike the BBC, features commercial advertising (Born, 2003) - and Wall to Wall Television, an independent British production company. Wall to Wall came to the U.S. and partnered with the New York public broadcasting station WNET to produce Frontier House and Colonial House. Outback House, originally a co-production with Wall to Wall, ended up as solely a production of ABC – the Australian Broadcasting Corporation – Australia's public broadcaster, and The Colony was a private-public international collaboration between production company Hilton Cordell and Irish public broadcaster Radio Telefís Éireann, The History Channel UK and the New South Wales Film and Television Office and its Regional Film Fund. Like the Australian productions, the Canadian "Quest" series were not part of the official "House" series, but were produced by independent production company Frantic Films, which made the programs in association with the cable channels History Television and the Life Network, with funding from the Canadian TV Fund and the Government of Manitoba. The connections with public broadcasters and public sources of funding meant that an educational mission and a nationalist orientation were elements of all the series.

These historical reality series achieved the somewhat rare combination of decent ratings with the prestigious glow of educational, quasi-documentary programming. Of course, these ratings must be interpreted relative to other programs with similar goals, that target similar audiences. Historical reality is certainly not in the same league of blockbuster reality hits like Survivor or Big Brother, but that has not been the genre's goal. These programs have sought to bring the reality format to a "quality" audience, to be popular within a particular niche, and to merge education and entertainment. However, ratings are important even for public broadcasters, and increasingly so. The ratings success was a major factor in the number of spin-off series inspired by the original Wall to Wall 1900 House production. While Variety dubbed 1900 House a "modest ratings success" for Channel Four, with its peak audience of 3.5 million viewers (Fry, 2000), it did double Channel Four's ratings in its time slot (Stanistreet, 1999). 1940s House did even better, with a peak primetime audience of 3.7 million viewers (Broadcast now, 2001), and Edwardian Country House also premiered at 3.6 million (Broadcast now, 2002). When 1900 House aired in the US, it earned an average of almost five million viewers per episode, more than doubling PBS' regular prime-time audience (Pereira, 2001), and making it the highest-rated multi-episode show of the season (Daily News, 2000). The American House spin-offs examined in this chapter, Frontier House and Colonial House, were also considered ratings successes for PBS. The audience for Frontier House was more than six million viewers (Baker, 2002), with Colonial House also bringing "tremendous viewership" according to a PBS executive (quoted in Jones, 2004).

Although the Canadian ratings appear small in comparison, Quest for the Bay's debut at 352,000 viewers made it one of History Television's (a Canadian cable channel) highest rated programs up to that point (Posner, 2002). In Australia, Outback House was a great ratings hope for public broadcaster ABC. While it started out strong at 1.2 million viewers, even beating Big Brother in Melbourne (Edmonds, Dennehy and Adams, 2005), its ratings dropped significantly in later weeks (Neill, 2005). Meanwhile The Colony aired on SBS - Australia's Special Broadcasting Service - which is tasked to reflect the diversity and multiculturalism of the country, and hence has lower expectations for ratings. The series debuted with 550,000 viewers, which was a record for the first episode of an Australian dramatic series on SBS, and was also the second highest-rated program of the year for the network (SBS Corporation, 2005).

As existing scholarship on the historical reality genre notes, these programs tend to be nationalist projects pursuing an ultimately elusive goal, a way to "know" the past, to transcend facts and dates and get inside the heads of historical subjects (Arrow, 2007; Edwards, 2007; Gapps, 2007; Rymsza-Pawlowska, 2007; Taddeo and Dvorak, 2007). It may be no accident that the original 1900 House series aired on the eve of the millennium, looking back almost 100 years to 1900 with intense curiosity, illustrating the truism that "the past is a foreign country" (Lowenthal, 1999), and trying to overcome that sense of foreignness through the reality formula. Certainly, to some extent, the series were successful. As period expert Daru Rooke comments on 1900 House, while replicas of Victorian houses are produced for museums, even experts are unsure of how things actually worked day-to-day. Joyce Bowler, the matriarch of the 1900 House, expresses her desire that the experience be authentic as possible, saying "Not that I'm play-acting it, but that I'm really living 1900" (1900 House, Episode 1).

Edwardian Country House seems to have been particularly successful in developing an Edwardian mindset among many of the participants. As housemaid Rebecca remarks, at the end of the series:

> It's been bizarre at times. Working alone in a room, suddenly look up and catch sight of myself in a mirror, in all my clothes, surrounded by all these Edwardian things, and thinking Edwardian. I'm not thinking anything whatsoever to do with my modern life. (*Edwardian Country House*, Episode 6)

Some of the "upstairs" family in the Edwardian House slip into their lives of privilege and status even more easily than the servants, such that Anna Olliff-Cooper comments by the end of the series that, "I actually now feel like mi'lady" (*Edwardian Country House*, Episode 6).

However, it quickly became clear that those who chose to participate paired the project of capturing the experiences and perspectives of people in the past with another, more personal project: to examine and come to know themselves. As much as these programs are nationalist projects, they are also projects of the self. Time travel presents a testing ground for getting to know the self, or explore the limits of the self, in a way that people perceive not being possible in their contemporary, everyday, unexamined lives. In this way the historical reality television genre has much in common with other "social experiment" reality programs, such as *Big Brother*, *The Real World*, and *Survivor*. As Mark Andrejevic (2004) has observed, the unreality of reality TV – its artificiality, its "social laboratory" premise, and the way it naturalizes constant surveillance – actually leads many participants to see the experience as an opportunity to be real, to be authentic, and to come to know the self in a way that is understood to be impossible in one's every day, "real" life.

There are commonalities across the national iterations of historical reality TV that can be attributed to the application of formula, which applies to those programs within the Wall to Wall *House* franchise as well as its imitators. Despite some of the different conditions of production, and the differing settings and national myths that the programs tap into, what is remarkable is their similarities, even for the series that aren't officially part of the *House* franchise, including: some kind of challenge that the participants must meet by the end of the series – particularly prominent in the Canadian, American and Australian versions; the use of experts to show the participants how to live in an authentically historical way; the use of anachronistic video cameras with which participants can record their "confessionals" or video diaries; a festival or party day; and a preoccupation with whether participants are cheating by living outside the historical rules of the day.

Beyond these elements in the genre's formula, there are similar dynamics of nationalism and the self that unfold across each of these series. Participants report taking part in historical reality series in order to test and come to know the true self. However, having to inhabit historical subject positions is frequently the catalyst for a crisis of the self, as people learn to live within the considerable constraints of the past. In particular, many find that they have to sacrifice their sense of individuality for the roles associated with their actual or assigned gender, race, class, or religious historical identities. In this way, the nationalist project embarked upon by the programs is threatened; the specificities of the volunteers' experiences expose the abstractions, even the fictional nature of a national identity. The project of bolstering national identity is further threatened by the unpleasant historical realities that the series deal with, particularly the realities of colonialism. However, the theme of sacrifice repairs these fissures, ultimately transforming these programs into commemorative rituals of nationalism, for both participants and viewers. The unreality of reality television, and in particular the ways in which the re-enacted sacrifices can never approximate the actual sacrifices of the past, actually strengthen the series' nationalist project.

One of the main uses of mass media is the brokering of group identities and sentiments, and given the national organization and regulation of much media, the group in question is often the nation. This chapter uses the analytic lens of ritual to understand how reality television participates in the production of group identities and sentiments. Marvin and Ingle (1999) define ritual as, "memory-inducing behavior that has the effect of preserving what is indispensable to the group" (p.129). Alexander (2004) writes, "Ritual effectiveness energizes the participants and attaches them to each other, increases their identification with the symbolic objects of communication, and intensifies the connection of the participants and the symbolic objects with the observing audience..." (p.527). As symbolic forms of communication that have as the implicit, and often explicit aim, a greater connection between members of the national group through "intensified" connections to the past, or at least to an imagined past, historical reality series are media rituals aimed at bolstering national sentiment.

Modern nations, in particular, must create a sense of groupness and belonging among people who are dispersed, diverse, and often fragmented along lines of ideology, identity, and access to power. The task of creating a sense of national connection to some kind of mythical center is a challenging one. Alexander (2004) writes, "Performances in complex societies seek to overcome fragmentation by creating flow and achieving authenticity. They try to recover a momentary experience of ritual, to eliminate or to negate the effects of social and cultural de-fusion" (p.548). As Benedict Anderson (1983) points out, it is through mass media that these performances are most often disseminated and experienced.

Complementing reflections on the importance of public rituals in group formation, Marvin and Ingle (1999) consider the role of representation in modeling origin moments of group "creation-sacrifice" (p.130). These moments of re-presenting or mimicking past sacrifices understood to have given birth to or strengthened the nation are commemorations that renew group identities for the time being, until the next commemoration or moment of creation-sacrifice. The reality show volunteers produce a commemorative ritual of sacrifice in part by mimicking the past – creating something that looks like or resembles past sacrifice – what Marvin and Ingle term "sympathetic magic" (ibid., p.130). But in addition they actually do engage in bodily sacrifice – the sacrifices of hard work, deprivation, suffering through difficult conditions, and sometimes even injury. The programs, then, contain elements of both "contagious magic" (actual bodily contact) and "sympathetic magic" (imitation) (ibid., p.130).

In sum, I consider the historical reality genre not as just a globalized reality formula, but as a cultural form whose logic and traction can be traced to the project, common to most national contexts, of producing or strengthening political identities through symbolic forms that hail a national self through the emotional appeal of ritual.

REALITY TV AND THE SELF

Although there are certainly historical insights that one can learn from these programs, they are more so about exploring and discovering the self. Here I agree with Malgorzata Rymsza-Pawlowska (2007), who wrote about *Frontier House*, that:

> Reality television….is not so much an account of the event, *but of the experience.* Documentary focuses on actions but reality television turns its attention to the actors. It is personalities that make reality shows memorable, as opposed to specific incidents. (emphasis in original, p.37)

For viewers, the accuracy of the historical setting is probably less important than the feeling that they are watching real people's reactions to unusual circumstances.

The participants describe a number of different motivations for taking part in these programs. Several people across the series cite their desire to escape the commercialism and distractions of modern life and technologies. A number of the families hope that they will experience more quality time with their children, and come together as a family unit. Some people's participation is inspired by particular ancestors or family members whose experiences they are hoping to honor through their own re-enactments. And some hope to recapture values that they associate with past eras, such as democracy, spirituality, connection to the land, and community.

In addition to these motivations, we often hear that the past provides a testing ground against which people can more clearly see or perceive their selves. Early in the *Quest for the Sea* series, the narrator asserts that, "Harold, like the others, has come to measure himself" (Episode 1). Joyce, the matriarch of the *1900 House*, looks back on the experience saying, "I've not only discovered lots about the period, about history, but about myself as well" (Episode 4). Similar to Joyce, Dan in *Outback House* reflects on his personal growth throughout the program: "This experience for me has become a rite of passage. This is much bigger than just giving something a go, or making a television show. For me this is testing what kind of man I am" (Episode 5). In a final example, Jonathan in *Colonial House*, after coming out as a gay man to the whole colony in Episode 4, reflects:

It's kind of ironic that I had to go back to the 17th century, an era of absolute intolerance, to really find myself, be honest with myself, and come out. And I'm going to take that back into the 21st century, and live with it for the rest of my life.

It's fascinating to see how many participants see a reality show in a historical setting as the opportunity to answer the question "who am I?" and "what am I capable of?" In this sense, historical reality television has much in common with other genres of reality television where, Mark Andrejevic (2004) argues, the "social experiment" and the constant surveillance are often viewed by participants not as intrusions or limits on their personal freedom, but as opportunities for a journey of self-discovery. Ironically, the panoptic experience becomes a guarantor or facilitator of authenticity, such that surveillance comes to be seen as a therapeutic condition, rather than an oppressive one. Although the re-enactment of a past historical era may seem patently artificial, much of reality television trades on the "reality of artifice" (Andrejevic, 2004, p.138). The more contrived the situation, the more of a "real" insight we are supposedly getting into the psyches of the reality show participants. In historical reality television, the fact that the scenario is contrived is never denied or hidden from view. Rather, it is the fidelity or authenticity of the contrivance – how "real" the artificial setting can be made – that is the only question. In fact, in *1900 House*, the first episode was devoted to the transformation of a 1999 house into a 1900 era house, and the struggles and challenges that the historical experts had in trying to re-construct the house with as much accuracy as possible. The construction of the artificial situation was offered as part of the viewing pleasure.

People may choose to take part in these programs to test their true mettle – to come to know the self better in a challenging, unusual, highly monitored situation. However, the programs fundamentally challenge modern ideas of the self organized by choice and self-determination by placing people in historical contexts where the self is organized by necessity, duty, and rigid social norms. No matter the particular historical era in question or the role being enacted, these reality show volunteers find their historical situation to be characterized by fewer choices, fewer freedoms, less room for expression of their individuality, and more constraints – both physical and social. Very soon into her stay in the *Edwardian Country House*, Antonia, who plays a servant, says:

> I think I'm going to cope with the independence thing with difficulty. I'm wanting to go downtown now, and explore, and go round the house. And I can't do it all. I just have to be told what to do, where to stand, when to go, when to get changed, what time I'm getting up, what time to have dinner. And I just want to go into town to the pub and have a bevvie [an alcoholic beverage], and I can't, and it's really strange. (Episode 1)

Participants across the series frequently have a crisis of the self at a point where the ways they are compelled to act and be, in order to re-enact the past, give them a sense of losing the self they thought they were. How much of our selves is essential or within us, and how much of our selves is produced by the contexts in which we find ourselves? Participants in historical reality programs routinely face this rather existential question. Sometimes the sense of transformation and turning into someone else is achieved through relatively small details, such as in *1900 House* when Joyce remarks that wearing 1900 dress makes her feel like "a completely different person" (Episode 4), or in *Edwardian Country House* when the maids observe that the class hierarchy is established very effectively and immediately through etiquette, clothes, and hair (Episode 1).

Modern participants, whose self-image generally demands that they respect the equality of persons, and whose impulse it is to try to make decisions that affect the entire group democratically, have considerable trouble acting out the values of the past organized by hierarchy and obedience. Scholar Michelle Arrow (2007) has drawn attention to the following exchange in Episode 2 of the Australian series *Outback House*, when Juli, who has been cast as squatter's wife (the squatter is at the top of the social hierarchy on the sheep station), tries to resolve a conflict with Carolina the cook, who is her employee. Carolina has been frustrated because the maids who are supposed to help her in the kitchen haven't been working hard enough. Because Juli doesn't run the house with a 19th century attitude towards hierarchy and authority, Carolina has taken on the task of discipline herself, to Juli's dismay:

> Juli: We are all equal—even myself.
> Carolina: In reality or in the life we're living now?
> Juli: No, no, we're still ourselves, and if we're going to make this a success we cannot order each other—we can ask. (*Outback House,* Episode 2)

Later in the series, squatter Paul finds himself in a conflict with his overseer Glen, who like Carolina, wishes Paul would enforce the 19th century hierarchy more rigidly. Finding himself in the relatively privileged position of overseer (which he believes he "deserves"), Glen wants Paul to enforce obedience and respect from the station hands. Paul explains to Glen:

> If we were to take the 1860s hierarchy, and replicate it with 21st century people and 21st century values, which we have, plus the options which everybody here has. We won't have anybody here. It would be you, me, and the chickens. (*Outback House,* Episode 2)

A theme across all the series is the ongoing struggle and negotiation that the participants have between their 21st century and historical selves. Often the standards of success for their historical selves – parameters set by the programs such as completing a journey, or producing profit, or accumulating enough goods for winter, or just living in a historically accurate way – are incompatible with their contemporary standards of personal success or integrity. Michelle Voorhees, in *Colonial House*, wants the colony to succeed, but doesn't want to have to attend Sabbath services, even if it leads to a poor evaluation by the inspectors. She says "We are not compromising our 21st century convictions about religion" (*Colonial House*, Episode 6). Sir John, in *Edwardian Country House*, is more willing to sacrifice certain convictions in the interests of historical accuracy, remarking about their period-appropriate fox hunt, "This hunt is part of the Edwardian project. Would I host a hunt in 2001? No I wouldn't. We do things in the house that conscience would not allow us to do in 2001" (Episode 4). Katherine, the eldest daughter in the *1900 House*, is conflicted about the maid-of-all-work that the family has decided to hire, in order to free themselves from the back-breaking housework. She explains:

> I know in 1900 that they would be employing somebody who was very poor and who needed the work, and needed the money, and that's the way we should be treating her. But I think, because we're actually a 1999 family, that it's not in our nature to be really horrible to anybody and make anybody do horrible skivvy jobs. (*1900 House*, Episode 3)

Although many historical reality program volunteers may see the experience as an opportunity to "learn about themselves," many also see that the experience requires them to act, behave, and even think in ways that contradict their values and 21st century identities.

CLASS, GENDER, RACE AND THE HISTORICAL SELF

Many participants in historical reality television have crises of the self in response to experiencing their class, gender, and racial identities in a very different way in their historical context than they do, or are aware of doing, in their contemporary lives. While participants bring their gender and racial identities with them into the past, their class status is assigned by the producers of the program. Class differences and the experience of social hierarchy are most dramatically represented in *Edwardian Country House*, explicitly organized around an "upstairs, downstairs" theme.[2] Perhaps because of this emphasis, this series comes across as the least celebratory of national identity. Indeed, the program is openly critical of the institutionalized inequalities that enabled the extravagant lifestyles of the Edwardian aristocracy, and frequently points out how unsustainable the system was, especially given changes in the economy and the onset of WWI. The ugly side of the master-servant dynamic is in evidence, both in terms of the contempt that the servants hold for the masters who work them too hard and treat them like second-class citizens, and the ease with which the upstairs family slip into their lives of luxury and blissful ignorance of the hard graft sustaining their lifestyles below stairs.

A common trope of every version of the historical reality genre examined is how women respond to their second-class status, limited freedoms, and lives of drudgery in past historical eras. For some of the series, particularly the *1900 House* and *1940s House* series, the programs end up being primarily about the experiences of women by virtue of focusing on the house itself – the domestic sphere where women reign and spend most of their time. In fact, women across the series begin to experience the eponymous "houses" as prisons, and many find creative ways to justify broadening their horizons. In *Frontier House*, participant Adrienne Clune describes the experience as a six-month labor camp (also observed by Edwards, 2007). Avril Anson, the upper-class spinster sister in *Edwardian Country House*, actually has a somewhat historically accurate nervous breakdown brought on by the enforced inactivity and loneliness of her social position, an intolerable contrast from her modern life of personal freedom and career success. She remarks, "I never for a moment thought I would be so constrained by this gilded cage" (*Edwardian Country House*, Episode 4). While Anson finds the house and its formalities to be a metaphorical cage, the women of *The Colony* quickly reject the physical caging they experience from their corsets and restrictive skirts and dresses, ditching them for more comfortable, but historically inaccurate, clothing that allows them to do the hard daily work on their farms.

It's striking how many of the women who join the program seeking a generalized connection with the past, or their ancestors, come away with a more particular perspective, informed by their connection to the women of the past and their *specific* experiences. The

---

[2] Upstairs, Downstairs was a British series depicting an Edwardian London House, which aired from 1971-1975 on ITV. It's a widely shared cultural reference point in Britain that participants, and likely audiences, used to interpret the program.

women volunteers sometimes find themselves torn between re-enacting the way woman actually lived, in order to demonstrate the lack of freedoms and suffering, and changing the situation to conform better to 21st century ideas about gender equality. When some of the women of *Colonial House* try to organize a more equitable share of domestic work with the men, Michelle Voorhees, who bitterly resents her second-class status, nonetheless exclaims, "We don't have the right to be pissed. We signed up for this trip, and we have to play it out" (*Colonial House*, Episode 2).

Other participants find that their ideas of having a generalized connection to the nation's "past" become disrupted by the particular racial or ethnic identities that they bring with them to the historical re-enactment. Some of the series aim for historical accuracy in how people of different races and ethnicities would have experienced life in these historical contexts. In *Edwardian Country House,* Reji Raj Singh is cast as an Indian tutor, and therefore experiences both the social possibilities and limits that his Indian heritage would have meant for him in early 20th century Britain. Like the upper-class Indians of the day who served as tutors to the British gentry, he occupies an awkward position in the house's social hierarchy. His race and position as employee place him below the family, but his education, status, and intimacy with the family place him above the servants, who come to resent the demands he places on them, and the fact that he may dine and socialize upstairs even though he is not a member of the family. When he says, "No matter how I try to be an Englishman, I'll *never* be accepted as one. I'm an Indian, and I'm proud of it," it is not clear if he's speaking about his Edwardian situation, his contemporary one, or both (*Edwardian Country House*, Episode 5).

In *The Colony*, set early in the 19th century, aboriginal people are positioned as part of a native clan who were living in the valley when the European settlers arrived. They are meant to live as their ancestors would have, and engage in historically accurate relations with the European families, in terms of trade and observing the laws and social taboos of the day. Ultimately, the aboriginal experience is very difficult to re-create. First, the bush foods that aboriginals of the early 19th century would have subsisted on are no longer available. Secondly, the aboriginals and Europeans enjoy spending time together and co-operating, reflecting their 21st century sensibilities rather than the 19th century rules. When the producers attempt to re-assert historical accuracy by issuing a Governor's proclamation that European families caught hosting aboriginals on their land will lose their government rations, and even face jail time, the aboriginal clan choose to leave the project for a time, and go on "walkabout." Even though much of their experience does not re-create the experience of their ancestors with much accuracy, they still feel the pain of exclusion and dispossession. They feel it, for example, when they are not invited to the big muster party towards the end of the project (a muster is an annual government inspection of the settlers' farms). A young male member of the Khoury clan, Jarlo, says, "Seeing the redcoats passing, and the horses, made us feel like it was back in 1800 and, man I felt like, we just weren't allowed to go to the muster, because the redcoats was there" (*The Colony*, Episode 5). On *Outback House* the aboriginal volunteers are more integrated with the whites on the sheep station, but station hand Malcolm has difficulty towards the end of the project when some of the other workers are making claims on parcels of land. Malcolm explains to the camera, "I belong to the land around here. It doesn't belong to me. I don't claim it. It claims me" (*Outback House,* Episode 5).

In the Australian and UK series, the programs include volunteers of color with a nod to historical accuracy, although ultimately the depictions are very much limited by modern sensibilities. In the North American series, racial and ethnic identities are dealt with much more obliquely. In *Frontier House,* there is an African-American cast member, Nate, who marries his fiancée who is white, Kristin, during the series - but for the most part his race and their inter-racial marriage are not treated in a historically accurate manner. In *Quest for the Bay,* one of the boat's crew, Ken, is Cree, but discussion of what his 19th century experience as the only Native American on the boat would have been is never broached. In fact, the Yorkboat and the project as a whole appear to be welcomed by the Cree communities they visit along the way. On *Colonial House,* "colonist" John Voorhees, who is part Native American, comments that re-living this part of American history is sometimes uncomfortable for him. Especially when members of the Passamaquoddy and Wampanoag tribes visit the colony, he feels he is "walking a fine line" between the two sides of his ancestry (*Colonial House,* Episode 1).

African-American Daniel Tisdale decides to take part in *Colonial House,* despite the historical inaccuracy of a person of color being in a 1628 New World colony, citing his identification and pride in being an American as his motivation. He explains, "I love the ideals of what this country was founded on" (*Colonial House,* Episode 1). However, his identification with American history in the abstract is challenged by the specificity of his experiences in the cultural hierarchies of 1628. Tisdale decides to leave the project before it is over, as he comes to realize that the demand for cheap labor and exploitation of indentured servants that he sees in the *Colonial House* setting (although he himself is assigned to the relatively privileged role of "freeman") are just the first steps towards the institution of slavery that was in place just fifty years later. Tisdale explains:

> This country was created on the backs of people who didn't have choices. As time progressed during the project I could see this idea of indentured servitude leading to slavery....For me it's a matter of conscience. How do I deal with it and what do I do about it? I've decided that it's time for me to leave. (*Colonial House,* Episode 5)

Tisdale's decision speaks to the stakes of the re-enactments in these historical reality series. As the participants start to inhabit their historical selves and see these historical periods play out, some of them grapple with the ethics of what they are re-presenting. Their struggles demonstrate the extent to which these performances are not just play. Whether the volunteers like it or not, their participation implies their approval and condoning of past circumstances. Tisdale comes to realize that his body is "emitting signs" (Foucault, 1995/1975, p.25) – appearing to resolve the tension between America's colonial past and his race – and not necessarily communicating the conflict he sees between the two. Like the aboriginals in *Outback House* and *The Colony,* he's caught between a project of visibility – being part of the historical reality series in order to assert his identity group's importance and contribution to the nation – and the pain of having to re-live and confront the racism of his nation in a different way than he experiences it today.

The historical reality genre arguably reveals the ways in which nations are truly "*imagined* communities," and therefore socially constructed rather than natural or "real" (Anderson, 1983). Ironically, it is the experience of historical re-enactment that raises the question of whether the participants' *contemporary* sense of national belonging is illusory, and by extension, our own. Is the idea of a national identity merely a story we tell ourselves that somehow secures our obedience and commitment to a national group? Are national identities fictions that obscure social inequalities defined by gender, race, and class? These are the questions that the historical reality genre raises quite powerfully in its various iterations. In attempting to revisit and honor the history of the nation, and in recruiting the kinds of people keen to participate in such an endeavor, the actual experiences of many of the participants undermine the story of national identity and unity that seemingly lie at the heart of the genre's premise.

RE-ENACTING COLONIALISM

The nationalist impulse of these programs is undermined by the way the historical reality genre challenges the idea of a unified group that gets to contribute and benefit equally from a national community. Additionally, many of these programs are set in periods where the early colonial context raises the question of whether nations should be celebrated when their foundation is the decimation and exploitation of other nations and peoples. Colonialism is arguably dealt with most consistently in Australia's *The Colony*, which includes four groups - an English family, an Irish family, a white Australian family, and an aboriginal clan - in order to represent the different groups that would have come into contact in Australia's early years. Similar to *Outback House,* the program does dramatize the ways in which aboriginal peoples were displaced from their land. After the Governor's proclamation makes it illegal for the aboriginal clan to visit the European farms, teenager Luana complains, "With this whole law thing, I just reckon this is really a bunch of crap. This is our land, they can't tell us to get off it. It's ours already" (*The Colony*, Episode 3). Deliberately socially-engineered to produce conflict, the program ends up having to explain most of the historical conflict and violence through voice-over, because the program's volunteers for the most part try to live according to 21st century values of cooperation and consensus. Further, as Australian scholar Michelle Arrow (2007) observes, colonialism is "a history that seems furthest from our grasp in terms of representation—it would be ethically impossible to accurately recreate such a history—to re-enact colonial violence, dispossession, and Indigenous resistance in a reality television program" (p.61). This is also why versions of the program such as "Antebellum in the South House" have not been produced, because viewers don't have the stomach for a re-enactment of slavery (Taddeo and Dvorak, 2007).

Edwardian Country House deals with colonialism to the extent that it includes Mr. Raj Singh as the tutor, who tries to teach young Master Guy about the British Empire, working against the grain of the privilege and entitlement that Guy is learning to adopt during his three month Edwardian experience. On the occasion of the "Empire Ball," meant to reproduce the celebrations that would have occurred when George V succeeded Edward VII, writer and social commentator Darcus Howe, a black Briton from Trinidad, visits the house in order to observe Edwardian power and privilege in action. The program invites us to feel the conflict between the patriotic elements of the Ball such as the pageant put on by children for the guests, and the lyrics of Rule Britannia which is sung with gusto – "Britons never never never shall be slaves" – and the experiences of Darcus Howe, his ancestors, and the millions of those who were colonized and enslaved around the world.

The US *Colonial House* worked with two Native nations, although the Native Americans' appearances and perspectives were incorporated in a less-sustained, integrated way than on *The Colony*. Indeed, the program's volunteers start to really get into the colonial spirit towards the end of the project - sending out map-making expeditions, making plans for expanding the colony, and creating propaganda to encourage other English people to join them in the New World. The ethics of what the colonists are re-enacting remain largely unexamined until the penultimate episode of the series, when the colony is visited by members of the Wampanoag Native American Nation, who in no uncertain terms condemn colonialism for its displacement and genocide of Native American peoples. The senior member of the tribe's visiting party, Ramona Peters, impresses on the colonists the seriousness of their intervention when she says, "Unlike yourselves we are wearing our best clothes in a sense, our traditional clothing, and we're all traditional people, so we're not playing" (*Colonial House*, Episode 7). Their visit puts the four month struggle of the colonists into a broader context, and puts into question the wisdom of buying into a 17th century mindset. After the Wampanoag's visit, California college professor (of anthropology, no less) and self-described liberal Carolyn Heinz explains:

> It suddenly sunk in, in a way that it hadn't until then, that I'm going along with being an imperialist. I think I'm going to go away, and people are going to say Carolyn, what did you think you were doing? But it didn't really sink in. That I'm re-enacting a whole system that I don't believe in and disapprove of, and yet, it's the roots of our own nation and of who we are! (*Colonial House*, Episode 7)

Heinz's response is similar to Tisdale's observations about how the roots of America are not just about courage, vision, and hard work, as has been emphasized throughout the series, but also about greed and exploitation. This observation is not unique to this series. The romance of going back into history, particularly in those series where participants must "live off the land," is often punctured by the realities of early capitalism. The lands they are living off were acquired through violence, and the profits they produce are often not for themselves, but for a larger corporate interest that exploits their labor – the fisheries in *Quest for the Sea*, the Hudson Bay Company in *Quest for the Bay*, the English investors in *Colonial House*, and the banks in *Outback House*.

Colonialism and capitalist imperialism are central to several of these historical reality series, something that would seem to undermine the nationalist spirit that arguably motivates them in the first place. However, for the most part acknowledgement of colonialism does limited damage to the overall narrative arcs of these programs. Sometimes discussion of colonialism is bounded in particular episodes or parts of the programs. For example, despite the intervention of the Wampanoag, the Plymouth Colony's assessment is carried out by white historical experts, using criteria that represent the interests of Britain. Or, when the relationships between colonists and colonized are depicted, they are, by necessity, such a pale imitation of the historical realities that they fail to make a substantial impact.

RITUAL REPAIR – RITES OF FERTILITY AND SACRIFICE

Within the historical reality genre, then, seemingly devoted to strengthening national identity through re-enactment, the fissures and myths of national belongingness are revealed. However, the series contain responses to this threat. They do this in part by emphasizing physical and genealogical connections between past and present, bringing attention to the contagious magic at work in these re-enactments. The ties of places, things, and even blood are used to anchor what otherwise might seem like tenuous connections between historical and contemporary national subjects. Danielle, the aboriginal maid servant in *Outback House*, remarks about the valley where the sheep station is located, "I've had so much déjà vu it's not funny, and I swear it's because I know people who are related to me have been here already" (*Outback House*, Episode 3). Like Danielle's experience, a number of the series include participants who can trace their families back to the actual people whose lives are being re-enacted. For example, in *Quest for the Bay*, one of the eight participants is Geoff Cowie, whose great-grandfather Isaac Cowie worked as a Yorkman in Manitoba in 1867, just as the crew is re-enacting. Geoff reads from his great-grandfather's diary along the way, and early in the journey they visit his grave, located on their route. Even though not all eight crew members can trace their families back to the actual people who made these treacherous journeys, Cowie's familial connection verifies the reality of the legacy they are honoring. In *Quest for the Sea*, all three men in the program are Newfoundlanders who can trace their families back to fishing villages similar to the one they are re-creating. Similarly, Mr. Edgar, who serves as Butler in the *Edwardian Country House* series, seeks to honor his own grandfather, who was in service early in the 20th century, as does housemaid Rebecca, who is following in her grandmother's footsteps. These more concrete connections between the present and the people of the past they seek to understand respond to the otherwise abstract connection of national identity over time. When this is not feasible, participants are invited to "touch" the past in other ways – by using objects or tools from the era, or in some of the series, meeting older people who lived lives similar to the ones they are re-enacting.

In addition to the contagious magic of physical or genealogical connections to the past, the series reinforce the pull of national identity on participants, and by extension on viewers, through rites of fertility and sacrifice. By fertility, I mean the *communitas* arising from group experiences, such as the high they get from feasts or parties, a regular feature of the series. Some series include actual fertility rituals in the form of weddings, which take place in both *Frontier House* and *Outback House*. The muster party in *The Colony* would have been an actual fertility ritual in the 19th century, as some people who attended were looking for potential husbands and wives.

Beyond that, the series encourage and celebrate familism, or the idea that the participants, initially strangers or separate family units, come together to forge a new quasi-family unit. This has varying success from series to series. In *Frontier House*, for example, by virtue of a structure where each incoming family works separately to prepare for winter, almost in competition with each other, the community never really gels as a larger family unit. In contrast, in series like *Edwardian Country House, Quest for the Bay, Outback House* and *Colonial House*, groups of people who start the projects as strangers report coming together "as a family" in response to their experiences, and in order to survive. On a small-scale, then, the series dramatize interdependence, the management of group conflict, and structured inequality, all characteristics of the past and present nation state. These families and "found families" are metonyms for the nation-as-family.

By and large the participants decide to take part in the programs in order to test themselves, but also, they report, to pay homage and do justice to the sacrifices of previous generations. Viewing the historical reality genre as a corpus, it becomes clear that the national self is a sacrificing, even sacrificial self. In an era where sacrifice may not have felt particularly immediate, the late 1990s and first half of the 2000s, modern reality TV

Reality Nations: An International Comparison of the Historical Reality Genre   273

volunteers got a taste of the sacrifices that are fetishized as giving birth to modern nations. Even as they complained and railed against the harshness and discomforts of their colonial, outport, frontier, outback, homefront, Victorian, or Edwardian lifestyles, the participants, and the programs, recognized these hardships as representing merely the tip of the iceberg of the actual human suffering that would have occurred in these various historical scenarios. Some of the sacrifices could be fully re-enacted, such as rationing in *1940s House*, where the whole family starts to go hungry, especially the women. In *Quest for the Bay*, the Yorkboat volunteers defy all odds by completing the infamous Robertson Portage, a challenge that is known to be back-breaking and dangerous. The program's narrator interprets this feat for us, saying:

> Two hundred years from their place in time, the 21st century crew can now stand beside their 19th century counterparts as equals. They have earned the right to call themselves Yorkmen, and have learned a basic truth of a Yorkman's life: the pain will pass, and the beauty remain. (*Quest for the Bay*, Episode 3).

For the most part though, the non-sacrificing modern self tries to inhabit the sacrificial historical self, and the embodiment is incomplete. Throughout all the programs, both participants and producers contemplate the gap between re-enactment and embodiment in actual historical circumstances. These gaps are particularly apparent when actual danger is near. The rules of the game, where participants are living in a period appropriate way, are broken or bent when people actually get sick, or are injured, or need to do things that violate contemporary perceptions of risk. Doctors and paramedics appear in present day clothes with 21st century technologies when people get sick or injured, as they do on both *Outback House* and *Quest for the Bay*. In *Quest for the Sea* and *Quest for the Bay* anachronistic safety gear is worn in dangerous waters. Ethics, good taste, and liability preclude re-enacting life as it would have actually been in most of these historical contexts, with its diseases, deaths resulting from the smallest accidents, such as an infected sliver, and armed conflicts between white settlers and indigenous peoples. The gap between past sacrifice and the program's re-enactment is also particularly apparent when historical dangers are simulated, such as the bombings during the blitz, in the *1940s House*. The family lives "as if" there are air raids, listening to recordings of bombs falling as they sit in their Anderson shelter.

Whereas the realism of reality television is normally a taboo topic to be raised within reality shows, in these series it becomes a central topic of conversation and worry. Producers challenge participants, and participants challenge themselves, to truly embody the experiences of their national forebears. However, the ways in which these historical experiments can never fully capture the subjectivities and experiences of the past is a constant source of reflection and anxiety. Examples of this range from the ridiculous to the sublime – from the *1900 House* family buckling and buying shampoo at the local store for their dried out, dirty hair, and then pouring it down the drain out of guilt for not living the full 1900 lifestyle, to the residents of the *1940s House* contemplating how the sacrifices they have endured over three months pale in comparison to the massive loss of life in WWII, on both the warfront and the homefront. This dynamic points to how these series function as a commemorative ritual of nation-creating bodily sacrifice. Here I use Carolyn Marvin's theoretical framework for understanding how nationalism and patriotism work.

According to Marvin and Ingle (1999), the nation is "the shared memory of blood sacrifice, periodically renewed," and further, the nation's power derives from the power to kill and sacrifice its own, which they label the "totem secret" (p.4). While Marvin and Ingle are concerned with creation-sacrifices of the highest order – soldiers sent to battle to die, and in so doing, renewing the nation with their blood – the sacrifices dealt with in historical reality television focus more on the everyday. They depict the physical sacrifices and hardships of pioneers, settlers, and the lower classes on whose backs nations were built. Marvin and Ingle (1999) insist that the "totem secret," the way that nations sacrifice their own citizens, is widely misrecognized, "concealed by the conviction that individualism is the defining myth of America, a way of thinking that seems far removed from any group idea" (p.2). While confronting the limits on individualism and the centrality of sacrifice might seem to puncture the strength of nationalism, Marvin and Ingle argue that it is central to its emotional power. They write, "Sacrifice disciplines groups....despite our conviction that violence is morally repugnant and should be eliminated, it creates the groups to which we feel the strongest attachments" (ibid., p.313). The historical reality genre's emotional power comes from its flirtation with the totem secret – that the nation is constructed out of violence and sacrifice. Danny Tisdale's reaction to the exploitation and violence at the roots of America, discussed earlier, is an example of this recognition. Participants in historical reality shows are often shocked by the hardships and sacrifices of historical subjects, but they are also in awe of them. For the most part, they leave with a sense of privilege and obligation to those who preceded them. The felt obligation to commemorate comes through in comments like this one, concluding *Quest for the Bay:* "The Yorkmen are gone, but their stories remain. They must. They are our legends" (Episode 5).

In Marvin and Ingle's framework, although the nation is periodically renewed through *actual* bodily sacrifice, in between past sacrifices must be recalled, even re-presented through ritual means, in order to extend their group-creating magic. Commemorative rituals "re-energize and re-dedicate sacred time and space created by acts of heroic predecessors" (Marvin and Ingle, 1999, p.134), and these increasingly occur through the mass media. They write:

> Media witness sacrifice and model it. Though they cannot perform real sacrifice, they scratch the itch in small ways and at regular intervals. They provide maintenance and memory until a big sacrifice comes again. Then they become the channel through which knowledge of sacrifice moves the nation. (ibid., p.141)

Although not on the scale of a media event such as a state funeral or Veteran's Day ceremony, historical reality series bear the logic of national connection that Marvin and Ingle explain, "scratching the itch" of national sentiment in a "small way."

HISTORICAL REALITY TV AS COMMEMORATIVE MEDIA RITUAL

The spectre of past sacrifices motivates historical reality television in the first place, providing the testing ground that participants crave. The programs expose the fissures and abstractions of national identity through the specificities of their re-enactments. Although the disconnect with the specificity of personal identity and experience threatens to expose the myth of national identity, the concreteness of past bodily sacrifices, whether actual deaths or the suffering resulting from hardships, re-asserts itself. The gap between the "reality" that can be shown in contemporary television series covered by insurance policies and limited by good taste, and the historical realities they propose to re-enact, create anxiety, just as all commemorative rituals create anxiety about their inability to re-create the ritual magic of actual creation-sacrifices. Marvin and Ingle (1999) contemplate this gap, inherent to all commemorative ritual:

> A defining feature of perfect creation-sacrifice, the utterly real, is that it has departed from the here and now. Existing only in memory, it is fragile. As a hedge against chaos, or forgetting, we can only imitate it. This anxious distance between model and reality provides the engine that sets the ritual cycle in motion. (p.135).

Like all reality television, the gap between the life-world reality and the mediascape reality is a central tension for producers, participants, and audiences (Lewis, 2004). However, this gap doesn't threaten the authenticity of the historical reality endeavor; rather, it verifies the solemnity of the series' undertaking. The participants bear witness by partially embodying and re-enacting the sacrifices of national forebears, and as viewers we bear witness to their witnessing. In this way, the series hail viewers as national subjects. In Mabel Berezin's (2001) terms, historical reality television series function as public political rituals that "dramatize political identity or felt membership" in the nation (p.84). The irony may be that, although they ostensibly set out to de-mystify the past by having modern people live like historical subjects, the programs of the historical reality genre largely end up re-mystifying the past, by drawing our attention and reverence to past sacrifices that are ritually experienced as birthing or renewing the modern nation.

REFERENCES

Alexander, J. C. (2004). Cultural pragmatics: Social performance between ritual and strategy. *Sociological Theory, 22*(4), 527-573.

Anderson, B. (1983). *Imagined communities: Reflections on the origins and spread of nationalism.* London: Verso.

Andrejevic, M. (2004). *Reality TV: The work of being watched.* New York: Rowman and Littlefield.

Arrow, M. (2007). "That history should not have ever been how it was": *The Colony, Outback House,* and Australian history. *Film and History, 37*(1), 54-66.

Baker, C. (2002, June 4). Public broadcasting increasingly businesslike; Competition forcing stricter attitude. *The Washington Times,* p.C08.

Berezin, M. (2001). Emotions and political identity: Mobilizing affection for the polity. In J. Goodwin, J.M. Jasper and F. Polletta (Eds.), *Passionate politics: Emotions and social movements* (pp.83-98). Chicago: University of Chicago Press.

Bignell, J. (2005). *Big brother: Reality TV in the 21st century.* New York: Palgrave Macmillan.

Born, G. (2003). Strategy, positioning, and projection in digital television: Channel Four and the commercialization of public service broadcasting at the UK. *Media, Culture and Society, 25*(6), 774-779.

Broadcast now. (2001, February 2). Older viewers lured into peaktime by *1940s House*. Retrieved from http://www.broadcastnow.co.uk/news/multi-platform/news/older-viewers-lured-into-peaktime-by-1940s-house/1169203.article

Broadcast now. (2002, May 20). Reviving Pet proves to be canny move for BBC1. Retrieved from http://www.broadcastnow.co.uk/news/multi-platform/news/reviving-pet-proves-to-be-a-canny-move-for-bbc-1/1143225.article

Brown, J. (2001). *Quest for the bay*. Frantic Films.

Brown, J. (2004). *Quest for the sea*. Frantic Films.

Burum, I. (2005). *Outback house*. Australian Broadcasting Corporation.

Daily News. (2000, December 4). Old-fashioned 'house' takes a trip west for PBS. p.92.

Edmonds, M., Dennehy, L., and Adams, C. (2005, June 16). Vets are on standby. *Herald Sun*, p.21.

Edwards, L. H. (2007). The endless end of frontier mythology: PBS's *Frontier House* 2002. *Film and History, 37*(1), 29-34.

Foucault, M. (1995/1975). *Discipline and punish: The birth of the prison.* (A. Sheridan Trans.). New York: Random House.

Fry, A. (2000, January 24). Showoff. *Variety*, p.47.

Gapps, S.(2007). Adventures in *The Colony*: *Big Brother* meets *Survivor* in period costume. *Film and History, 37*(1), 67-72.

Gardam, T. (2003, December 15). End of risk TV? *The Guardian*, p.8.

Graham, A. (2000). *1940s house*. Channel Four Television Corporation and Wall to Wall Television.

Graham, A., and Willis, E. (2002). *Edwardian country house*. Channel Four Television Corporation and Wall to Wall Television.

Hilton, C. (2005). *The colony.* Film Finance Corporation Australia Limited, New South Wales Film and Television Office and Hilton Cordell Productions.

Hoppe, B. (2000). *The 1900 house*. London: Channel Four Television Corporation and Wall to Wall Television.

Hoppe, B. (2002). *Frontier house*. Educational Broadcasting Corporation and Wall to Wall Television.

Hoppe, B. (2004). *Colonial house*. Educational Broadcasting Corporation and Wall to Wall Television.

Jones, J. (2004). PBS evolves, keeps its traditions intact. *Television Week, 23*(21), 20-23.

Lewis, J. (2004). The meaning of real life. In S. Murray and L. Ouellette (Eds.), *Reality TV: Remaking television culture* (pp.288-302). New York: New York University Press.

Lowenthal, D. (1999). *The past is a foreign country*. Cambridge, UK: Cambridge University Press.

Marvin, C., and Ingle, D. W. (1999). *Blood sacrifice and the nation: Totem rituals and the American flag*. Cambridge, UK: Cambridge University Press.

Neill, R. (2005, July 9). Broadcast blues – troubles with our AUNT. *Weekend Australian*, p.R01.

Outright Distribution. (N.D.) *The History House*. Retrieved from http://www.outrightdistribution.com/programme.aspx?program=709

Pereira, J. (2001, January 26). Rating reality. *Report on Business Magazine*, p.23.

Posner, M. (2002, January 11). Reality bites back. *The Globe and Mail*, p.R12.

Rymsza-Pawlowska, M. (2007). *Frontier House*: Reality television and the historical experience. *Film and History, 37*(1), 35-42.

SBS Corporation. (2005). *SBS Annual Report 2004-2005*. Retrieved from http://media.sbs.com.au/sbscorporate/documents/777101_front_section.pdf

Schlesinger, P. (2008). Cosmopolitan temptations, communicative spaces, and the European Union. In D. Hesmondhalgh and J. Toynbee (Eds.), *The media and social theory* (pp. 75-92). New York: Routledge.

Stanistreet, M. (1999, October 17). Big Brother is watching, and now you can too. *Sunday Express*. Retrieved from LexisNexis database.

Taddeo, J. A., and Dvorak, K. (2007). The PBS historical house series: Where historical reality succumbs to reel reality. *Film and History, 37*(1), 18-28.

Volcic, Z. (2008, December). *That's Me*: Nationalism, power, and identity on Balkan reality TV. Presentation at *Real worlds: Global perspectives on the politics of reality television*. Annenberg School for Communication, Philadelphia, PA.

ABOUT THE CONTRIBUTORS

Mark Andrejevic is an Associate Professor in the Department of Communication Studies at the University of Iowa and a Postdoctoral Research Fellow at the University of Queensland's Centre for Critical and Cultural Studies. He is the author of two books: Reality TV: The Work of Being Watched (2004) and iSpy: Surveillance and Power in the Interactive Era (2007). He is also the author of numerous book chapters and articles on surveillance, popular culture, and reality TV.

Helena Bilandzic holds a doctorate from Ludwig-Maximilians-Universität München in Germany and works as a Professor at the Zeppelin University, Friedrichshafen, Germany. Her research interests include media use, cultivation, narrative persuasion, qualitative and quantitative methodology.

Sevilay Celenk was born in 1967 in Turkey. She attended Ankara University, where she received her Bachelor's degree in International Relations and a Master's degree in Communications. She also received her Doctorate degree from the same university in 2003. For the last twelve years, she has been teaching television criticism, media criticism, creative writing, studies of everyday life, popular culture, semiotics of television, visual culture and ideology at the Faculty of Communications, Ankara University. She has published several books, articles and essays in the same area.

Dror Abend-David is the Chair of the English program at Ohalo College in Israel, and teaches Communications, Translation and Literature at Tel-Aviv University, Bar Ilan University, and the Interdisciplinary Center in Herzliya – all in the same country. He previously taught in the United States, Turkey and Northern Cyprus. Dror graduated with a doctorate in Comparative Literature from New York University in spring 2001. His book, based on his dissertation, was published in 2003 by Peter Lang. It is titled: 'Scorned my Nation:' A Comparison of Translations of The Merchant of Venice into German, Hebrew, and Yiddish. In addition to his work on Hebrew and Yiddish Literatures, he also published a number of articles about Media, Cultural Studies and Translation Theory, Modern Poetry and Drama.

Norah Dunbar has a Ph.D. in Communication from the University of Arizona (2000), an M.A. from California State University Chico, and a B.A. from the University of Nevada Reno. Her research interests include interpersonal deception and nonverbal expressions of power and dominance in interpersonal relationships. Methodologically, she uses behavioral observation techniques to examine verbal and nonverbal communication displays. Her recent publications include articles in Interpersona: An International Journal on Personal

Relationships, Journal of Social and Personal Relationships, Journal of Family Communication, and Communication Reports. She has also written chapters on nonverbal dominance and influence in The Persuasion Handbook, Beyond Words: A Sourcebook of Methods for Measuring Nonverbal Cues, and The Sage Handbook of Nonverbal Communication. Currently, she is working on several projects including a study on the detection of deception in mediated interactions. She teaches undergraduate and graduate classes in Interpersonal Communication, Nonverbal Communication, Deception, and Communication Theory.

Matthias R. Hastall (M.A., Dresden University of Technology, Germany) works as research fellow at the Zeppelin University, Friedrichshafen, Germany. His research interests include selective exposure research, health communication, and narrative persuasion.

Erich Hayes is a Ph.D. candidate at the University of Oklahoma. He has an M.A. from Missouri State University (2008) as well as B.A. (2004) and B.J. (2005) from the University of Missouri-Columbia. He took part at the 2009 NCA doctoral honors seminar hosted at West Virginia University. His writing and research focuses on interpersonal communication. Currently, he is working on finishing his Ph.D. in communication and teaches undergraduate courses in public speaking and fundamentals of communication.

Amir Hetsroni is an Associate Professor of Communication at Ariel University Center in Israel. He holds a Ph.D. in communication from the Hebrew University of Jerusalem. His research concerning popular TV programming has appeared in highly acclaimed scholarly journals such as Journal of Communication, Journal of Advertising, Communication Monographs, Sex Roles, Social Science Quarterly and others. He is co-author of the book Sex differences: Summarizing more than a century of scientific research (published by Routledge in 2008).

Jürgen Grimm has been a professor for communication at the University of Vienna since 2004. Apart from his teaching activities, he carries out intensive empirical research especially in the area of media effects and the resulting consequences for media acts. He is also heading the Forum for Methods at the Faculty of Social Sciences of the University of Vienna. Before he came to Vienna, Professor Grimm taught and did research at several German universities such as Mannheim, Münster, Augsburg and Siegen. Since 1994, he is a member of the board of trustees of the Freiwillige Selbstkontrolle Fernsehen ("Organization for the Voluntary Self-Regulation of Television"). He is also chairman of the "Association for the Promotion of Media Research" in Austria and Germany. He has published widely on depictions of media violence, on media effects in politics, on news processing, on reality TV, talk shows and other aspects of media entertainment. At present, he studies war and crisis journalism, media effects concerning migration and integration, patriotism and historical consciousness. He is also working on a project concerning the effects of advertising.

Derek Lackaff has a Ph.D. from the University at Buffalo and is now a Postdoctoral Fellow and Lecturer in the Department of Radio-Television-Film at the University of Texas at Austin. His research focuses on social media, social network analysis and methodologies, and the social psychology of communication technology use.

Tania Lewis is a Senior Research Fellow in Sociology at La Trobe University in Melbourne, Australia. She is the author of Smart Living: Lifestyle Media and Popular Expertise (Peter Lang, New York: 2008), editor of TV Transformations: Revealing the Makeover Show (Routledge, London: 2009), and co-editor (with Emily Potter) of Ethical Consumption: A Critical Introduction (Routledge, forthcoming 2010). Her current research is on sustainable lifestyles, green citizenship and ethical consumption. She is also a chief investigator on a four year (2010-2013) comparative study of lifestyle advice television in Asia funded by the Australian Research Council.

Oren Livio is a doctoral candidate at the Annenberg School for Communication at the University of Pennsylvania. His research focuses on cultural constructions of the relationships between nationalism, militarism, and citizenship in popular and everyday discourse. His work has appeared in Journal of Communication, Journalism: Theory, Practice and Criticism, The Communication Review, and Mass Communication & Journalism Quarterly. He received his Master's degree in communication from the University of Haifa, and his bachelor's degree in communication and psychology from the Hebrew University in Jerusalem.

Motti Neiger earned his Ph.D. from both the Communication department and Hebrew Literature department at the Hebrew University, Jerusalem (2000). Today he is a senior lecturer at the School of Mass Communication in Netanya Academic College and a visiting professor at Universidad Complutense de Madrid. His research interests are focused on issues associated with media and culture: The role of Culture Mediators: literary supplements, books publishers and literary periodicals; Cultures of journalism and journalism practice during violent conflict; Popular culture, collective memory and identity. Between the years 2006-2009 he was the chairman of the Israel Communication Association (ISCA, the umbrella organization for all Israeli Media Scholars) and the founding editor of "Media Frames: Israel Journal of Communication".

Maria Raicheva-Stover is Assistant Professor of Journalism and New Media at Washburn University, Kansas. Her research interests encompass the study of Eastern European media and the social impact of new communication technologies. Raicheva-Stover's work has appeared in the Howard Journal of Communications and the International Journal of Communication. She is also the co-author of a chapter in Negotiating Democracy: Media Transformations in Emerging Democracies.

Devan Rosen has a Ph.D. from Cornell University and works now as an Assistant Professor of Speech, University of Hawaii at Manoa. He has published on topics including social network analysis, decentralized networks and self-organizing systems, flock theory, and computer-mediated communication. He has also developed network analytic methods for the structural and content analysis of online communities and virtual worlds.

Michael A. Stefanone has a Ph.D. from Cornell University and works as an assistant professor in the Department of Communication at the State University of New York at Buffalo. His research focuses on the intersection of people, organizations, and technology. His current research explores the relationship between traditional mass media and new media use and how people's social context influence technology adoption and use.

Zala Volcic is a Postdoctoral Fellow at the Centre for Critical and Cultural Studies, University of Queensland, Australia. She is interested in the cultural consequences of nationalism, capitalism, and globalization, with a particular emphasis on international communication and media identities. She published a book on Identity and Media (2008). Her work has appeared in numerous journals, including *Social Semiotics, Canadian Journal of Communication,* and *International Communication Journal, Gazette.*

Emily West is an Assistant Professor in the Department of Communication at the University of Massachusetts Amherst, USA. In addition to a focus on media and nationalism, her areas of research include consumer culture, emotion in mediated culture, and ritual. Her past publications on media and nationalism appear in International Journal of

Communication, European Journal of Communication, and Critical Studies in Media Communication.

INDEX

9

9/11, 172

A

abstraction, 212
abuse, 230
access, 8, 49, 59, 91, 220, 263
accessibility, 41, 223, 257, 260
accommodation, 109, 167
achievement, 74, 138, 178, 221
acquaintance, 132
actuality, 191, 197, 199
adaptability, 53, 59
adaptation, x, xi, 2, 69, 84, 124, 135, 138, 165,
    198, 211
adaptations, 3, 4, 124, 138, 139, 140, 144, 147,
    225, 226
adjustment, 194, 216
adolescents, 43, 99, 109, 230, 232, 256, 257
advertising, viii, 32, 53, 55, 65, 66, 67, 68, 70,
    72, 74, 82, 83, 89, 151, 152, 161, 172, 187,
    193, 202, 261, 280
aesthetics, 192, 197, 203, 205, 206, 207
affirming, 30, 70, 93, 206
Africa, 13, 14, 85, 92, 185, 260
age, 3, 12, 13, 15, 16, 18, 21, 33, 35, 39, 44, 51,
    56, 57, 61, 99, 104, 105, 106, 126, 132, 133,
    135, 143, 208, 254, 258
aggression, 41, 42, 56, 142, 229, 249
aggressive behavior, 256
aggressiveness, 28, 43
alternatives, 71, 74, 138, 220, 221
American culture, 2, 133
American Psychological Association, 227, 253
anger, 97, 121, 154, 158, 217, 249
ANOVA, 19, 38, 105
anxiety, 67, 73, 90, 107, 157, 217, 258, 273, 275

applications, 27, 30, 38, 65, 215, 239
Arab countries, 173
Arab world, 256
Argentina, 49
arousal, 97, 245, 253, 255, 258
articulation, 141, 173, 181, 204, 260
Asia, 190, 201, 206, 207, 208, 281
assertiveness, 216, 217, 226, 229, 232, 233, 237
assessment, 8, 129, 233, 252, 271
assimilation, 87, 139
assumptions, 168, 207, 216, 252
attachment, 73, 130, 142, 253
attitudes, 27, 28, 33, 75, 108, 117, 219, 227, 245,
    249, 250
attractiveness, 11, 39, 235, 246
audition, 40, 169, 176, 177
Australia, x, 3, 14, 47, 61, 79, 85, 88, 91, 169,
    189, 190, 191, 195, 199, 204, 206, 208, 259,
    261, 270, 276, 280, 281
Austria, xi, 3, 211, 221, 225, 230, 231, 232, 234,
    235, 236, 239, 243, 244, 280
authenticity, 24, 49, 70, 74, 75, 84, 85, 87, 90, 91,
    124, 153, 154, 159, 180, 213, 220, 221, 263,
    265, 275
authoritarianism, 227, 228, 230, 232, 234, 235,
    236, 237, 238, 249, 253
authorities, 66, 73, 75, 157, 218, 220, 227
authority, 70, 178, 184, 220, 221, 223, 226, 227,
    228, 232, 234, 266
authors, 2, 3, 28, 82, 154
autobiographical memory, 216, 231, 254
autonomy, 24, 185, 227, 235
awareness, 147, 219, 224

B

background, 52, 67, 72, 86, 87, 121, 127, 130,
    145, 171, 172, 174, 182, 241
bankruptcy, ix, 115, 120

banks, 120, 271
barriers, 223, 250
beauty, 56, 70, 157, 158, 202, 203, 273
beer, 58, 83, 85
behavior, vii, 7, 9, 27, 29, 30, 31, 32, 33, 35, 37,
38, 39, 40, 44, 51, 102, 110, 130, 146, 168,
177, 217, 218, 219, 227, 228, 231, 232, 236,
237, 244, 246, 248, 249, 253, 263
behaviors, 12, 22, 27, 29, 30, 33, 38, 39, 97, 104,
185, 231, 250, 254
Belgium, 49
beliefs, 28, 198
belongingness, 260, 271
benign, 82
bias, 10, 99, 232, 241, 250
Big Brother, v, vii, viii, x, xi, 2, 3, 7, 28, 47, 48,
49, 50, 51, 52, 53, 54, 55, 56, 57, 58, 59, 60,
61, 62, 63, 79, 84, 85, 86, 88, 92, 93, 95, 97,
98, 99, 100, 101, 102, 103, 104, 105, 106, 107,
108, 109, 110, 115, 119, 120, 121, 135, 138,
140, 151, 152, 154, 155, 156, 157, 159, 161,
162, 168, 174, 187, 189, 191, 192, 195, 196,
197, 207, 211, 212, 213, 214, 218, 220, 221,
245, 253, 254, 255, 256, 260, 261, 262, 276,
277
birth, vii, viii, 25, 65, 69, 212, 214, 258, 263,
273, 276
blog, 63, 87, 91, 92, 115, 116, 119, 120, 129
blogs, 26, 27, 29, 32, 43, 80, 87, 119, 122, 126
blood, 120, 177, 272, 274
body weight, xi, 189
bonds, 20, 143, 147
Bosnia, 83
Brazil, xi, 3, 49, 211, 221, 225, 228, 231, 234,
235, 237, 240, 241, 243, 256
Britain, 2, 3, 75, 195, 196, 197, 208, 231, 267,
268, 271
Bulgaria, v, 2, 47, 51, 52, 53, 54, 55, 56, 58, 59,
60, 61, 63, 128
bullying, 71, 73
Bush, President George W., 172
business model, 43

C

cable television, viii, 79
campaigns, 82, 91, 95, 204, 214, 254
Canada, x, 3, 131, 165, 169, 170, 171, 172, 177,
180, 184, 185, 187, 259, 260
candidates, ix, 50, 52, 84, 100, 102, 103, 104,
106, 107, 115, 127, 129, 202
capital flows, 82
capitalism, x, 66, 67, 77, 125, 128, 135, 165, 166,
204, 271, 281

case study, 2, 48, 202, 207, 259
cast, 13, 56, 84, 85, 86, 87, 88, 90, 91, 127, 128,
156, 169, 170, 266, 268, 269
casting, 50, 52, 55, 58, 87, 116, 119, 121, 218,
222, 223, 245
catalyst, 56, 107, 108, 263
catastrophes, 212, 215
causality, 107, 121
CBS, 8, 14, 15, 97
CEE, 82, 84
censorship, 50, 152
Census, 15, 23, 24
Central Europe, 61, 80, 83, 91
certificate, 129, 173
challenges, 7, 8, 11, 12, 13, 14, 16, 17, 18, 21, 22,
63, 73, 87, 124, 126, 127, 223, 225, 235, 242,
252, 265, 270
channels, viii, 1, 53, 54, 79, 106, 109, 121, 124,
138, 139, 142, 151, 198, 202, 261
chaos, 221, 275
character, 32, 73, 80, 82, 85, 97, 130, 136, 160,
172, 178, 212
childhood, 132, 133, 221, 244
children, 13, 74, 76, 127, 132, 143, 193, 197,
221, 223, 224, 225, 226, 227, 228, 229, 230,
231, 232, 233, 234, 235, 239, 241, 244, 248,
249, 250, 251, 254, 257, 264, 270
China, 14, 186, 195, 201, 212, 219, 221, 228
circulation, 141, 153, 259
citizenship, 175, 190, 192, 193, 194, 195, 202,
204, 206, 208, 257, 260, 281
civilization, 148, 225
class struggle, 147
classes, 146, 213, 223, 226, 274, 280
classification, 97, 194
clients, 144, 221, 223, 240, 251
climate, 182, 220, 228, 229, 235, 249
clinical psychology, 234
CMC, 25, 26, 27
CNN, 61, 139
coding, 99, 100, 153, 231, 234
cognition, 216, 235, 237, 242
cognitive theory, 27, 28, 30, 33, 38, 41
college students, 29, 33, 41, 125, 136
colonization, 140, 166, 198, 224, 250
commerce, viii, 65
commercials, 61, 132, 146
commodity, 48, 66, 68, 69, 73, 75, 214
communication, ix, 3, 10, 23, 24, 26, 27, 29, 30,
40, 41, 95, 96, 97, 98, 100, 102, 103, 104, 105,
106, 107, 108, 109, 110, 158, 166, 168, 184,
185, 214, 216, 218, 219, 231, 237, 254, 263,
279, 280, 281

communication technologies, 26, 281
community, ix, 1, 48, 49, 80, 87, 90, 123, 126,
   127, 135, 139, 142, 154, 171, 175, 191, 193,
   199, 200, 201, 202, 204, 254, 264, 272
competence, 12, 23, 217, 253
competition, ix, x, 24, 49, 88, 123, 124, 134, 137,
   138, 143, 144, 147, 152, 165, 169, 173, 175,
   176, 177, 179, 180, 243, 272
competitor, 52, 88, 138, 143, 144, 145, 241
competitors, viii, 12, 21, 22, 60, 73, 79, 88, 138,
   139, 143, 144, 145, 223
complexity, 126, 135, 140, 234
components, 26, 132, 134, 143, 167, 168, 173,
   175, 176, 230, 234
comprehension, 182
compression, 166
computer technology, 41
computer-mediated communication (CMC), 25
confession, 189, 192, 194, 199, 220, 251
confessions, 26, 118, 213
conflict, 8, 9, 11, 126, 128, 129, 130, 131, 133,
   134, 135, 200, 216, 220, 224, 230, 233, 246,
   250, 254, 266, 269, 270, 272, 281
conformity, 66, 69, 70, 73
confrontation, 3, 213, 244, 247
consciousness, 117, 252, 280
construction, vii, 25, 81, 82, 90, 92, 118, 125,
   141, 159, 168, 171, 174, 177, 265
consulting, 82, 251
consumers, 26, 28, 38, 42, 73, 98, 181, 204
consumption, 2, 24, 26, 27, 28, 33, 34, 35, 38, 40,
   43, 67, 73, 76, 82, 89, 96, 181, 183, 185, 186,
   189, 190, 193, 198, 199, 200, 201, 204, 205,
   206, 207, 208, 256, 281
consumption patterns, 24, 256
content analysis, x, xi, 3, 28, 151, 153, 211, 231,
   235, 248, 257, 281
control, viii, 12, 47, 82, 83, 96, 97, 147, 215, 217,
   221, 224, 225, 227, 228, 229, 230, 233, 235,
   237, 238, 239, 241, 246, 247, 248, 249, 250,
   257
convention, 185, 186, 187, 253
convergence, 83, 191, 193
conversion, 129, 132
conviction, 176, 179, 274
Cook Islands, 12, 14, 21
cooking, 55, 88, 116, 202, 205, 206, 222
cooling, 142
coping strategies, 213, 220
correlation, 2, 35, 105, 106, 158, 232
correlation coefficient, 105, 106
correlations, 36, 104, 105, 106
costs, 26, 38, 69, 124, 138

counseling, 221, 223, 225, 231, 232, 234, 236,
   237, 242, 248, 251, 252
credibility, 218, 221
credit, 115, 242
crime, 199, 212, 222, 237, 241
crisis management, 213
critical period, 141
criticism, 98, 122, 140, 147, 154, 155, 156, 157,
   158, 160, 174, 177, 178, 227, 230, 232, 235,
   250, 251, 279
Croatia, 47, 83
cross-cultural differences, 3
cultivation, 28, 39, 42, 43, 279
cultural differences, 181, 182
cultural heritage, 87
cultural identities, 126, 139, 168, 183
cultural imperialism, 166, 167, 168
cultural stereotypes, 9, 21
cultural studies, vii, 3, 136, 166, 183, 184, 187
cultural tradition, 56, 198
cultural values, 139, 204, 206
curiosity, 100, 104, 108, 215, 245, 250, 253, 262
currency, 189, 190, 195
Cyprus, 279
Czech Republic, 47

D

danger, 223, 273
database, 110, 154, 277
dating, 3, 28, 42, 81, 123, 125, 182, 185, 212, 224
death, 116, 200, 214, 217, 230, 253
deaths, 63, 273, 275
decisions, vii, 7, 12, 220, 232, 240, 266
defense, 121, 220
definition, 1, 32, 40, 58, 102, 116, 117, 119, 120,
   141, 156, 175, 179, 200, 212, 213, 219, 220,
   227, 233, 234, 236, 237
democracy, 3, 63, 80, 81, 179, 182, 187, 213,
   229, 235, 237, 238, 243, 264
democratization, 191, 192
demographic characteristics, 21
demographics, 13, 15, 33
demonstrations, 229, 232
Denmark, 49
dependent variable, 35, 37, 38, 40
deregulation, 82, 83, 191, 202
detection, 280
developed nations, 139
deviation, 97
dialogues, 72, 159, 160
differentiation, 218, 225, 227, 230
diffusion, 30, 225
digital cameras, 30, 43

digital television, 276
directors, 50, 155, 160
disaster, 70, 246, 247, 248
discipline, 86, 226, 227, 266
disclosure, 26, 29, 31, 40, 194
discourse, x, 1, 28, 67, 107, 108, 137, 139, 140, 141, 142, 145, 146, 147, 149, 151, 156, 162, 166, 171, 173, 174, 175, 176, 179, 180, 190, 191, 201, 207, 221, 231, 259, 281
distance learning, 42
distress, 157, 224, 248
diversification, 56, 241, 242, 256
diversity, 13, 15, 21, 29, 82, 96, 98, 118, 134, 166, 172, 174, 192, 197, 261
dominance, x, 12, 23, 24, 53, 66, 75, 139, 165, 201, 218, 260, 279
downward comparison, 98, 218
drawing, 60, 80, 84, 116, 124, 190, 275
dream, 92, 119, 124, 128, 130, 143, 149, 179, 180, 184, 201
dreams, 42, 67, 135, 138, 143, 180, 181, 183, 215
drug use, 157, 228
dynamics, ix, 21, 42, 59, 123, 168, 229, 263

E

earth, 76, 131, 180, 248, 255
East Asia, 256
Eastern Europe, 47, 51, 80, 82, 84, 281
economic crisis, 158
economic development, 190
economic policy, 80
editors, 81, 121, 155, 156
Education, 36, 37, 57, 63, 102, 104
Egypt, 228
emotion, 70, 201, 216, 217, 237, 240, 242, 245, 247, 249, 255, 281
emotional distress, 247
emotional responses, 69, 197
emotions, 107, 197, 245, 247, 249
empathy, 72, 239, 246, 247, 254
empirical studies, 237
employment, 1, 28
empowerment, 237, 238
energy, vii, 25, 204
energy consumption, 204
engagement, 197
England, xi, 211, 232, 234, 236, 241
English language program, 202
entrepreneurs, 69
environment, 7, 8, 26, 27, 28, 30, 31, 32, 33, 49, 91, 118, 133, 136, 191, 197, 215, 216, 218
environmental influences, 30
estrangement, 67, 139, 140

ethics, 103, 269, 271
ethnic background, 22, 33, 172
ethnic groups, 175
ethnic minority, 72
ethnicity, 21, 175
etiquette, 194, 224, 266
EU, 81, 142
Eurasia, 61
Europe, v, 2, 45, 61, 84, 89, 109, 138, 148, 149, 155, 157, 161, 190, 191, 226, 254
European Union, 277
evening, 58, 83, 127, 202, 249
evening news, 58, 83
evolution, 222, 248
exaggeration, 142, 146, 219
examinations, 168, 260
exclusion, 141, 142, 175, 223, 268
expertise, 69, 72, 155, 193, 208
exploitation, 269, 270, 271, 274
exports, 260
exposure, 31, 39, 40, 96, 97, 98, 99, 101, 104, 105, 107, 108, 109, 110, 157, 230, 280
external locus of control, 246, 250
external relations, 30, 32

F

Facebook, 2, 26, 27, 29, 30, 32, 38, 40, 41, 44, 72, 251
failure, 158, 190, 213, 242, 243, 244, 246, 250, 251
faith, 67, 75, 142
family conflict, 246, 250
family members, 26, 30, 97, 240, 241, 264
family therapy, 221
family units, 272
fear, 11, 215, 217, 224, 229, 247, 258
fears, 83, 214, 217
feelings, 26, 49, 72, 75, 96, 97, 98, 145, 173, 230, 260
Fiji, 13, 14, 21
films, 144, 152, 154, 155, 156, 158
formula, vii, x, 3, 7, 47, 58, 59, 165, 169, 182, 215, 262, 263, 264
foundations, 41, 42, 126, 128
France, 49, 88, 156, 220, 221, 254
franchise, 51, 152, 156, 168, 169, 170, 173, 262
freedom, 194, 218, 222, 225, 249, 257, 265, 267
friendship, 23, 24, 29, 30, 32
frustration, 226, 249
fuel, 107, 219

Index

287

G

Gabon, 13, 14
garbage, 102, 154, 155, 161
Gaza Strip, 116
gender, 8, 35, 39, 88, 136, 156, 174, 192, 196,
    241, 243, 250, 263, 267, 268, 269
gender differences, 39
gender equality, 268
gender role, 156
genres, vii, xi, 2, 28, 33, 40, 117, 154, 190, 191,
    192, 198, 207, 211, 259, 265
Georgia, 174
Germany, xi, 3, 49, 59, 88, 95, 107, 211, 215,
    220, 221, 225, 227, 230, 231, 232, 234, 241,
    243, 244, 259, 279, 280
girls, 90, 124, 125, 126, 127, 128, 130, 131, 152,
    155
global communications, 185
global mobility, 195, 206
global village, 4, 183
globalization, x, 48, 80, 87, 89, 90, 135, 136, 139,
    165, 166, 168, 170, 181, 182, 183, 184, 185,
    186, 222, 225, 227, 250, 281
Globalization, ix, 62, 123, 124, 136, 166, 168,
    186, 187
goals, 26, 81, 175, 181, 203, 214, 257, 261
goods and services, viii, 65, 67, 68
gossip, 32, 98, 100, 104, 105, 106, 107, 108
governance, 81, 90, 193, 194, 222, 225
government, vii, 47, 50, 53, 60, 81, 82, 89, 121,
    194, 203, 204, 205, 206, 227, 268
Great Britain, 2, 47, 84, 260
Greece, 49, 156
group identity, 260
group membership, 98
group size, 102
groups, 10, 23, 29, 30, 98, 99, 108, 110, 141, 157,
    216, 219, 231, 245, 246, 270, 272, 274
growth, ix, 123, 138, 196, 225, 264
Guatemala, 14
guidance, 4, 189, 196, 201, 213, 218, 235, 237,
    238, 249, 253
Guinea, 82

H

hair, 205, 227, 266, 273
happiness, 76, 194, 241
harmony, 70, 250
Hawaii, 25, 44, 281
health, 12, 20, 66, 143, 157, 193, 194, 197, 207,
    280

health problems, 20, 143
hegemony, 147, 173, 174, 191
high school, 13, 22, 99, 202
higher education, 108, 244
Hispanics, 15
historical reality genre, xi, 259, 262, 264, 267,
    269, 270, 271, 272, 274, 275
holding company, 61
home ownership, 199
homework, 226, 249
Hong Kong, 256
host, 8, 21, 28, 30, 69, 70, 71, 74, 127, 171, 172,
    177, 178, 179, 200, 202, 203, 266
hostilities, ix, 115, 120
House, xi, 91, 198, 259, 260, 261, 262, 264, 265,
    266, 267, 268, 269, 270, 271, 272, 273, 275,
    276, 277
human rights, 88, 239
Hungary, 47, 59
husband, 32, 126, 127, 200
hybrid, xi, 48, 87, 133, 138, 140, 143, 166, 167,
    173, 203, 212, 259
hybridity, 59, 167
hybridization, 48, 133, 167, 168, 184, 186, 191,
    207
hypothesis, 35, 95, 96, 235

I

Iceland, 49
icon, 185, 187
ideal, 33, 71, 74, 75, 107, 118, 127, 131, 135,
    144, 193, 206, 222, 226
ideal forms, 222
idealization, 216
ideals, 86, 146, 194, 227, 269
identification, viii, 68, 71, 79, 82, 84, 87, 89, 90,
    98, 133, 134, 181, 183, 207, 217, 239, 263,
    269
ideology, viii, 65, 93, 125, 141, 158, 184, 263,
    279
Idol, vi, x, 3, 34, 40, 51, 53, 54, 84, 91, 138, 165,
    168, 169, 170, 171, 172, 173, 174, 175, 176,
    177, 178, 179, 180, 181, 182, 183, 184, 185,
    186, 187, 260
illusion, 58, 125, 135, 225, 227
image, 13, 24, 32, 40, 71, 79, 82, 87, 130, 142,
    145, 168, 171, 173, 226, 250
imagery, 67, 79, 125, 202, 212, 218, 255
images, 31, 39, 67, 121, 145, 184, 187, 203, 204,
    212, 238
imagination, x, 119, 142, 183, 189, 237, 256
imitation, 31, 178, 264, 271
immunity, 7, 8, 12, 16, 17, 18, 21, 22

imperialism, 166, 187, 271
implementation, 213, 225, 235
incarceration, 115, 119, 120
inclusion, 1, 97, 141, 193, 241
inclusiveness, 172, 174
income, 152, 250
indentured servant, 269
independence, 80, 81, 91, 224, 265
independent variable, 35
India, 187
Indians, 268
indication, 75, 84, 158
indicators, 16, 18, 141, 231
indices, 37
indigenous, 138, 166, 273
indigenous peoples, 273
individual differences, 254
individual personality, 71
individualism, 81, 166, 175, 194, 201, 206, 274
individuality, 3, 70, 75, 176, 177, 178, 182, 263, 265
individualization, 224, 226
indoctrination, 218, 223
industrial relations, 140
industrialization, 87
industry, 9, 40, 68, 76, 153, 155, 159, 160, 189, 190, 191, 192, 193, 195, 196, 197, 202
inequality, 167, 272
information seeking, 109, 216
information technology, 43
innovation, 43, 65
inoculation, 218, 257, 258
insight, 85, 203, 251, 265
institutions, 138, 145, 167, 168, 185, 187, 222, 223, 225
integration, 28, 81, 280
integrity, 10, 175, 266
intentions, 74, 129, 181, 217, 218, 249, 251
interaction, vii, 25, 26, 27, 30, 32, 40, 140, 220, 223, 230, 231, 239, 241, 244, 249
interactions, 8, 12, 27, 33, 40, 118, 124, 143, 214, 217, 229, 240, 280
interdependence, 187, 272
intermediaries, 69, 74, 193
international communication, 187, 281
international trade, 204
internet, vii, 1, 15, 25, 27, 28, 39, 90, 98, 100, 105, 108, 253
interpersonal communication, viii, 2, 3, 26, 38, 95, 96, 97, 98, 107, 108, 110, 111, 280
interpersonal relations, 8, 23, 30, 192, 279
interpersonal relationships, 8, 23, 30, 279
interrogations, 191

intervention, 75, 217, 220, 221, 222, 223, 230, 232, 233, 236, 237, 242, 246, 251, 271
interview, 82, 87, 127, 161, 223, 224, 234, 258
intimacy, 27, 30, 66, 72, 96, 192, 268
introspection, 74
intrusions, 265
invasion of privacy, 156
investment, 29, 74, 75, 83
investors, 83, 271
Iran, 116, 174
Iraq, 139, 141, 145, 146, 148, 172, 214
irony, 275
isolation, 42, 215
Italy, 49, 52, 130, 220

J

Japan, 185, 187, 204, 228
Jews, 130, 156, 162, 173, 174, 175, 227
jobs, 162, 266
journalism, 62, 117, 119, 191, 280, 281
journalists, 81, 107
judges, 169, 171, 172, 174, 175, 177, 178, 179, 220
judgment, 178, 214, 219, 220, 240, 248

K

Korea, 228

L

labor, 267, 269, 271
labour, 68, 192, 209
land, 130, 134, 145, 161, 264, 268, 270, 271
landscape, viii, 25, 26, 30, 47, 60, 62, 80, 81, 87, 171
Landscape, 81, 190
language, 72, 81, 85, 128, 141, 144, 145, 172, 182, 202, 203, 228, 234
languages, 52, 182
Latin America, 167, 187, 228
laughing, 219, 245
leadership, 228, 254
leadership style, 228, 254
learning, 31, 41, 42, 215, 218, 219, 220, 231, 235, 237, 238, 239, 240, 246, 249, 250, 270
legend, 183
leisure, 67, 75, 125, 131, 196
leisure time, 67
lens, 90, 206, 263
liberalisation, 201
liberation, 222, 225
licenses, 61, 72, 140, 152, 156
life experiences, 3, 134

lifestyle, xi, 2, 50, 66, 67, 68, 69, 71, 72, 73, 74, 75, 76, 117, 130, 189, 190, 191, 192, 193, 194, 195, 196, 197, 198, 199, 200, 201, 202, 203, 204, 205, 206, 207, 208, 209, 211, 219, 234, 273, 281
likelihood, 34, 35, 39
line, 9, 17, 39, 48, 50, 51, 53, 72, 73, 75, 97, 142, 146, 147, 154, 155, 157, 159, 160, 161, 178, 193, 204, 221, 231, 238, 242, 269
linkage, 125
links, 28, 82, 130, 132, 192, 245
listening, 273
living conditions, 215, 218
living environment, xi, 211, 215, 216, 218
localization, x, 165, 170, 181, 182, 184
locus, 215, 237, 246
loneliness, 96, 267
longitudinal study, 43
love, 51, 52, 96, 107, 123, 124, 125, 126, 128, 132, 133, 136, 145, 155, 179, 218, 269
loyalty, 10, 84, 89, 91
lying, 10, 131

M

Macedonia, 51, 59, 260
magazines, 32, 58, 100, 105, 198, 208
maintenance, 27, 30, 152, 274
majority, 2, 8, 10, 12, 15, 16, 17, 18, 20, 21, 33, 34, 68, 82, 83, 95, 107, 175, 202, 223, 245, 248, 250
management, 11, 66, 83, 92, 110, 152, 192, 193, 216, 217, 218, 235, 237, 240, 242, 245, 247, 249, 250, 253, 254, 255, 256, 272
Mandarin, 204, 205
manipulation, 10, 12, 27, 58
manners, 66, 239, 240
market, ix, x, 29, 43, 48, 51, 54, 60, 70, 72, 74, 81, 83, 87, 89, 123, 124, 165, 189, 190, 192, 195, 196, 198, 204, 206, 235
market economy, 81
market share, 54, 124
marketing, xi, 28, 40, 68, 73, 82, 90, 91, 92, 102, 162, 168, 181, 191, 211, 222, 223, 225, 235
marketing strategy, 73, 91
marketplace, 69
markets, vii, 48, 51, 84, 185, 190, 191, 192, 195, 196, 198, 204, 205, 206, 208, 256
marriage, 127, 202, 214, 269
married couples, 244
mass communication, 27, 95, 109, 110, 242, 248
mass customization, 80, 90
mass media, 2, 27, 28, 38, 40, 87, 95, 96, 97, 109, 110, 125, 168, 171, 185, 263, 274, 281

measurement, 110, 170
measures, 18, 35, 56, 100, 105, 237, 239, 240, 242
membership, 142, 275
memory, xi, 119, 121, 216, 256, 257, 259, 260, 263, 274, 275, 281
men, 39, 50, 52, 56, 101, 103, 107, 117, 126, 143, 196, 197, 202, 241, 242, 243, 244, 250, 268, 272
mental health, 56
messages, 2, 27, 28, 31, 38, 39, 52, 59, 65, 67, 69, 73, 143, 160
metaphor, 127
methodology, 3, 279
Mexico, 49, 191, 234
middle class, 69, 71, 80, 130, 131, 192, 193, 194, 199, 201, 205, 207, 208
Middle East, v, x, 3, 113, 189, 256
military, vii, 25, 50, 144, 145, 195
minorities, 21, 56, 156, 160, 162
minority, 21, 157, 175
mobility, 3, 90, 179, 180, 182, 189, 192, 195, 201, 206, 253
model, vii, 3, 7, 16, 19, 22, 26, 27, 28, 30, 31, 33, 35, 40, 43, 56, 68, 73, 82, 86, 148, 152, 157, 181, 192, 193, 196, 206, 213, 224, 227, 234, 235, 237, 239, 241, 242, 244, 249, 274, 275
modeling, 30, 32, 263
modelling, 199, 203, 206, 207
models, 3, 18, 26, 31, 33, 35, 37, 39, 40, 41, 73, 75, 109, 160, 166, 189, 190, 193, 194, 195, 200, 202, 204, 205, 206, 207, 218, 219, 220, 230, 231, 255
modern society, 130
modernity, x, 82, 110, 165, 168, 183, 184, 192, 195, 202, 204, 206, 207, 254
modernization, 81, 226, 227
money, x, 13, 31, 67, 75, 83, 102, 107, 120, 121, 124, 126, 156, 180, 189, 266
Montana, 260
mood, 146, 178, 251
moral code, 152
moral development, 194
morality, 209, 225, 258
Moscow, 51, 61, 62
mothers, 144, 219, 226, 241, 242, 243, 250
motion, 156, 220, 275
motivation, 9, 22, 32, 39, 100, 181, 231, 269
motives, xi, 3, 96, 97, 98, 99, 100, 104, 105, 107, 108, 110, 212, 215, 217, 238, 244, 245, 248
movement, 116, 198
multiculturalism, 261
multidimensional, 254

music, viii, 58, 69, 79, 82, 91, 98, 126, 138, 146, 166, 167, 169, 172, 173, 174, 185, 186
music industry, 169
MySpace, 29, 30, 40
mythology, 68, 198, 276

N

narratives, 27, 66, 67, 68, 75, 84, 134, 185
nation, viii, x, 4, 41, 77, 79, 80, 81, 82, 84, 85, 89, 92, 93, 129, 130, 135, 141, 142, 145, 147, 148, 165, 166, 168, 171, 173, 174, 176, 183, 184, 205, 208, 209, 233, 260, 263, 268, 269, 270, 271, 272, 273, 274, 275, 276
nation building, 79, 82
national character, 91, 138, 181
national community, 90, 270
national culture, ix, 48, 80, 81, 123, 186
national identity, x, 3, 79, 80, 81, 83, 84, 85, 87, 88, 89, 91, 93, 142, 143, 146, 148, 165, 171, 174, 175, 176, 177, 178, 181, 182, 259, 260, 263, 267, 269, 271, 272, 274
national origin, 172
nationalism, 80, 81, 82, 83, 89, 90, 91, 134, 138, 140, 141, 142, 145, 148, 181, 183, 186, 187, 204, 256, 263, 273, 274, 275, 281
nationality, 82, 224
nation-building, 80, 81, 173
Native Americans, 270
native population, 168
NCA, 280
negotiation, 11, 12, 48, 184, 266
neoliberalism, 204, 206, 208
Netherlands, 48, 84, 148
network, 20, 26, 29, 30, 32, 34, 35, 37, 38, 39, 40, 61, 83, 140, 169, 197, 198, 200, 262, 281
new media, viii, 30, 47, 54, 60, 69, 109, 281
New South Wales, 261, 276
New Zealand, 169, 259
news coverage, 3, 115, 116
newspapers, 100, 105, 120, 145, 153
next generation, 43, 117
Nobel Prize, 159
North America, v, 2, 5, 133, 176, 268
Norway, 49, 185
nostalgia, viii, 79, 93, 207

O

obedience, 226, 227, 230, 266, 269
obesity, 158, 194, 198, 204
obligation, 10, 213, 274
observational learning, 30, 31, 32
observed behavior, 30, 31, 32

obstacles, 125, 217
Oklahoma, 7, 280
older people, 104, 272
openness, 28, 218, 222, 235, 237, 252
openness to experience, 28
opportunities, 59, 60, 90, 139, 265
orientation, 26, 82, 97, 136, 199, 215, 216, 218, 219, 220, 221, 222, 234, 235, 238, 240, 241, 244, 245, 246, 249, 250, 252, 259, 261
outsourcing, 225
overweight, x, 189, 193, 199
ownership, 89, 91

P

Pacific, 15
pain, 70, 268, 269, 273
Pakistan, 260
Panama, 12, 14, 21
parameters, 142, 266
parental control, 227, 239, 253
parental influence, 221
parenting, xi, 197, 211, 221, 224, 225, 226, 227, 228, 229, 230, 231, 232, 233, 234, 235, 236, 237, 238, 239, 240, 241, 242, 243, 244, 248, 249, 250, 251, 253, 254, 256, 257
parenting styles, 226, 227, 228, 229, 231, 232, 234, 236, 237, 254, 256
parents, 4, 68, 85, 128, 131, 132, 133, 197, 221, 223, 224, 225, 226, 227, 228, 229, 230, 231, 232, 233, 234, 235, 236, 237, 238, 239, 240, 241, 242, 243, 245, 248, 249, 250, 251, 253, 254, 257
Parliament, 116, 148
particles, 212, 252
patriotism, 134, 141, 144, 146, 172, 273, 280
pedagogy, 203, 251
perceptions, 3, 28, 42, 43, 125, 256, 273
performance, x, 11, 41, 72, 151, 172, 173, 174, 181, 182, 193, 228, 235, 249, 252, 275
performers, 169, 174, 217
permeability, 213
personal identity, vii, 25, 274
personal life, 191, 193, 200
personality, viii, 9, 12, 17, 21, 28, 41, 42, 65, 69, 133, 146, 200, 224, 227, 228, 246, 253
personality test, 246
personality traits, 28, 42
personhood, 209
persuasion, 218, 219, 258, 279, 280
phenomenology, xi, 211
photographs, 33, 34, 35, 127
planning, 51, 73, 82, 133
platform, 32, 34, 41, 276

Index

291

plausibility, 237

pleasure, 97, 134, 158, 174, 193, 265

pluralism, 91, 134, 166, 220, 222

Poland, 2, 47, 50, 53, 59, 185

polarization, 136

police, 65, 66, 215, 222, 224

politics, 56, 76, 90, 92, 141, 158, 162, 184, 186, 187, 208, 213, 225, 246, 275, 277, 280

poor, 9, 133, 195, 237, 266

popular vote, 260

population, 15, 26, 33, 49, 51, 81, 99, 107, 108, 139, 143, 172, 173, 174, 175, 201, 254

porous borders, 175

Portugal, 49, 157

positive correlation, 105, 234

positive relation, 38, 40

positive relationship, 38, 40

poverty, 81, 143

power, vii, x, 9, 12, 20, 23, 24, 25, 39, 42, 71, 72, 73, 74, 96, 123, 125, 133, 156, 165, 166, 167, 168, 176, 179, 192, 193, 198, 208, 209, 217, 220, 223, 224, 225, 226, 228, 232, 241, 250, 256, 263, 270, 274, 277, 279

power relations, x, 73, 165, 167, 168

predictor variables, 17, 18

predictors, 16, 17, 19, 24, 35

preference, 86, 167, 252

prejudice, vii, 25, 253

pressure, 67, 189, 191, 196, 237, 238, 249

prices, 70, 203

primacy, 32, 70, 133, 224

primary function, 12

primetime slot, viii, 79, 200

prisoners, 119, 120

privacy, 11, 27, 29, 31, 41, 42, 135, 153, 155, 159, 160, 230

private sector, 80, 90

probability, 217, 221, 244

problem solving, 216, 217, 221, 233, 241, 242, 243

production, 15, 31, 32, 38, 48, 49, 51, 55, 82, 83, 84, 86, 117, 124, 127, 130, 134, 138, 139, 140, 146, 152, 153, 158, 159, 169, 182, 191, 196, 261, 262, 263

production costs, 48, 51, 84, 124, 138, 152

professionalism, 179, 180, 184

profit, 81, 83, 90, 251, 266

profits, 82, 271

project, x, xi, 3, 8, 20, 22, 80, 81, 87, 165, 173, 175, 176, 185, 193, 201, 208, 211, 222, 227, 231, 255, 262, 263, 264, 266, 268, 269, 271, 280

proliferation, 40, 152, 153, 154

propaganda, 53, 58, 80, 90, 151, 181, 214, 218, 271

psychoanalysis, 122

psychological processes, 43

psychology, 23, 24, 41, 73, 136, 253, 281

public broadcasting, 93, 138, 261

public discourse, x, 107, 151, 156, 174

public education, 197, 204, 205

public health, 194

public life, 92, 108

public opinion, 181

public service, 53, 65, 72, 75, 191, 196, 198, 261, 276

public television, 81, 83, 161

punishment, 221, 224, 226, 227, 228, 229, 231, 232, 254

pupil, 224, 226

Pyszczynski, 253

R

race, vii, 8, 12, 15, 16, 18, 19, 21, 23, 25, 136, 174, 263, 268, 269

racial minorities, 13

racism, 174, 269

radio, 42, 49, 58, 62, 100, 105, 115, 119, 120, 202

range, 13, 22, 27, 29, 30, 32, 37, 38, 39, 40, 58, 83, 85, 96, 166, 168, 174, 190, 191, 193, 195, 196, 197, 199, 201, 204, 220, 238, 241, 248, 273

rape, 230

ratings, viii, 9, 11, 23, 51, 52, 56, 59, 60, 79, 80, 83, 84, 85, 89, 92, 95, 124, 152, 169, 170, 183, 195, 198, 202, 204, 261

rationality, 218, 220, 225

reactions, 107, 116, 177, 215, 264

realism, 66, 68, 70, 75, 218, 247, 273

reason, 8, 16, 17, 20, 74, 89, 97, 99, 117, 118, 133, 138, 140, 146, 154, 156, 157, 159, 183, 191, 212, 224, 243, 247, 255

reception, 2, 47, 59, 80, 160, 166, 182, 195, 206, 214, 216, 244, 255

recognition, 119, 274

recommendations, iv, xi, 82, 211, 221, 222, 223, 225, 231, 233, 234, 236, 237, 248, 249, 251

reconcile, 128

reference frame, 216, 218

reflection, 1, 69, 73, 90, 107, 181, 231, 233, 238, 273

reflexivity, 222, 223

reforms, 81

regenerate, 225

region, 13, 15, 21, 80, 84, 201, 205

regression, 16, 17, 18, 35, 66
regression analysis, 16, 17, 18
regression equation, 18
regulation, 50, 72, 152, 197, 217, 227, 249, 263
reinforcement, 257
relationship, viii, 10, 11, 13, 24, 27, 29, 33, 34, 35, 38, 39, 40, 65, 87, 90, 91, 95, 107, 119, 128, 132, 134, 145, 192, 194, 206, 213, 221, 222, 226, 229, 235, 239, 249, 281
relevance, 118, 160, 216, 217, 221, 222
reliability, 86, 153, 221, 231, 252
religion, 3, 128, 132, 182, 215, 256, 266
religiousness, 128, 256
reproduction, 32, 80, 87, 141, 142, 175, 241
reputation, 9, 29, 246
resistance, 140, 142, 177, 182, 187, 214, 218, 219, 225, 237, 249, 257, 258, 270
resources, 29, 34, 41, 83, 166, 216, 224, 225, 230, 248
respect, 47, 85, 88, 126, 132, 181, 227, 266
restructuring, 221, 229
returns, 23, 128, 131
rewards, 7, 8, 12, 21, 23, 179
rhetoric, 131, 158
rhythm, 55, 118
rights, iv, 41, 50, 88, 239
risk, 81, 143, 178, 181, 192, 216, 223, 230, 250, 251, 273, 276
Romania, 47, 59
routines, 82, 141, 214, 215, 216, 223, 229, 246, 252
Russia, 2, 47, 50, 52, 59, 174, 185

S

Saddam Hussein, 214
safety, 90, 157, 217, 232, 246, 273
sales, 15, 80, 89
sanctions, 226, 227
Saudi Arabia, 185, 219
scholarship, 4, 154, 190, 262
school, 99, 100, 101, 102, 104, 107, 130, 225, 226, 228, 229, 230, 234, 237, 250, 258
search, 4, 56, 124, 135, 172, 183, 186, 202
season finale, viii, 47, 145, 169, 176
secondary schools, 228
second-class citizens, 267
security, 65, 90, 255
self-concept, 234
self-control, 97, 221, 224, 225, 250
self-discipline, 224
self-discovery, 265
self-doubt, 70

self-efficacy, 215, 217, 218, 222, 230, 237, 238, 239, 242
self-empowerment, 214, 217, 219, 221, 225, 237, 238, 239, 248
self-enhancement, 98
self-esteem, 69, 74, 97, 98, 108, 110, 228, 255, 256
self-expression, 33
self-identity, 85, 254
self-image, 266
self-improvement, 198, 206
self-irony, 219
self-presentation, 220
self-reflection, 134, 223, 249
self-regulation, 194, 204, 225, 227, 229, 249, 254
self-worth, 70, 74, 217, 218, 246
semiotics, 279
sensation, 135, 215, 246
sensation seeking, 215, 246
sensitivity, 147, 246, 247
Serbia, 47, 51, 89
sex, 12, 13, 16, 18, 21, 27, 42, 43, 99, 104, 105, 106, 110, 126, 153, 161, 162, 185
sex differences, 185
sexual abuse, 230
sexual harassment, 157, 162
sexual orientation, 56
sexuality, 56, 108, 174, 219, 226
shape, 22, 70, 108, 116, 129, 140, 237
shaping, 209, 217, 249
shares, 52, 56, 83
sharing, 20, 27, 29, 30, 32, 33, 35, 39, 42, 71, 97, 144
sheep, 261, 266, 268, 272
siblings, 143, 144
signalling, 88
signs, 66, 67, 68, 69, 70, 71, 88, 132, 141, 269
simulation, 28, 40, 76, 252
Singapore, 3, 122, 190, 195, 201, 202, 204, 205, 206, 207
singers, ix, 56, 137, 138, 143, 147, 156, 172
skills, 8, 10, 13, 22, 27, 33, 68, 69, 88, 192, 193, 197, 202
slavery, 269, 270
Slovakia, 59
smoking, 27, 43, 44, 218, 257
SMS, 52, 143
SNS, 27, 29, 32, 33, 34, 35, 36, 37, 39, 40, 41
soccer, 56, 216
social attributes, 246
social behavior, 27, 217, 223
social behaviour, 257
social capital, 9, 12, 20, 21, 22, 23, 29

Index

293

social change, 87, 118
social class, 52, 69, 118
social comparison, 97, 98, 103, 108, 109, 111, 245
social conflicts, 125, 256
social construct, 28
social constructivism, 28
social context, 28, 30, 281
social control, 222
social desirability, 246
social development, xi, 211, 256
social environment, 26, 27, 216, 223, 244, 249, 250
social exchange, vii, 25
social group, 10, 56, 98, 100, 183
social hierarchy, 130, 266, 267, 268
social identity, 110, 203, 205
social identity theory, 110
social inequalities, 269
social learning, 30, 251
social learning theory, 30
social life, 31, 86, 136
social movements, 275
social network, 2, 13, 26, 29, 30, 32, 33, 34, 35, 38, 39, 40, 41, 42, 43, 74, 280, 281
social norms, 97, 98, 107, 265
social perception, 43
social psychology, 111, 217, 256, 280
social relations, ix, 8, 29, 30, 67, 95, 97, 98, 133, 166, 168, 177
social relationships, ix, 8, 29, 30, 67, 95, 97, 98
social roles, ix, 95
social services, 225, 251
social skills, 10, 13, 22
social status, 98, 130
social structure, 29
social support, 29
social theory, 92, 260, 277
social welfare, 197
socialization, 224, 257
software, 43
soil, 52, 67, 130
solidarity, 3, 10, 144, 171, 172, 176, 225
South Africa, 49
South Korea, 41
sovereignty, 82, 249
Soviet Union, ix, 123
space, 3, 4, 29, 74, 75, 82, 139, 141, 175, 186, 190, 192, 203, 274
Spain, xi, 3, 49, 211, 221, 225, 228, 231, 232, 234, 241, 243, 256, 257, 259
specialists, 194
spectrum, 30, 200

speech, 119, 147, 178, 185
spin, xi, 49, 54, 58, 98, 169, 202, 259, 261
spin-offs, xi, 49, 58, 98, 259, 261
spirituality, 131, 264
Sri Lanka, 82, 92
stability, 23, 73, 231
standard deviation, 36, 99
standardization, 166
standards, 86, 126, 149, 182, 266
stars, 33, 76, 83, 84, 169, 185, 218, 219, 223
state of emergency, 214, 219
statistics, 20, 36, 59
stereotypes, 28, 42, 43, 70, 85, 131, 134, 141, 156, 160
strategies, 8, 10, 11, 24, 27, 68, 80, 171, 181, 182, 184, 213, 217, 218, 239, 241
strategy, 12, 66, 68, 72, 82, 95, 98, 110, 146, 196, 275
strategy use, 12
stratification, 167, 185
strength, 8, 12, 88, 130, 134, 147, 216, 220, 233, 239, 274
stress, 130, 221, 247
structuring, 28, 181
students, 15, 39, 56, 153, 202, 203
subjectivity, 194, 207
success rate, 242, 243
suicide, 117, 122
summer, 116, 133, 180, 183
Sun, 224, 258, 276
Supernanny, xi, 3, 194, 197, 198, 211, 221, 222, 223, 224, 225, 227, 229, 230, 231, 232, 234, 235, 237, 239, 240, 241, 242, 243, 244, 245, 246, 247, 248, 249, 250, 251, 254, 255, 257
supernatural, 71
supervision, 226, 228, 231
supervisor, 242
suppliers, 200
supply, 241
surplus, 214
surveillance, vii, x, 1, 31, 47, 49, 65, 97, 100, 116, 151, 194, 197, 199, 214, 215, 221, 222, 260, 262, 265, 279
survival, xi, 7, 13, 20, 22, 67, 74, 211, 217, 221
Survivor, v, vii, x, xi, 2, 7, 8, 9, 10, 11, 12, 13, 14, 15, 16, 17, 18, 20, 21, 22, 23, 24, 27, 32, 40, 42, 51, 54, 115, 117, 118, 120, 124, 135, 140, 151, 152, 154, 155, 190, 195, 199, 202, 203, 204, 205, 206, 212, 217, 221, 259, 261, 262, 276
suspects, 31, 214
suspense, 118, 120, 134, 220
Switzerland, 49

symbolic associations, 67
symbolism, viii, 2, 65
symbols, 66, 67, 68, 70, 91, 125, 126, 133, 141, 142
sympathy, 13, 120, 139
system analysis, 231
systematic processing, 38

T

Taiwan, 205
takeover, 196, 209
talent, ix, 3, 53, 55, 137, 138, 143, 147, 169, 176, 179, 212, 219
teaching, 31, 131, 193, 197, 199, 200, 205, 206, 249, 279, 280
technology transfer, 166
teens, 28, 199
television stations, 83, 99
television viewing, 9, 27, 28, 33, 35, 37, 38, 39, 40
tenants, 89, 98
tension, x, 3, 120, 128, 137, 139, 140, 144, 147, 172, 175, 250, 269, 275
tensions, x, 123, 125, 126, 134, 165
territory, 234, 254
terror management theory, 255
testing, xi, 2, 33, 189, 211, 262, 264, 274
Thailand, 14, 49
The Biggest Loser, x, 168, 189, 193, 194, 195, 198, 199, 201, 203, 204
The Farm, viii, 2, 79, 80, 86, 87, 89, 90, 91, 92
theatre, 212, 214
therapeutic process, 249
therapists, 75
therapy, 66, 77, 224, 237, 251
thinking, 55, 67, 70, 89, 193, 195, 204, 262, 274
Third World, 166, 168, 184, 185
thoughts, 26, 27, 31, 49, 97, 103, 123, 159, 174
threat, 12, 13, 17, 20, 21, 67, 87, 108, 126, 137, 173, 246, 271
threats, 20, 237
tourism, 82, 186
trade, 10, 130, 140, 191, 248, 268
tradition, xi, 67, 88, 118, 119, 126, 128, 132, 133, 173, 176, 191, 196, 199, 201, 224, 243, 259
traditional gender role, 28
traditionalism, 193
traditions, xi, 81, 88, 89, 119, 166, 167, 182, 197, 202, 206, 207, 211, 234, 276
traffic, 59, 191, 198
training, 50, 193, 195, 197, 229, 231
traits, 12, 28
transcendence, 42, 119

transformation, 26, 53, 56, 66, 67, 68, 70, 71, 73, 74, 75, 76, 192, 197, 198, 199, 200, 207, 208, 209, 242, 251, 265
transformations, 66, 68, 74, 80, 81, 109, 142, 191, 198, 199, 203, 208, 209
transition, viii, 47, 55, 130, 142, 180, 191
transitions, 67, 191
trends, viii, xi, 20, 26, 33, 48, 65, 82, 118, 122, 139, 140, 160, 192, 204, 211
tribes, 7, 8, 11, 12, 21, 118, 269
trust, 10, 20, 22, 71, 74, 218, 224, 229
Turkey, ix, 49, 88, 117, 137, 138, 139, 141, 142, 143, 144, 145, 146, 147, 148, 279
Turkish Armed Forces, 141
Turks, 146

U

Ukraine, 52
uncertainty, 96, 208, 220, 226, 252
unemployment, 219, 221
United Kingdom, x, 49, 91, 165
United States, x, 15, 47, 48, 59, 85, 89, 91, 116, 126, 131, 132, 165, 185, 187, 260, 279
upward mobility, 198, 200
urban life, 42, 75

V

Vanuatu, 14
variables, 15, 16, 18, 19, 34, 35, 36, 37, 38, 40, 231
variance, 35, 39, 40, 228
variations, 192, 241
vein, viii, 56, 79, 103
Vietnam, 227
village, 4, 87, 230, 260, 261
violence, 27, 41, 42, 88, 208, 228, 229, 237, 248, 254, 255, 270, 271, 274, 280
violent crime, 116
vision, 49, 215, 271
voice, 25, 72, 107, 175, 176, 204, 231, 270
voters, 179
voting, 2, 8, 9, 10, 16, 17, 18, 20, 21, 22, 97, 168, 174, 175, 178, 179, 181
voting record, 8

W

Wales, 261
war, 121, 214, 227, 280
waste, 71, 102
wealth, 33, 40, 92
wear, 50, 69, 129, 234

Index 295

web, 26, 29, 30, 32, 38, 43, 49, 51, 54, 58, 59, 60, 61, 115
websites, 8, 16, 28, 29
weight loss, 193, 195
welfare, 90, 224, 234, 251, 257
welfare state, 90, 224
Western countries, 47
Western Europe, 84, 191
winning, 12, 13, 18, 20, 21, 22, 23, 85, 88, 155, 173, 179, 182, 204
winter, 99, 266, 272
women, 13, 39, 51, 52, 56, 68, 69, 99, 101, 103, 107, 125, 127, 128, 143, 156, 157, 158, 196, 198, 199, 208, 241, 242, 243, 244, 250, 267, 273
workers, 68, 69, 88, 159, 268

worldview, 182, 256
worry, 132, 135, 273
writing, 1, 14, 40, 158, 160, 169, 190, 202, 279, 280

Y

yellow journalism, 157
Yemen, 174
young women, 123, 126, 127
Yugoslavia, 80, 83, 93, 260

Z

zeitgeist, vii, 25, 227
zero sum game, 246